Contents

Response of Upper Gulf Coast Estuaries to Holocene Climate Change and Sea-Level Rise

Edited by

John B. Anderson
Department of Earth Science
Rice University
6100 Main Street, MS 126
Houston, Texas 77005
USA

Antonio B. Rodriguez
Institute of Marine Sciences
University of North Carolina at Chapel Hill
3431 Arendell St.
Morehead City, North Carolina 28557
USA

THE
GEOLOGICAL
SOCIETY
OF AMERICA®

Special Paper 443

3300 Penrose Place, P.O. Box 9140 ▪ Boulder, Colorado 80301-9140, USA

2008

Copyright © 2008, The Geological Society of America (GSA). All rights reserved. GSA grants permission to individual scientists to make unlimited photocopies of one or more items from this volume for noncommercial purposes advancing science or education, including classroom use. For permission to make photocopies of any item in this volume for other noncommercial, nonprofit purposes, contact The Geological Society of America. Written permission is required from GSA for all other forms of capture or reproduction of any item in the volume including, but not limited to, all types of electronic or digital scanning or other digital or manual transformation of articles or any portion thereof, such as abstracts, into computer-readable and/or transmittable form for personal or corporate use, either noncommercial or commercial, for-profit or otherwise. Send permission requests to GSA Copyright Permissions, 3300 Penrose Place, P.O. Box 9140, Boulder, Colorado 80301-9140, USA. GSA provides this and other forums for the presentation of diverse opinions and positions by scientists worldwide, regardless of their race, citizenship, gender, religion, or political viewpoint. Opinions presented in this publication do not reflect official positions of the Society.

Copyright is not claimed on any material prepared wholly by government employees within the scope of their employment.

Published by The Geological Society of America, Inc.
3300 Penrose Place, P.O. Box 9140, Boulder, Colorado 80301-9140, USA
www.geosociety.org

Printed in U.S.A.

GSA Books Science Editors: Marion E. Bickford and Donald I. Siegel

Library of Congress Cataloging-in-Publication Data

Response of upper Gulf Coast estuaries to Holocene climate change and sea-level rise / edited by John B. Anderson, Antonio B. Rodriguez.
p. cm. — (The Geological Society of America special paper ; 443)
Includes bibliographical references.
ISBN 978-0-8137-2443-0 (pbk.)
1. Climatic changes—Environmental aspects—Gulf Coasts (U.S.) 2. Global warming—Environmental aspects—Gulf Coasts (U.S.) 3. Climatic changes—Environmental aspects—Gulf States. 4. Sea level—Gulf Coasts (U.S.) 5. Estuaries—Gulf Coasts (U.S.) I. Anderson, John B., 1944–. II. Rodriguez, Antonio B.

QC984.G82R47 2008
551.45′80976—dc22

2008022309

Cover: Digital elevation map of the northern Gulf Coast highlighting the fluvial valleys that were flooded to create modern estuaries. The inset is of Galveston Bay.

10 9 8 7 6 5 4 3 2 1

Preface

The International Panel on Climate Change (IPCC) has proposed three different scenarios for sea-level rise in the next century that vary between 15 cm to nearly a meter of rise (Church and White, 2006). Given its low gradient, the U.S. Gulf Coast is particularly vulnerable to inundation by sea-level rise. In Louisiana, coastal subsidence rates are estimated to be somewhere between 15 cm/century (Penland and Ramsey, 1990) and 150 cm/century (Shinkle and Dokka, 2004). This is significantly greater than the average rate for the past 4000 yr (Milliken et al., this volume). The cause of this increase in relative sea-level rise is therefore likely anthropogenic, most logically due to oil and gas extraction (Morton et al., 2006). The present U.S. Gulf Coast is also characterized by a strong climatic gradient; precipitation varies between 50 cm/yr and 200 cm/yr from east to west across the region (Fig. 1). This strong gradient makes the region especially vulnerable to climate change. Climate change and associated changes in river discharge and vegetation cover within a drainage basin could, in turn, result in significant and rapid (decadal to century time scale) changes in the sediment supply of rivers to the coast.

Currently, the impact of accelerated relative sea-level rise is manifest as high rates of coastal erosion and wetlands loss. As the rate of rise increases this century, the magnitude and extent of coastal change will increase. However, the exact response of coastal systems remains uncertain, mainly because the responses of different coastal environments to rising sea level and/or diminishing sediment supply are not well understood. One way to improve our understanding is to study the geological record of coastal response to climate change and rising sea level.

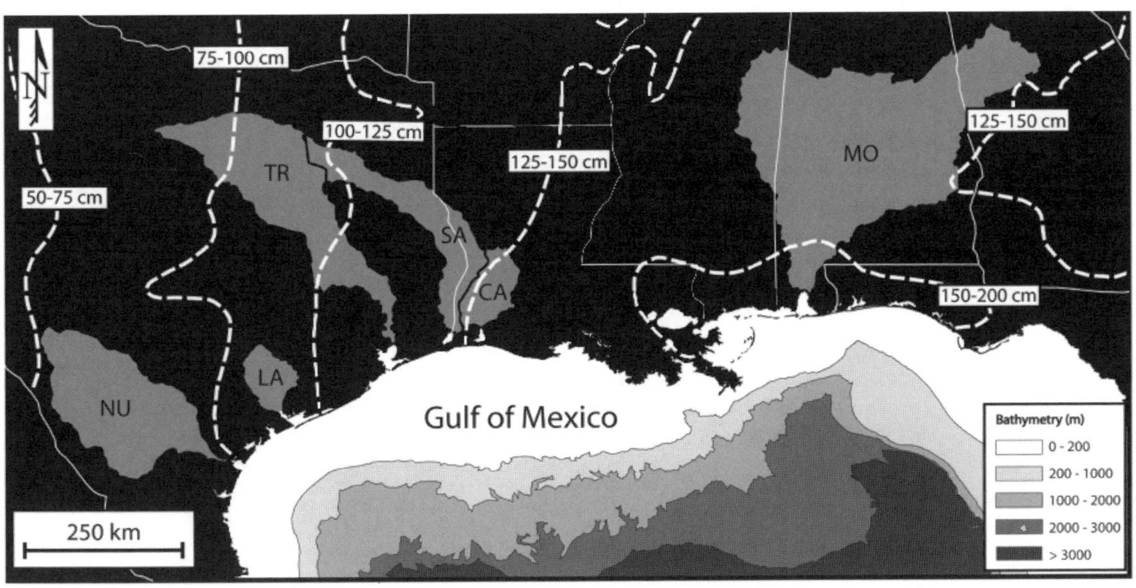

Figure 1. Regional study area map of the northern Gulf of Mexico margin depicting the drainage basins of the main rivers that flow into the estuaries studied (NU—Nueces River, LA—Lavaca River, TR—Trinity River, SA—Sabine River, CA—Calcasieu River, and MO—Mobile-Tensaw River), isopleths (dashed lines) of mean annual precipitation (in cm), and shelf bathymetry.

The papers presented in this volume provide documentation for rapid flooding and associated environmental change in Gulf Coast estuaries in the past. The approaches used to study these estuaries were similar so that comparisons could be made between them. The overall approach involved initial collection and analysis of high-resolution seismic data used to map the incised valleys occupied by the estuaries and the sediment fill of these valleys. This was followed by the acquisition of drill cores along the axis of each estuary and sedimentological and paleontological examination of these cores. These combined data were used to characterize and map seismic facies, lithofacies, and biofacies and to reconstruct paleoenvironments.

One key to understanding the response of estuaries to sea-level rise is a precise sea-level curve for the Holocene. A composite sea-level curve for the northern Gulf of Mexico was constructed using published and new sea-level datums derived from basal peats and foreshore deposits (Milliken et al., this volume). The curve shows rapid and possibly episodic sea-level rise during the early Holocene, followed by slower more continuous rise during the middle and late Holocene.

The history of all of the estuaries we studied was punctuated by rapid flooding events, during which there were major reorganizations of the environments. The primary manifestation of rapid inundation of an estuary is a flooding surface, which is a stratigraphic surface indicative of an increase in water depth and/or marine influence. A robust radiocarbon chronostratigraphy for each estuary was used to constrain the age and rates of flooding events. The radiocarbon reservoir for each estuary was constrained using published data and by dating paired shell and wood or peat samples collected at the same stratigraphic level. Using these corrections, we argue that it is possible to constrain the timing of flooding surfaces to within a century or two in areas where the surface is sampled and dated in more than one location.

The largest flooding events resulted in landward shifts of bayhead deltas by tens of kilometers and increases in estuary area up to 30% in a century or two. Most of these flooding events occurred prior to 7500 yr ago, when sea level was rising at an average rate of 4.2 mm/yr. However, there is also evidence for rapid change in several Gulf Coast estuaries after ca. 7500 yr, when the rate of sea-level rise was actually decreasing. The largest of these events impacted only east Texas and western Louisiana estuaries and is believed to have resulted from a reduction in sediment supply as the region transitioned out of the cool, wet glacial to warmer and drier conditions.

The stratigraphic record is quite clear. When the supply of sediment to an estuary falls below the critical level needed to maintain a bayhead delta during sea-level rise, the response is one of rapid landward retreat of the delta and dramatic reorganization of estuarine environments. Regardless of the way in which these different scenarios for sea-level rise play out by the end of this century, humans have already exerted a significant, probably irreversible, impact on estuaries and wetlands by altering sediment supply through construction of dams, land-use changes, and modification of river channels and by increasing rates of coastal subsidence with subsurface fluid extraction. Increasing global sea-level rise will only exacerbate these human impacts. The results from this study tell us that rapid and dramatic reorganization of estuarine environments will likely occur in the next few centuries.

John B. Anderson
Antonio B. Rodriguez

REFERENCES CITED

Church, J.A., and White, N.J., 2006, A 20th century acceleration in global sea-level rise: Geophysical Research Letters, v. 33, L01602, doi: 10.1029/2005GL024826.

Milliken, K.T., Anderson, J.B., and Rodriguez, A.B., 2008, this volume, A new composite Holocene sea-level curve for the northern Gulf of Mexico, *in* Anderson, J.B., and Rodriguez, A.B., eds., Response of Upper Gulf Coast Estuaries to Holocene Climate Change and Sea-Level Rise: Geological Society of America Special Paper 443, doi: 10.1130/2008.2443(01).

Morton, R.A., Bernier, J.C., and Barras, J.A., 2006, Evidence of regional subsidence and associated wetland loss induced by hydrocarbon production, Gulf Coast region, USA: Environmental Geology, v. 50, p. 261–274, doi: 10.1007/s00254-006-0207-3.

Penland, S., and Ramsey, K.E., 1990, Relative sea-level rise in Louisiana and the Gulf of Mexico 1908–1988: Journal of Coastal Research, v. 6, p. 323–342.

Shinkle, K.D., and Dokka, R.K., 2004, Rates of Vertical Displacement of Benchmarks in the Lower Mississippi Valley and the Northern Gulf of Mexico: National Oceanic and Atmospheric Administration (NOAA) Technical Report NOS/NGS 50, 135 p.

The Geological Society of America
Special Paper 443
2008

A new composite Holocene sea-level curve for the northern Gulf of Mexico

K.T. Milliken
John B. Anderson
Department of Earth Science, Rice University, 6100 Main Street, MS 126, Houston, Texas 77005, USA

Antonio B. Rodriguez
Institute of Marine Sciences, University of North Carolina at Chapel Hill, 3431 Arendell Street, Morehead City, North Carolina 28557, USA

ABSTRACT

A composite Holocene sea-level curve for the northern Gulf of Mexico coastal region was constructed using basal or bayline peat and swash-zone deposits determined from careful facies analysis. Initial work involved an assessment of published sea-level datums for the region, which show wide scatter. Reevaluation of individual data points based on sea-level–specific criteria required exclusion of most of these data from the composite curve. Additionally, variations in the age of the radiocarbon reservoir across the region were determined, and site-specific corrections were applied to the sample ages, further increasing the accuracy of the final sea-level curve, which is used in our reconstruction of the evolution of the bays of the region in companion papers herein.

The refined regional sea-level curve indicates that site-to-site variations in relative sea-level rise across the coastal plain spanning central Texas to Alabama are minimal. The data indicate rapid and possibly episodic rise during the early Holocene, followed by slower and more continuous rise during the middle and late Holocene. In addition, the new composite regional curve unequivocally plots sea level from −10 to −3 m below present from 8000 cal B.P. to 4000 cal B.P., which precludes the occurrence of a middle Holocene highstand above present sea level in this region. Last, a comparison of our composite curve to Caribbean sea-level curves suggests that coastal subsidence within the study area over the past several thousand years has been minimal. A comparison of the late Holocene record of sea-level rise to satellite altimetry and tide-gauge records indicates that rates of rise have increased by an order of magnitude over the past century.

Keywords: Quaternary, peat, incised valley, subsidence, eustasy.

Milliken, K.T., Anderson, J.B., and Rodriguez, A.B., 2008, A new composite Holocene sea-level curve for the northern Gulf of Mexico, *in* Anderson, J.B., and Rodriguez, A.B., eds., Response of Upper Gulf Coast Estuaries to Holocene Climate Change and Sea-Level Rise: Geological Society of America Special Paper 443, p. 1–11, doi: 10.1130/2008.2443(01). For permission to copy, contact editing@geosociety.org. ©2008 The Geological Society of America. All rights reserved.

INTRODUCTION

Previous studies have shown that rates of shoreline retreat along the Texas coast during the Holocene were episodic (Rodriguez et al., 2004). This volume presents results that show dramatic changes in the evolution of Gulf Coast estuaries during the past 10,000 yr. These changes are manifest as flooding surfaces, which mark landward shifts in depositional environments, typically separating upper estuary (e.g., bayhead delta) sedimentary facies below from marine and open-estuary sedimentary facies above. The rate at which some of these flooding events occurred appears to have been quite rapid, resulting in significant migration of estuarine and coastal environments. In some cases, bayhead deltas shifted up-valley tens of kilometers within centuries. In other examples, the areas of estuaries increased by 20% to 30% within a few centuries. Causes of flooding events include inundation of irregular antecedent topography, variations in the rate of sea-level rise, and changes in sediment supply. Of these, sea-level rise has been the overriding control on estuary evolution. The effects of antecedent topography and sediment supply vary across the study area and are addressed in the individual papers in this volume. This paper presents the composite sea-level curve used to evaluate sea-level rise as a cause for flooding events during the Holocene. This curve is based on a careful screening of published sea-level data from the northern Gulf of Mexico and integration into a database with 138 new data points from our study sites (Fig. 1).

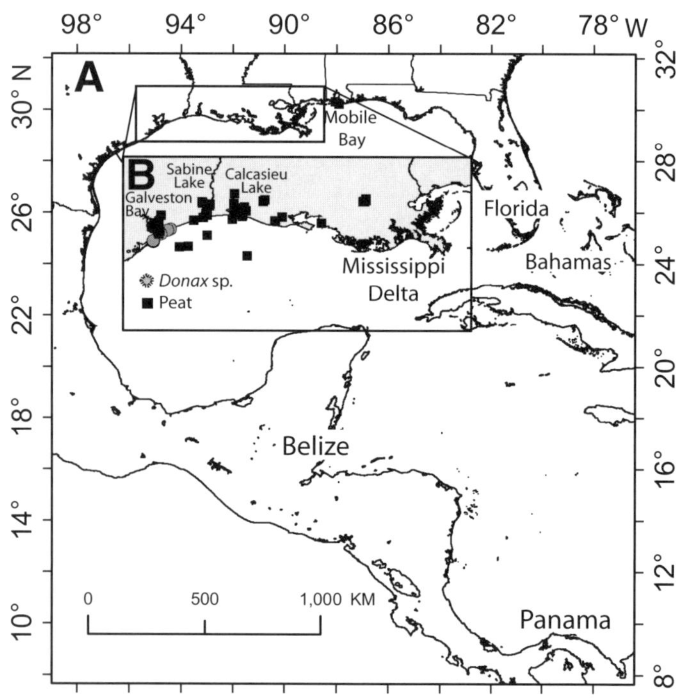

Figure 1. Location map of Texas–Louisiana Gulf Coast area showing individual sampling locations.

Several studies document dramatically increased rates of subsidence along the Louisiana and Texas Gulf Coast within historical times, but they disagree on the causal mechanism (Penland et al., 1989; Shinkle and Dokka, 2004; Morton et al., 2006). Regardless of the source of increased subsidence rates, this area will likely be most impacted by predicted accelerated sea-level rise during the next century (Penland et al., 1989; Roberts et al., 1994; White et al., 2002; Morton et al., 2006). Our goal is to allow coastal scientists to relate changes in bay evolution to variations in the rate of sea-level rise of the past 10,000 yr, and to other factors such as changes in sediment supply. A second objective is to compare long-term rates of eustasy and subsidence to modern rates, necessitating the separation of the eustatic record from subsidence. It is our hope that these results will be used to predict future coastal change and/or shoreline migration resulting from accelerated sea-level rise over the next century and beyond, and to increase public awareness of the vulnerability of the low-gradient Gulf Coastal Plain to rapid landward translation of both gulf and estuarine shorelines.

METHODS

The accuracy of the composite sea-level curve is dependent on the paleoenvironmental specificity of the sea-level indicators used and on the accuracy of the radiocarbon ages. The following is a summary of the methods used to select and refine each of the data sets.

Conventional radiocarbon ages (yr B.P.) compiled from the regional literature were obtained on organic carbon and calcium carbonate material according to the methods outlined in Stuiver and Polach (1977). Conventional radiocarbon age implies use of the Libby half-life (5568 yr), use of oxalic acid as a standard, the year 1950 as the base year, normalization of $\delta^{13}C$ to −25‰ to account for isotopic fractionation, and assumption of constant atmospheric ^{14}C levels during the past. In fact, it has been shown that atmospheric ^{14}C production has fluctuated over time (Stuiver, 1980; Stuiver and Braziunas, 1989; Turnbull et al., 2007), resulting in significant corrections needed to convert conventional radiocarbon ages to calendar years using dendrochronological data and U-Th–dated coral records (Stuiver, 1982; Bard et al., 1990; Stuiver and Reimer, 1993; Stuiver et al., 1998). In addition, many river-estuary-marine organisms utilize CO_2 that has not been in contact with the atmosphere for extended periods (centuries to millennia; e.g., Hogg et al., 1998). This reservoir effect must be accounted for in order to compare radiocarbon dates between marine and terrestrial organisms.

Previous regional studies have investigated potential marine reservoir effects utilizing charcoal-wood pairs, particularly in the Trinity River–Galveston Bay system (Fig. 1) (Aten, 1983; Kibler, 1999). In an attempt to determine the ^{14}C reservoir influence on radiocarbon ages of estuary fauna across the study area, additional radiocarbon dates were obtained on wood-shell pairs sampled from the same stratigraphic interval. The discrepancy between conventional radiocarbon ages in which the calcium carbonate shell material was older than wood or peat was taken to be the reservoir age for that particular fluvial-estuary system (Fig. 2).

Preliminary data plotted in Figure 2 suggest that each estuary has its own unique reservoir effect. There is also a noticeable trend whereby the reservoir effect decreases in age from west to east. The measured reservoir trend corresponds to measured estuary bicarbonate levels (Aten, 1983). Bicarbonate levels diminish from southwest to northeast; averages are 269 mg/L in the Lavaca-Matagorda river-estuary system, 113 mg/L in the Trinity-Galveston river-estuary system, and 37.5 mg/L in the Sabine-Neches river-estuary system (Aten, 1983). The trend of diminishing reservoir from west to east likely correlates to the age and source of river-estuary bicarbonate and has at least two explanations. One explanation involves differences in the geology of the fluvial drainage basins, especially the extent of exposures of old carbonate rocks in contact with water in the basins. An alternative explanation involves increased aridity toward the west.

The modern climate grades from semiarid in the west, which has average annual precipitation of 500 mm (20 in.) received primarily during the spring, to subhumid in the east, which has average annual precipitation of 1500 mm (55 in.) distributed more evenly throughout the year. Average annual temperature is warmest in the southwest, at an average of 24 °C (75 °F), compared to 18 °C (65 °F) in the northeastern portion of the drainage basins. The climatic conditions translate to broad vegetation patterns of shrub and grasslands in the west to woodlands in the east. Soil characteristics broadly follow climate and show greater depth to carbonate in the east as compared to the west (Retallack, 1990; Royer, 1999; Nordt et al., 2006).

The age of pedogenic carbonate could be on the order of several thousand years, related to the age of soil development. Carbonate nodules and calcrete within the soil zone could provide a source of carbon reservoir from the groundwater that flows through the soils and discharges to the rivers and estuaries. Alternatively, the Nueces and Trinity Rivers drain areas of Cretaceous-age rocks, which are composed of significant amounts of carbonates. The Lavaca River system does not drain Cretaceous carbonates but does flow through Tertiary and Pleistocene sediments on which carbonaceous soils have developed in the semiarid landscape. The Sabine-Neches River system drains very little Cretaceous-age limestone, is located in a subhumid landscape with carbonate-leached soils, and has a lower carbon reservoir than the river-estuary systems to the west. Little reservoir data exist for the Calcasieu River system, but climate and geologic data suggest that a very low carbon reservoir should exist in this system.

A literature review of radiocarbon dates from 1956 to present indicated that prior to the early 1980s, analysis and/or reporting of sample $\delta^{13}C$ values and correction for isotopic fractionation was not standard practice. In order to most accurately utilize the extensive radiocarbon data set, all dates have been corrected to modern $\delta^{14}C$ fractionation using the relationship of $Fm = e^{-0.1245x}$, where Fm = fractionation modern and x = yr B.P. or radiocarbon age (Stuiver and Polach, 1977). The fractionation modern values were then normalized to a $\delta^{13}C$ value of –25‰ (assuming a $\delta^{13}C$ value of –27‰) using $Fm_{\delta^{13}C} = Fm\{([1 - 0.025]/[1 + \delta^{13}C/1000])^2\}$. The radiocarbon age was calculated according to the formula, $-8033 \times \ln(Fm_{\delta^{13}C})$ (Stuiver and Polach, 1977). For samples for which the $\delta^{13}C$ value was not reported, a $\delta^{13}C$ range of 10‰ was applied. The resultant age range was added to the error and generally resulted in, at most, a few hundred years error, which is shown in the resulting plots. Applicable calibration curves from the Calib05 program were utilized for various materials depending upon the reported depositional environment. A northern hemisphere terrestrial calibration data set, IntCal05.14C NH, was used for plant and terrestrial material whereas a "global" marine calibration data set, Marine05.14C, was used for shell material. The marine calibration data set has an average reservoir of 370 ± 40 yr built in (Stuiver and Braziunas, 1993). Additional reservoir values were incorporated into the final age calculation using the Calib05 program (Stuiver et al., 2005).

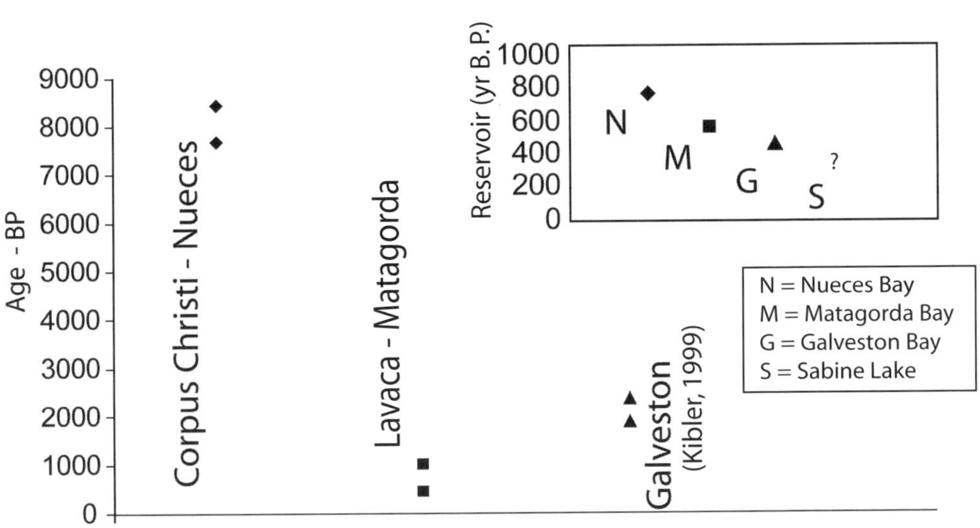

Figure 2. Conventional wood/peat–shell pair ages from various river-bay systems (Maddox et al., this volume; Simms et al., this volume). The age differences, or reservoir effect for each locality, are shown in the inset graph.

Our study included 138 new potential sea-level datums acquired from cores, grab samples, and trenches within Galveston Bay, Sabine Lake, Calcasieu Lake, Mobile Bay, and surrounding coastal environments (Fig. 1). The sampling and analytical methods for the data are presented in companion papers in this volume.

RESULTS

The comprehensive regional sea-level database started out with 138 new samples and ~561 potential sea-level indicators published in previous studies from the Gulf of Mexico shelf and Texas and Louisiana coastal areas and adjacent continental shelf. The data were sorted into broad marine/beach (149 samples), estuary (254 samples), and marsh (288 samples) depositional environments according to observed facies. Age-depth relationships of samples obtained from offshore marine environments showed the widest scatter, likely due to greater rates of subsidence and uncertainties as to the exact environments of deposition. These data were subsequently excluded, with the exception of the *Donax* sp. samples obtained from the swash-zone depositional environment. To facilitate analysis and data display, Figure 3A shows an edited data set in which most marine and beach samples were excluded. The excluded marine data points showed even greater scatter than that observed in Figure 3A. Despite the relatively large scatter in the resulting data, calibrated radiocarbon age-depth relationships for the complete Gulf of Mexico data set show a nonlinear trend, in which decreasing age coincides with

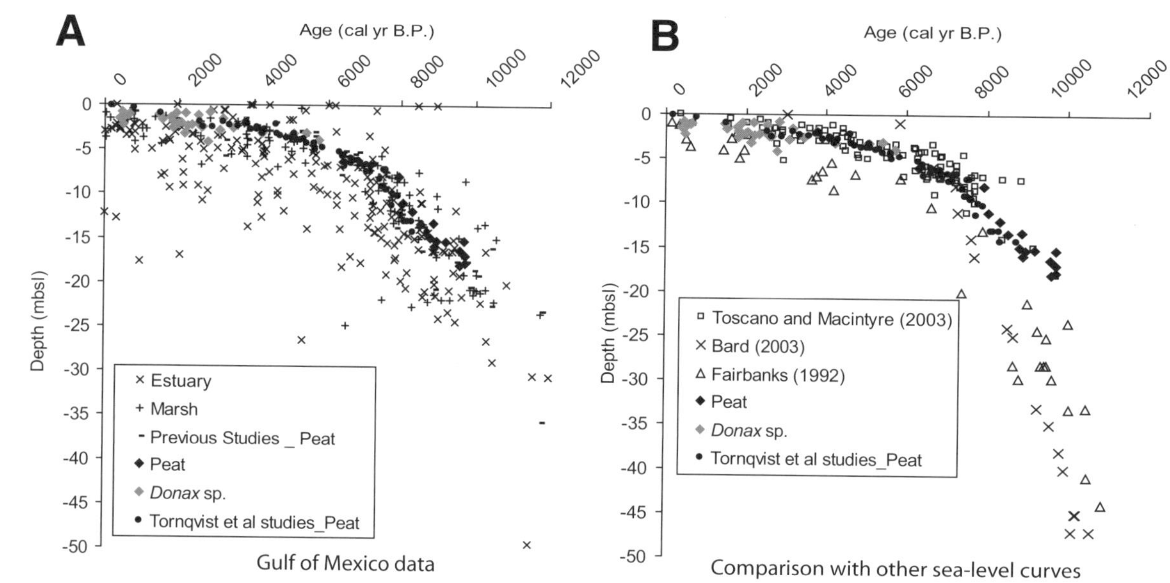

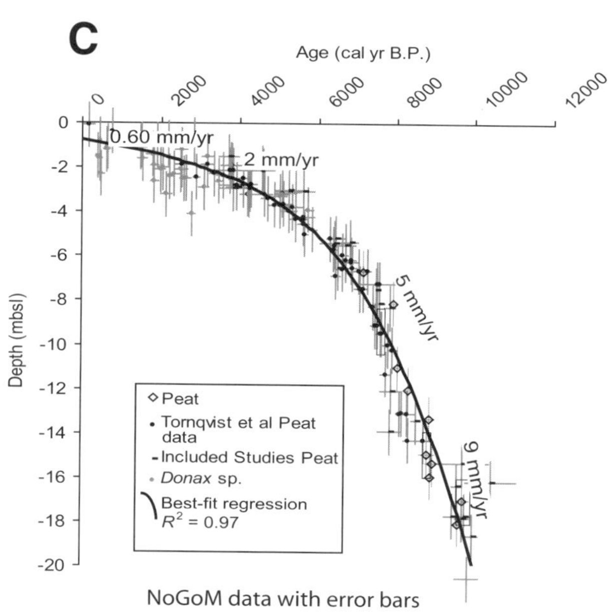

Figure 3. (A) Age-depth relationships of estuary, marsh, and swash-zone depositional environments for the northern Gulf of Mexico (mbsl—meters below sea level), illustrating that peat and swash-zone depositional environments are more useful as sea-level datums than bay samples. (B) Age-depth relationship of northern Gulf of Mexico sea-level indicators (peat and *Donax* sp.), along with sea-level curves from the peripheral Gulf of Mexico, Barbados, Tahiti, and the Sunda Shelf. Comparison of the northern Gulf of Mexico with other sea-level curves shows the similarity of the data from 8000 yr B.P. to present. (C) Age-depth relationship of peat and swash-zone samples and associated horizontal and vertical errors. The best-fit nonlinear regression is drawn (Törnqvist et al., 2004, 2005, 2006). NoGoM—northern Gulf of Mexico.

decreasing depths. A third-order polynomial nonlinear regression was performed on these data, resulting in a goodness of fit statistical value $R^2 = 0.45$. Estuary and marsh samples also show wide scatter due to uncertainties in depth of deposition and degree of compaction. In particular, marsh samples embedded in compactable Holocene deposits commonly plot below the nonlinear trend. A number of estuary and marsh samples yielded anonymously old ages for their depths, suggesting reworking, and so they were also excluded.

For basal peat, selection criteria included core location, core lithofacies and biofacies, and documentation of peat immediately overlying a pedogenically dewatered and compacted Pleistocene surface (Table 1). Nonbasal (166 ages) as well as 13 basal peat ages were excluded, leaving 108 ages from reliable peat samples in the final plot. For shells, Rodriguez et al. (2004) obtained continuous cores from beach ridges on Galveston Island and Bolivar Peninsula and identified facies indicative of swash-zone conditions. Swash-zone deposits are characterized by gently dipping laminations and alternating light (quartz) and heavy mineral layers, and by an indigenous fauna dominated by *Donax* sp. The probable elevation for the swash-zone or region between high and low water levels is at most 1 to 2 m, which is comparable to the basal peat sea-level datum. In addition, swash-zone deposits are interbedded with organic-rich clay that formed in freshwater marshes within the swales between beach ridges. These modern freshwater marshes occur within 2 m of sea level. Rodriguez et al. (2004) used articulated shells for dating wherever possible (Table 2). Thirty-four *Donax* sp. shell ages were incorporated into the sea-level database. Elevations of marsh peat developed on precompacted Pleistocene deposits (inferred basal peat) should be most accurate. As expected, the basal peat and swash-zone samples show the least scatter (Fig. 3B). Nonlinear regression of the final data set with a third-order polynomial best fit gave an R^2 value of 0.97 (Fig. 3C).

DISCUSSION

Numerous studies, beginning with Shepard (1956), have documented relative sea-level rise on the Gulf of Mexico shelf since the Last Glacial Maximum (Shepard, 1956; Gould and McFarlan, 1959; Curray, 1960; McFarlan, 1961; Coleman and Smith, 1964; McEwen, 1969; Rehkemper, 1969; Nelson and Bray, 1970; Frazier, 1974; Pearson et al., 1987; Penland et al., 1989; Thomas, 1990; Anderson et al., 1991; Morton et al., 2000; Törnqvist et al., 2004, 2005, 2006). These studies utilized sea-level datums from offshore, estuary, beach and chenier ridge, marsh, and intertidal swash-zone deposits to construct the history of sea-level change. In this study, we critically examined the published data sets and selected a subset of basal peat samples, corrected for isotopic fractionation where appropriate. If some of the basal peat samples in a particular published data set were suspect, i.e., we could not definitively determine that the age was obtained from peat as compared to organic-rich clay, we did not include the entire data set in the northern Gulf of Mexico curve (Gould and McFarlan, 1959; Rehkemper, 1969; Penland et al., 1989). In addition to basal peat deposits, foreshore (swash-zone) deposits represent the intertidal zone and thus theoretically provide the best sea-level constraints, especially along the Louisiana and Texas coasts, where the spring tidal range is less than 60 cm. However, they are not always easily distinguished from upper shoreface deposits, and *Donax*, which is used to recognize and date foreshore deposits, is commonly reworked into the upper shoreface. This probably accounts for some of the outliers in the *Donax* ages shown in Figure 3B. Radiocarbon-dated mollusks from estuary and shallow-marine deposits provide lower limits to sea level, while shell material extracted from beach and chenier ridges is commonly anomalously old, having been reworked into the subaerial ridges (Gould and McFarlan, 1959; Morton et al., 2000). Basal peat forms within a few meters of mean sea level,

TABLE 1. BASAL AND BAYLINE PEAT DATES

Sample	Lab no.	Ref†	$\delta^{13}C$ (‰)	Conventional ^{14}C age (yr, 2σ)	Calibrated ^{14}C age (yr B.P., 2σ)		Depth‡ (m)	Type	Location	Long (°W)	Lat (°N)
					Lower	Upper					
04cl07_25_30_26	OS-46751	1	–24.5	6240 ± 30	7170	7210	–6.7	Peat	Calcasieu Lake, LA	93.28	30.20
04sl02_30_35_18_21	OS-46747	1	–28.2	7110 ± 45	7860	9020	–8.1	Peat	Sabine Lake, LA	93.78	29.95
04cl05_40_45_8	OS-46750	1	–25.3	7190 ± 40	7890	8110	–11.0	Peat	Calcasieu Lake, LA	93.29	30.00
04cl06_40_45_106	OS-46749	1	–27.3	7540 ± 35	8270	8460	–12.0	Peat	Calcasieu Lake, LA	93.27	29.91
04sl03_50_55_86t	OS-46746	1	–28.2	7910 ± 45	8550	8930	–14.9	Peat	Sabine Lake, LA	93.83	29.87
04cl05_45_50_83	OS-46981	1	–27.4	7980 ± 30	8730	9010	–13.3	Peat	Calcasieu Lake, LA	93.29	30.00
04sl03_50_55_131	OS-46763	1	–26.0	8070 ± 35	8840	9170	–15.3	Peat	Sabine Lake, LA	93.83	29.87
04cl01_60_64_96	OS-46748	1	–26.8	8620 ± 35	9500	9640	–19.2	Peat	Calcasieu Lake, LA	93.29	29.78
04sl01_60_64_10_15	OS-46681	1	–29.0	8720 ± 40	9500	9840	–17.0	Peat	Sabine Lake, LA	93.82	29.77
MS-04-4 (591 cm)	OS-44720	1		5470 ± 50	6210	7310	–5.9	Peat	Mobile Bay, AL		
MB-03-4 (1514 cm)	OS-42942	1		8270 ± 60	9200	9320	–15.1	Peat	Mobile Bay, AL		
SB93_7_431	BETA65497	2	–20.2	8677 ± 270	8460	8640	–15.3	Peat	Sabine Bank	93.82	29.37
HB_93_7_330	BETA65496	2	–15.1	8570 ± 70	9480	9560	–16.3	Peat	Heald Bank	94.20	29.15
SWL00_4_160	AA39388	2	–1.2	2999 ± 41	2720	2800	–1.6	Taselus	Galveston Island, TX	94.89	29.26
TIV_1_4_1340–1342	AA39403	3	–28.1	7738 ± 60	8445	8520	–13.4	Wood	Galveston Bay, TX	94.73	29.77
GB_01_1_1781–1785	AA44773	3	–28.6	8702 ± 61	9550	9740	–17.8	Wood	Galveston Bay, TX	94.87	29.53

Note: LA—Louisiana, AL—Alabama, TX—Texas.
†Published reference: 1—new data; 2—Rodriguez et al. (2004); 3—Rodriguez et al. (2005).
‡Meters relative to mean sea level.

TABLE 2. *DONAX* SP. DATES

Sample	Lab no.	$\delta^{13}C$ (‰)	Conventional ^{14}C age (yr, 2σ)	Calibrated ^{14}C age (yr B.P., 2σ) Lower	Upper	Depth† (m)	Type	Location	Long (°W)	Lat (°N)
BP_00_6_157	AA39407	−1.4	523 ± 35	110	170	−1.6	*Donax*	Bolivar Peninsula, TX	94.73	29.39
BP_00_6_234	AA39401	−0.4	581 ± 38	200	240	−2.3	*Donax*	Bolivar Peninsula, TX	94.73	29.39
BP_00_4_115	AA43895	−0.7	764 ± 42	360	410	−1.2	*Donax*	Bolivar Peninsula, TX	94.73	29.40
BP_00_1_106	AA43897	−0.6	783 ± 40	400	440	−1.1	*Donax*	Bolivar Peninsula, TX	94.74	29.40
BP_00_3_106	BETA163926	−0.8	1610 ± 60	1100	1170	−1.1	*Donax*	Bolivar Peninsula, TX	94.74	29.40
PIT1-2_1/4	AA44793	−0.5	1669 ± 42	1200	1240	−1.6	*Donax*	Bolivar Peninsula, TX	94.67	29.45
PIT2-1_1/2	AA44789	−0.9	1885 ± 57	1370	1440	−1.4	*Donax*	Bolivar Peninsula, TX	94.64	29.46
BP_00_13_261	AA39377	−0.4	1926 ± 37	1430	1480	−2.6	*Donax*	Bolivar Peninsula, TX	94.59	29.47
BP_00_3_191	AA39396	0.1	1932 ± 37	1430	1480	−1.9	*Donax*	Bolivar Peninsula, TX	94.74	29.40
PIT1-3	AA44780	−0.4	2000 ± 36	1520	1560	−1	*Donax*	Bolivar Peninsula, TX	94.67	29.45
PIT3-4	AA44779	0.4	2044 ± 45	1560	1620	−2.1	*Donax*	Bolivar Peninsula, TX	94.55	29.49
BP_00_3_202	AA43896	−1.0	2115 ± 44	1630	1690	−2	*Donax*	Bolivar Peninsula, TX	94.74	29.40
PIT3-2	AA44782	−0.2	2150 ± 40	1690	1740	−3.2	*Donax*	Bolivar Peninsula, TX	94.55	29.49
SWL00_7_93	AA43893	0.6	2206 ± 47	1750	1810	−0.9	*Donax*	Galveston Island, TX	94.88	29.24
SWL00_7_244	AA39398	−0.2	2237 ± 38	1800	1850	−2.4	*Donax*	Galveston Island, TX	94.88	29.24
BP_00_12_230	AA39394	−0.2	2342 ± 34	1910	1950	−2.3	*Donax*	Bolivar Peninsula, TX	94.59	29.47
SWL00_6_209	AA39389	0.6	2457 ± 35	2050	2100	−2.1	*Donax*	Galveston Island, TX	94.88	29.24
SWL00_6_123	AA43887	0.0	2476 ± 40	2060	2120	−1.2	*Donax*	Galveston Island, TX	94.88	29.24
SWL00_1_250	AA39405	0.5	2537 ± 38	2160	2220	−2.5	*Donax*	Galveston Island, TX	94.88	29.25
BP_00_2_252	AA39380	0.3	2577 ± 37	2210	2260	−2.5	*Donax*	Bolivar Peninsula, TX	94.74	29.40
SWL00_1_407	AA43888	0.3	2673 ± 47	2310	2360	−4.1	*Donax*	Galveston Island, TX	94.88	29.25
BP_00_5_79	AA43898	−0.4	2745 ± 43	2350	2430	−0.8	*Donax*	Bolivar Peninsula, TX	94.73	29.39
BP_00_1_209	AA39392	0.6	2948 ± 35	2710	2750	−2.9	*Donax*	Bolivar Peninsula, TX	94.74	29.40
SWL00_2_146	AA39306	0.0	3013 ± 40	2740	2780	−1.5	*Donax*	Galveston Island, TX	94.88	29.25
SWL00_2_256	AA39397	0.6	3241 ± 38	2990	3050	−2.6	*Donax*	Galveston Island, TX	94.88	29.25
SWL00_5_185	AA43891	0.5	3513 ± 93	3290	3400	−1.9	*Donax*	Galveston Island, TX	94.88	29.26
SWL00_4_319	AA39382	0.2	3828 ± 40	3720	3780	−3.2	*Donax*	Galveston Island, TX	94.89	29.26
SWL00_3_328	AA39408	−4.9	4450 ± 43	4540	4620	−3.3	*Donax*	Galveston Island, TX	94.89	29.26
SWL00_5_316	AA39387	−0.4	4486 ± 44	4610	4670	−3.2	*Donax*	Galveston Island, TX	94.88	29.26
BP_00_4_314	AA43892	−0.6	4756 ± 43	4900	4980	−3.1	*Donax*	Bolivar Peninsula, TX	94.73	29.40
SWL00_4_387	AA43894	1.6	5052 ± 52	5330	5400	−3.9	*Donax*	Galveston Island, TX	94.89	29.26

Note: TX—Texas. Published reference: Rodriguez et al. (2004, 2005).
†Meters relative to mean sea level.

accumulates on exposure surfaces, and subsequently becomes altered, compacted, and consolidated (e.g., Törnqvist et al., 2004). Peat types in this study range from freshwater to brackish water swamp and marsh environments (Chmura et al., 1987).

Continuous versus Episodic Rise

It is important to note that error bars on the curve limit resolution to about ±1 m. Lower-magnitude sea-level fluctuations could potentially produce dramatic changes on the low-gradient coast. This being said, there are some key problems that can be addressed with the northern Gulf of Mexico curve (Figs. 3C and 4). One of these is a long-standing debate about whether Holocene sea-level rise in the northern Gulf of Mexico was episodic or continuous (e.g., Curray, 1960; Nelson and Bray, 1970; Frazier, 1974; Pearson et al., 1987; Penland et al., 1989; Thomas and Anderson, 1994). Geologic evidence for these rapid rises in sea level should occur contemporaneously across the study area, since a rapid rise will transcend hydro-isostatic and tectonic effects (Simms et al., 2007). Furthermore, the timing of these events should coincide with episodes of significant ice-sheet decay, although the details of ice-sheet retreat history are still poorly constrained.

We found several flooding events in the sedimentological record during the early Holocene that appear to have occurred contemporaneously across the different study areas, at least within the limits of radiocarbon errors and ^{14}C reservoir variability across the northern Gulf Coast region. These flooding events occurred at ca. 9500–9800 cal yr B.P., ca. 8500–8900 cal yr B.P., ca. 8000–8400 cal yr B.P., and ca. 6800–7400 cal yr B.P. (Fig. 4). All of these events occurred when sea level was rising rapidly and may have been rising episodically (Fig. 4). The most recent event, at ca. 6800–7400 cal yr B.P., is also known to have coincided with a major climate transition in the region, which amplified the environmental changes that occurred in estuaries of the east Texas coast (Bryant and Holloway, 1985; Meltzer, 1991; Toomey et al., 1993; Humphrey and Ferring, 1994; Nordt et al., 1994, 2002).

The early Holocene was indeed a time of episodic retreat of ice sheets in both hemispheres, although the record of ice-sheet retreat is still too sparse to allow direct correlation between specific sea-level events and ice-sheet history (e.g., Blanchon and Shaw, 1995; Alley et al., 1997; Anderson et al., 2002; Carlson et al., 2007; Heroy and Anderson, 2007). In general, the rapid rise that ended ca. 9600 cal B.P. coincided with retreat of the West Antarctic Ice Sheet from the inner continental shelf of the Amundsen Sea and Bellingshausen Sea (Anderson et al., 2002; Lowe and Anderson, 2002) and retreat of the Antarctic Peninsula Ice Sheet from the inner continental shelf (Heroy and Anderson, 2007). The ca. 8200 cal B.P. event coincided with a rapid sea-level rise that was interpreted by Törnqvist et al. (2005) to have been associated with mass wasting of the Laurentide Ice Sheet (Alley et al., 1997). Additionally, an event related to the final collapse of the Laurentide Ice Sheet is age constrained to ca. 7600 cal B.P. (Blanchon and Shaw, 1995; Carlson et al., 2007).

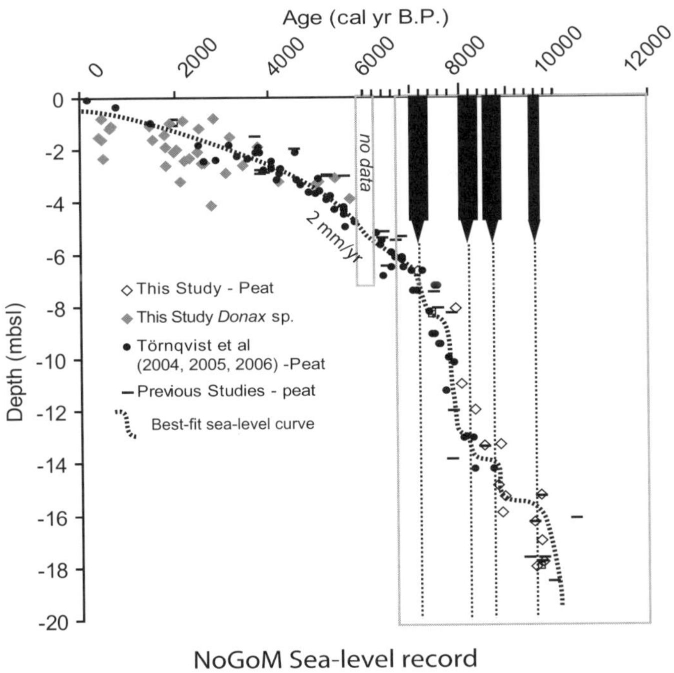

Figure 4. Age-depth relationships of peat and *Donax* sp. ages compared to previously published studies (mbsl—meters below sea level). A nonlinear curve is drawn that takes into account vertical and horizontal error bars, which are not shown in this figure. Given the paucity of data points for the early Holocene portion of the curve, an episodic rise in sea level is indicated for this time interval. Early Holocene flooding events that occur in all studied estuaries are marked with black arrows. The width of the arrow indicates uncertainty in timing of a flooding interval. NoGoM—northern Gulf of Mexico.

Middle Holocene Highstand

Another ongoing controversy concerns the interpretation of a mid-Holocene sea-level highstand above its present level (Morton et al., 2000; Blum et al., 2001; Törnqvist et al., 2004). The new northern Gulf of Mexico curve (Fig. 3C) plots sea level between −10 m and −3 m from 8000 cal B.P. to 4000 cal B.P. Given the density of quality-controlled data used to establish this record, the maximum possible rapid rise and fall during any as-yet-unresolved sea-level events would be less than 2 m during this interval. Thus, the new composite sea-level curve for the region records no occurrence of a mid-Holocene highstand in this area of the northern Gulf of Mexico. Although there is a data gap between ca. 5900 and 6300 cal B.P., four centuries is hardly enough time to raise and lower sea level ~6 m, which would call for complete deglaciation of West Antarctica followed by reestablishment of the ice sheet.

Subsidence versus Eustasy

The age-depth plot for the study area melds subsidence and eustasy and thus provides a cumulative measure of relative sea level. Our approach to separating eustasy from subsidence was to relate the regional curve for the northwestern Gulf of Mexico to the widely accepted eustatic curve from Barbados (Fairbanks, 1989, 1992; Bard et al., 1990). We also included the widely cited sea-level curves derived from Tahiti and the Sunda Shelf for comparison (Bard et al., 1996; Hanebuth et al., 2000; Bard, 2003). Data obtained from the periphery of the Gulf of Mexico basin include sites from Florida, Belize, and Panama (Toscano and Macintyre, 2003). The peripheral Gulf of Mexico data set is particularly pertinent as a means to compare our composite sea-level record to sites located on slowly subsiding carbonate platforms (Toscano and Macintyre, 2003). Note that the Barbados data typically plot up to 5 m below the other data in Figure 3B (Fairbanks, 1989, 1992). This reflects the depth range of the corals used to acquire these data points.

In making the comparison between the northern Gulf of Mexico curve and generally accepted eustatic curve from Barbados (Fairbanks, 1992; Bard et al., 1996; Bard, 2003), it is important to consider isostatic effects, which can result in significant regional differences in the magnitude of sea-level change (Lambeck and Chappell, 2001). The northern Gulf of Mexico is subject to both loading from the Mississippi Delta and fan complex and expansion and contraction of the Laurentide Ice Sheet (Clark et al., 1978; Lambeck and Chappell, 2001; Ivins et al., 2005, 2007; Simms et al., 2007). The results obtained from numerical modeling studies suggest that Holocene water loading has minimal effects on modern subsidence as compared to Holocene and Pleistocene deltas loading the shelf, which produced 1.0–1.5 mm/yr long-term subsidence on the inner shelf (Ivins et al., 2005, 2007). Simms et al. (2007) showed that the western Louisiana and east Texas coasts lie beyond the limits of down-loading by the Holocene Mississippi Delta complex. They also found that the effects of water loading on the Gulf of Mexico shelf due to deglaciation were minimal after ca. 8000 cal B.P.

Utilizing the observation that eustatic rise ranged from 0.2 to 1.2 mm/yr during the past 4000 yr (Clark et al., 1978; Fairbanks, 1992; Bard et al., 1996; Bard, 2003; Toscano and Macintyre, 2003), the sinking and burial of marsh and swash-zone deposits at higher rates will largely be a function of subsidence. Our data for the past 4000 yr show scatter in the peat, *Donax* depths of <3 m, and age errors on the order of two centuries (Fig. 5A). Statistical treatments of data from the northern Gulf of Mexico compared to the data published by Toscano and Macintyre (2003) and Törnqvist et al. (2004, 2005, 2006) are shown in Figure 5B. Peat datums from previous studies (referenced previously) are also shown as bars in Figure 5A, but they were not included in the linear regressions. The dark-gray line, defined by the equation, depth (mm) = (0.37 ± 0.14) × age (cal B.P.) + (1181 ± 314), shows an average rate of 0.4–0.6 mm/yr and an extrapolated depth of 1.18 m for swash-zone environments (Fig. 5B). The light-gray line, depth (mm) = (0.55 ± 0.2) × age (cal B.P.) + (254 ± 586), illustrates the linear regression of marsh and peat deposits of Toscano and Macintyre (2003) from the periphery of the Gulf of Mexico basin, i.e., Florida

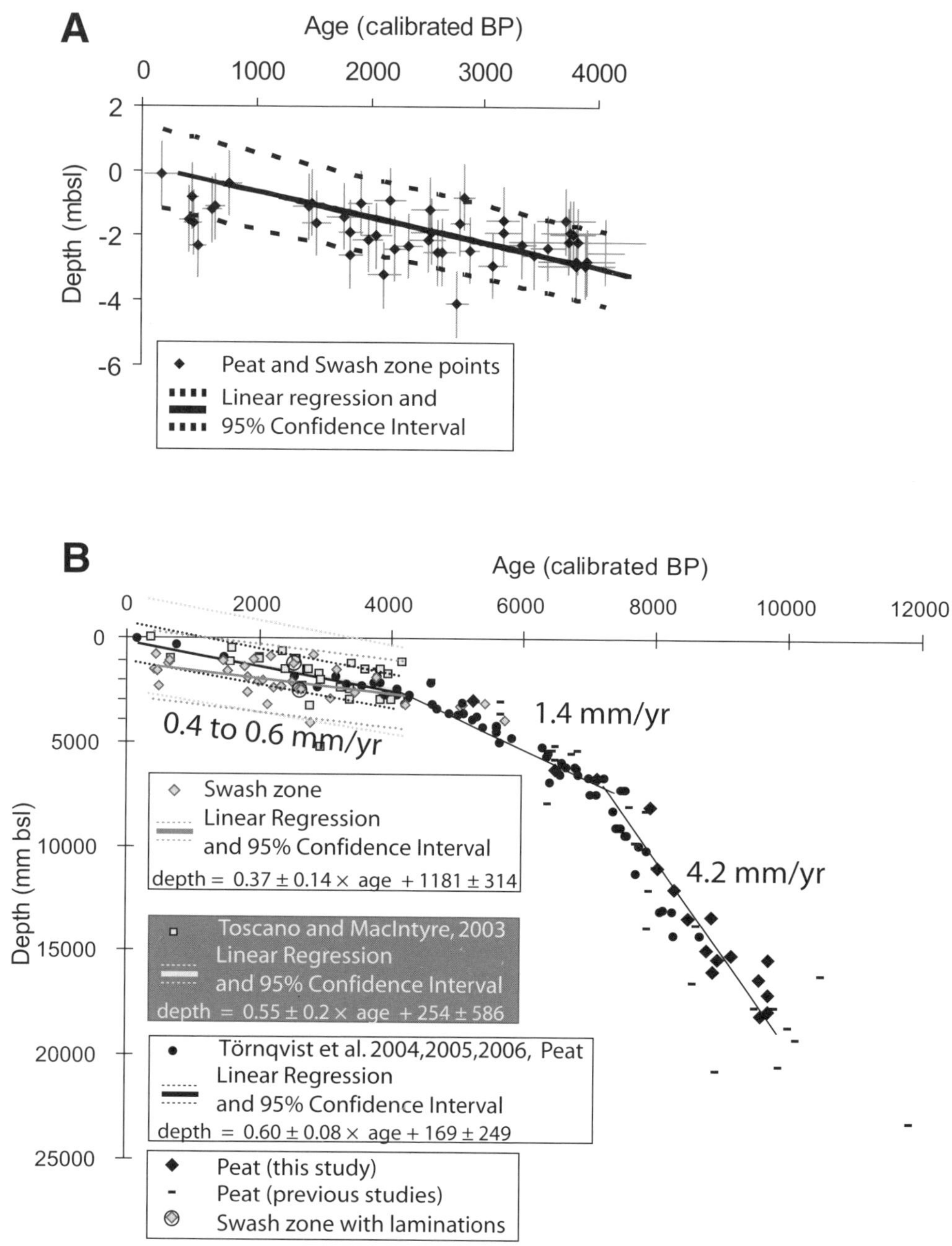

Figure 5. (A) Linear regression and 95% confidence interval brackets for peat and swash-zone samples showing the vertical resolution of this data set (mbsl—meters below sea level). (B) Linear relationship of the past 4000 yr of data for the northern Gulf of Mexico data (gray diamonds, black circles, black diamonds, dashes) compared to peripheral Gulf of Mexico data (gray squares) published by Toscano and Macintyre (2003). The similarity of the linear regressions constrains the subsidence rates along the northern Gulf of Mexico shoreline over the past 4000 yr.

Keys, Belize, and Panama. The black line, depth (mm) = (0.60 ± 0.08) × age (cal B.P.) + (169 ± 249), shows the linear regression obtained using the peat data published by Törnqvist et al. (2004, 2005, 2006) as a comparison. The sea-level rise around the periphery is similar to the relative sea-level rise of the study area within a 95% confidence interval. These three data sets illustrate the similarity of relative sea-level rise at all stable locations around the Gulf of Mexico basin. Therefore, subsidence along the northern Gulf Coast from ca. 4000 to 100 cal yr B.P. was a minor component of the observed relative sea-level rise for that time interval.

Comparison of Holocene to Modern Sea-Level Rise

The relative sea-level rise rate (0.4–0.6 mm/yr) derived from the sedimentary record for the time interval between 4000 and 100 cal B.P. is significantly less than tide gauge-derived rates of 2–5 mm/yr (Gornitz and Lebedeff, 1987; Paine, 1993; Gornitz, 1995; Penland and Ramsey, 1990; Douglas, 1991, 1997; Zervas, 2001; Miller and Douglas, 2004; Morton et al., 2006), and it is an order of magnitude less than the 2005 National Geodetic Survey (NGS) releveling/benchmark-derived rates (10–15 mm/yr) of Shinkle and Dokka (2004). Thus, tide gauge– and geodetic survey–derived subsidence rates cannot be reconciled with millennial scale subsidence along the northern Gulf Coast. The decadal- to century-scale tide and geodetic survey records reveal high-frequency perturbations superimposed on otherwise very slow subsidence, extremely high modern sea-level rise rates, or a very recent phenomenon related to anthropogenic activity. None of the mechanisms can be ruled out by our data set, but high-frequency oscillations of ~1.5 m/century (Shinkle and Dokka, 2004) seem untenable given the regional extent of present-day subsidence. It is more plausible that the current regional phenomenon of rapid modern subsidence is a result of large-scale fluid extraction from the subsurface (Morton et al., 2006). In any case, accelerating rates of relative sea-level rise along the west Louisiana and east Texas coasts, coupled with reduced sediment supply due to dams and other water management projects, will heavily impact the coast in the coming decades, particularly because decimeter-scale rapid rises along low-gradient coasts could result in major lateral shifts in coastal environments.

CONCLUSIONS

A new composite sea-level curve for the northern Gulf of Mexico shows rapid, and possibly episodic, rise during the early Holocene followed by a slow and continuous rise during the middle and late Holocene. This regionally consistent sea-level history provides a framework to assess whether differential sea-level rise was a controlling factor for shoreline configuration and estuary evolution, or whether differences in shoreline and bayline retreat and advance can be ascribed to other forcing mechanisms, such as climate, sediment supply, and/or antecedent topography. The results indicate that rapid sea-level rise dominated estuary evolution during the early Holocene and that apparently contemporaneous flooding episodes across the northern Gulf Coast may have been caused by episodic sea-level events. The changes that occurred during the middle and late Holocene were diachronous and localized and resulted from variations in antecedent topography and sediment supply. The composite sea-level curve shows that sea level was between −10 and −3 m below present during the middle Holocene (8000 cal B.P. to 4000 cal B.P.) and did not rise to modern levels during this time interval.

The influence of relative sea-level rise (subsidence and eustasy) on Gulf of Mexico shoreline and bayline configuration over the past 4000 yr can be constrained by comparison of regional and global sea-level curves. Late Holocene differential relative sea-level rise across the northern Gulf of Mexico was minimal and on the order of ~0.4–0.6 mm/yr. Although modern relative sea-level rise rates (2–15 mm/yr) are significantly greater than these millennial-scale rates, the errors associated with measurement of geologic rates cannot rule out low-frequency, high-amplitude periods of enhanced relative sea-level rise. With this said, the current rate of rise appears to be unprecedented for the past 4000 yr. If current rates continue over several decades, resulting changes to future shoreline and bayline configurations will be dramatic.

ACKNOWLEDGMENTS

We thank Chip Anderson, Jessica Maddox, Patrick Taha, Alex Simms, and Craig Meyers for field assistance. Funding was provided by the National Science Foundation EAR-0107650 grant and the David Worthington research fund. The manuscript benefited from reviews by M. Toscano and H.R. Wanless.

REFERENCES CITED

Alley, R.B., Mayewski, P.A., Sowers, T., Stuiver, M., Taylor, K.C., and Clark, P.U., 1997, Holocene climatic instability: A prominent widespread event 8200 yr ago: Geology, v. 25, p. 483–486, doi: 10.1130/0091-7613(1997)025<0483:HCIAPW>2.3.CO;2.

Anderson, J.B., Siringan, F.P., Taviani, M., and Lawrence, J., 1991, Origin and evolution of Sabine Lake, Texas–Louisiana: Transactions of the Gulf Coast Association of Geological Societies, v. 41, p. 12–16.

Anderson, J.B., Shipp, S.S., Lowe, A.L., Wellner, J.S., and Mosola, A.B., 2002, The Antarctic Ice Sheet during the Last Glacial Maximum and its subsequent retreat history: A review: Quaternary Science Reviews, v. 21, p. 49–70, doi: 10.1016/S0277-3791(01)00083-X.

Aten, L.E., 1983, Indians of the Upper Texas Coast: New York, Academic Press, 370 p.

Bard, E., 2003, Tahiti Deglacial Relative Sea-level Reconstruction: Boulder, Colorado, IGBP Pages/World Data Center for Paleoclimatology Data Contribution Series 2003-028, NOAA/NGDC Paleoclimatology Program, ftp://ftp.ncdc.noaa.gov/pub/data/paleo/paleocean/relative_sea_level/tahiti/bard1996.txt (last accessed 30 May 2008).

Bard, E., Hamelin, B., Fairbanks, R.G., and Zindler, A., 1990, Calibration of the ^{14}C timescale over the past 30,000 years using mass spectrometric U-Th ages from Barbados corals: Nature, v. 345, p. 405–409, doi: 10.1038/345405a0.

Bard, E., Hamelin, B., Arnold, M., Montaggioni, L., Cabioch, G., Faure, G., and Rougerie, F., 1996, Sea-level record from Tahiti corals and the timing of deglacial meltwater discharge: Nature, v. 382, p. 241–244, doi: 10.1038/382241a0.

Blanchon, P., and Shaw, J., 1995, Reef drowning during the last deglaciation: Evidence for catastrophic sea-level rise and ice-sheet collapse: Geology, v. 23, no. 1, p. 4–8, doi: 10.1130/0091-7613(1995)023<0004:RDDTLD>2.3.CO;2.

Blum, M.D., Misner, T.J., Collins, E.C., Scott, D.B., Morton, R.A., and Aslan, A., 2001, Middle Holocene sea-level rise and highstand at + 2 m, central Texas coast: Journal of Sedimentary Research, v. 71, p. 581–588, doi: 10.1306/112100710581.

Bryant, V.M., and Holloway, R.W., 1985, The Late Quaternary paleoenvironmental record of Texas, *in* Bryant, V.M., and Holloway, R.W., eds., Pollen Records of Late Quaternary North American Sediments: Dallas, Texas, American Association of Stratigraphers and Palynologists, p. 39–70.

Carlson, A.E., Clark, P.U., Raisbeck, G.M., and Brook, E.J., 2007, Rapid Holocene deglaciation of the Labrador sector of the Laurentide Ice Sheet: Journal of Climate, v. 20, p. 5126–5130, doi: 10.1175/JCLI4273.1.

Chmura, G.L., Aharon, P., Socki, R.A., and Abernethy, R., 1987, An inventory of ^{13}C abundances in coastal wetlands of Louisiana, USA: Vegetation and sediments: Oecologia, v. 74, no. 2, p. 264–271, doi: 10.1007/BF00379369.

Clark, J.A., Farrell, W.E., and Peltier, W.R., 1978, Global changes in postglacial sea level: A numerical calculation: Quaternary Research, v. 9, p. 265–287, doi: 10.1016/0033-5894(78)90033-9.

Coleman, J.M., and Smith, W.G., 1964, Late Recent rise of sea level: Geological Society of America Bulletin, v. 75, p. 833–840, doi: 10.1130/0016-7606(1964)75[833:LRROSL]2.0.CO;2.

Curray, J.R., 1960, Sediments and history of Holocene transgression, continental shelf, northwest Gulf of Mexico, *in* Shepard, F.P., Phleger, F.B., and van Andel, T.H., eds., Recent Sediments, Northwest Gulf of Mexico: Tulsa, Oklahoma, American Association of Petroleum Geologists, p. 221–266.

Douglas, B.C., 1991, Global sea-level rise: Journal of Geophysical Research, v. 96, p. 6981–6992, doi: 10.1029/91JC00064.

Douglas, B.C., 1997, Global sea rise: A redetermination: Surveys in Geophysics, v. 18, p. 279–292, doi: 10.1023/A:1006544227856.

Fairbanks, R.G., 1989, A 17,000-year glacio-eustatic sea-level record: Influence of glacial melting rates on the Younger Dryas event and deep-ocean circulation: Nature, v. 342, p. 637–642, doi: 10.1038/342637a0.

Fairbanks, R.G., 1992, Barbados Sea Level and Th/U ^{14}C Calibration: Boulder, Colorado, IGBP PAGES/World Data Center for Paleoclimatology Data Contribution Series 92-020: NOAA/NGDC Paleoclimatology Program, ftp://ftp.ncdc.noaa.gov/pub/data/paleo/contributions_by_author/fairbanks1990/barbados_1990.txt (last accessed 30 May 2008).

Frazier, D.E., 1974, Depositional Episodes: Their Relationship to the Quaternary Stratigraphic Framework in the Northwest Portion of the Gulf Basin: Austin, Texas, Bureau of Economic Geology, Geologic Circular 74-1, 28 p.

Gornitz, V., 1995, Sea-level rise: A review of recent past and near-future trends: Earth Surface Processes and Landforms, v. 20, p. 7–20, doi: 10.1002/esp.3290200103.

Gornitz, V., and Lebedeff, S., 1987, Global sea level changes during the past century, *in* Nummedal, D., Pilkey, O.H., and Howard, J.D., eds., Sea-Level Fluctuation and Coastal Evolution: Society for Economic Paleontologists and Mineralogists (SEPM) Special Publication 41, p. 3–16.

Gould, H.R., and McFarlan, E., 1959, Geologic History of the chenier plain, southwestern Louisiana: Gulf Coast Association of Geological Societies Transactions, v. 9, p. 261–272.

Hanebuth, T., Stattegger, K., and Pieter, G., 2000, Rapid flooding of the Sunda Shelf; a late-glacial sea-level record: Science, v. 288, no. 5468, p. 1033–1035, doi: 10.1126/science.288.5468.1033.

Heroy, D.C., and Anderson, J.B., 2007, Radiocarbon constraints on Antarctic Peninsula ice sheet retreat following the Last Glacial Maximum: Quaternary Science Reviews, v. 26, p. 3286–3297, doi: 10.1016/j.quascirev.2007.07.012.

Hogg, A.G., Higham, F.G., and Dahm, J., 1998, ^{14}C dating of modern marine and estuarine shellfish: Radiocarbon, v. 40, no. 2, p. 975–984.

Humphrey, J.D., and Ferring, C.R., 1994, Stable isotopic evidence for latest Pleistocene and Holocene climate change in north-central Texas: Quaternary Research, v. 41, p. 200–213, doi: 10.1006/qres.1994.1022.

Ivins, E.R., Blom, R.G., Wu, X., and Dokka, R., 2005, Late Pleistocene sedimentation increase in the Gulf of Mexico and modeling of time-dependent long-wavelength flexural subsidence: Eos (Transactions, American Geophysical Union), v. 86, no. 52, Fall meeting supplement, abstract G21D-05.

Ivins, E.R., Dokka, R.K., and Blom, R.G., 2007, Post-glacial sediment load and subsidence in coastal Louisiana: Geophysical Research Letters, v. 34, p. L16303, doi: 10.1029/2007GL030003.

Kibler, K.W., 1999, Radiocarbon dating of Rangia cuneata: Correction factors and calibrations for the Galveston Bay area: Texas Archaeological Society Publication, v. 70, p. 457–466.

Lambeck, K., and Chappell, J., 2001, Sea-level change through the Last Glacial cycle: Science, v. 292, p. 679–686, doi: 10.1126/science.1059549.

Lowe, A.L., and Anderson, J.B., 2002, Late Quaternary advance and retreat of the West Antarctic Ice Sheet in Pine Island Bay: Antarctica: Quaternary Science Reviews, v. 21, p. 1879–1897, doi: 10.1016/S0277-3791(02)00006-9.

Maddox, J., Anderson, J.B., Milliken, K.T., Rodriguez, A.B., Dellapenna, T.M., and Giosan, L., 2008, this volume, The Holocene evolution of the Matagorda and Lavaca estuary complex, Texas, USA, *in* Anderson, J.B., and Rodriguez, A.B., eds., Response of Upper Gulf Coast Estuaries to Holocene Climate Change and Sea-Level Rise: Geological Society of America Special Paper 443, doi: 10.1130/2008.2443(07).

McEwen, M.C., 1969, Sedimentary facies of the modern Trinity Delta, *in* Lankford, R.R., and Rogers, J.J.W., eds., Holocene Geology of the Galveston Bay Area: Houston, Houston Geological Society, p. 53–77.

McFarlan, E., 1961, Radiocarbon dating of late Quaternary deposits, south Louisiana: Geological Society of America Bulletin, v. 72, p. 129–158, doi: 10.1130/0016-7606(1961)72[129:RDOLQD]2.0.CO;2.

Meltzer, D.J., 1991, Altithermal archaeology and paleoecology at Mustang Springs, on the southern High Plains of Texas: American Antiquities, v. 56, p. 236–267, doi: 10.2307/281417.

Miller, L., and Douglas, B.C., 2004, Mass and volume contribution to twentieth-century global sea-level rise: Nature, v. 428, no. 6981, p. 406–409, doi: 10.1038/nature02309.

Morton, R.A., Paine, J.G., and Blum, M.D., 2000, Responses of stable bay-margin and barrier-island systems to Holocene sea-level highstands, western Gulf of Mexico: Journal of Sedimentary Research, v. 70, no. 3, p. 478–490, doi: 10.1306/2DC40921-0E47-11D7-8643000102C1865D.

Morton, R.A., Bernier, J.C., and Barras, J.A., 2006, Evidence of regional subsidence and associated wetland loss induced by hydrocarbon production, Gulf Coast region, USA: Environmental Geology, v. 50, p. 261–274, doi: 10.1007/s00254-006-0207-3.

Nelson, H.F., and Bray, E.E., 1970, Stratigraphy and history of the Holocene sediments in the Sabine–High Island area, Gulf of Mexico, *in* Morgan, J.P., ed., Deltaic Sedimentation Modern and Ancient: Society for Economic Paleontologists and Mineralogists (SEPM) Special Publication 15, p. 48–77.

Nordt, L.C., Boutton, T.W., Hallmark, C.T., and Waters, M.R., 1994, Late Quaternary vegetation and climate change in central Texas based on isotopic composition of organic carbon: Quaternary Research, v. 41, p. 109–120, doi: 10.1006/qres.1994.1012.

Nordt, L.C., Boutton, T.W., Jacob, J.S., and Mandel, R.D., 2002, C_4 plant productivity and climate-CO_2 variations in south-central Texas during the late Quaternary: Quaternary Research, v. 58, p. 182–188, doi: 10.1006/qres.2002.2344.

Nordt, L., Orosz, M., Driese, S., and Tubbs, J., 2006, Vertisol carbonate properties in relation to mean annual precipitation: Implications for paleoprecipitation estimates: The Journal of Geology, v. 114, no. 4, p. 501–510, doi: 10.1086/504182.

Paine, J.G., 1993, Subsidence of the Texas coast: Inferences from historical and late Pleistocene sea levels: Tectonophysics, v. 222, p. 445–458, doi: 10.1016/0040-1951(93)90363-O.

Pearson, C.E., Kelley, D.B., Weinstein, R.A., and Gagliano, S.M., 1987, Archaeological investigations on the outer continental shelf: A study within the Sabine River valley, offshore Louisiana and Texas: Reston, Texas, Coastal Environments Inc., a report for Minerals Management Service, U.S. Department of the Interior, 338 p.

Penland, S., and Ramsey, K.E., 1990, Relative sea-level rise in Louisiana and the Gulf of Mexico: 1908–1988: Journal of Coastal Research, v. 6, no. 2, p. 323–342.

Penland, S., Ramsey, K.E., McBride, R.A., Moslow, T.F., and Westphal, K.A., 1989, Relative Sea Level in Louisiana and the Gulf of Mexico: Baton Rouge, Louisiana Geological Survey, Coastal Geology Technical Report No. 3, 65 p.

Rehkemper, L.J., 1969, Sedimentology of Holocene Estuarine Deposits, Galveston Bay, Texas [Ph.D. dissertation]: Houston, Rice University, 61 p.

Retallack, G.J., 1990, Soils of the Past: An Introduction to Paleopedology: London, Unwin Hyman, 512 p.

Roberts, H.H., Bailey, A., and Kuecher, G.J., 1994, Subsidence in the Mississippi River Delta—Important influences of valley filling by cyclic deposition, primary consolidation phenomena, and early diagenesis: Gulf Coast Association of Geological Societies Transactions, v. 44, p. 619–629.

Rodriguez, A.B., Anderson, J.B., Siringan, F.P., and Taviani, M., 2004, Holocene evolution of the east Texas coast and inner continental shelf: Along-strike variability in coastal retreat rates: Journal of Sedimentary Research, v. 74, p. 405–412, doi: 10.1306/092403740405.

Royer, D.L., 1999, Depth to pedogenic carbonate horizon as a paleoprecipitation indicator? 1999: Geology, v. 27, no. 12, p. 1123–1126, doi: 10.1130/0091-7613(1999)027<1123:DTPCHA>2.3.CO;2.

Shepard, F.P., 1956, Late Pleistocene and Recent history of the central Texas coast: The Journal of Geology, v. 64, no. 1, p. 56–69.

Shinkle, K.D., and Dokka, R.K., 2004, Rates of Vertical Displacement at Benchmarks in the Lower Mississippi Valley and the Northern Gulf Coast: Silver Spring, Maryland, National Oceanic and Atmospheric Administration (NOAA) Technical Report NOS/NGS 50, NOAA, 135 p.

Simms, A.R., Lambeck, K., Purcell, T., Anderson, J.B., and Rodriguez, A.B., 2007, Sea level within the Gulf of Mexico since the Last Glacial Maximum with implication for the melting history of the Laurentide Ice Sheet: Quaternary Science Reviews, v. 26, p. 920–940.

Simms, A.R., Anderson, J.B., Rodriguez, A.B., and Taviani, M., 2008, this volume, Mechanisms controlling environmental change within an estuary: Corpus Christi Bay, Texas, USA, *in* Anderson, J.B., and Rodriguez, A.B., eds., Response of Upper Gulf Coast Estuaries to Holocene Climate Change and Sea-Level Rise: Geological Society of America Special Paper 443, doi: 10.1130/2008.2443(08).

Stuiver, M., 1980, Solar variability and climatic change during the current millennium: Nature, v. 286, p. 868–871, doi: 10.1038/286868a0.

Stuiver, M., 1982, A high-precision calibration of the A.D. radiocarbon time scale: Radiocarbon, v. 24, no. 1, p. 1–26.

Stuiver, M., and Braziunas, T.F., 1989, Atmospheric ^{14}C and century-scale solar oscillations: Nature, v. 338, p. 405–408, doi: 10.1038/338405a0.

Stuiver, M., and Braziunas, T.F., 1993, Modeling atmospheric ^{14}C influences and ^{14}C ages of marine samples back to 10,000 B.C.: Radiocarbon, v. 35, p. 137–189.

Stuiver, M., and Polach, H.A., 1977, Discussion: Reporting of ^{14}C data: Radiocarbon, v. 19, p. 355–363.

Stuiver, M., and Reimer, P.J., 1993, Extended C-14 database and revised Calib 3.0 C-14 age calibration program: Radiocarbon, v. 35, no. 1, p. 215–230.

Stuiver, M., Reimer, P.J., and Braziunas, T.F., 1998, High-precision radiocarbon age calibration for terrestrial and marine samples: Radiocarbon, v. 40, no. 3, p. 1127–1151.

Stuiver, M., Reimer, P.J., and Reimer, R.W., 2005, CALIB 5.0: http://calib.qub.ac.uk/calib/ (last accessed 29 April 2008).

Thomas, M.A., 1990, The Impact of Long-Term and Short-Term Sea-Level Changes on the Evolution of the Wisconsinan-Holocene Trinity/Sabine Incised Valley System, Texas Continental Shelf [Ph.D. dissertation]: Houston, Rice University, 314 p.

Thomas, M.A., and Anderson, J.B., 1994, Sea-level controls on the facies architecture of the Trinity/Sabine incised-valley system, Texas continental shelf, *in* Dalrymple, R.W., Boyd, R., and Zaitlin, B.A., eds., Incised-Valley Systems: Origin and Sedimentary Sequences: Society for Economic Paleontologists and Mineralogists (SEPM) Special Publication 51, p. 63–82.

Toomey, R.S., Blum, M.D., and Valastro, S., 1993, Late Quaternary climates and environments of the Edwards Plateau, Texas: Global and Planetary Change, v. 7, p. 299–320, doi: 10.1016/0921-8181(93)90003-7.

Törnqvist, T.E., Gonzalez, J.L., Newsom, L.A., van der Borg., K., de Jong, A.F.M., and Kurnik, C.W., 2004, Deciphering Holocene sea-level history on the U.S. Gulf Coast: A high-resolution record from the Mississippi Delta: Geological Society of America Bulletin, v. 116, p. 1026–1039.

Törnqvist, T.E., Bick, S.J., Gonzalez, J.L., van der Borg, K., and de Jong, A.F.M., 2005, Tracking the sea-level signature of the 8.2 ka cooling event: New constraints from the Mississippi Delta: Geophysical Research Letters, v. 31, L23309, p. 1–4.

Törnqvist, T.E., Bick, S.J., van der Borg, K., and de Jong, A.F.M., 2006, How stable is the Mississippi Delta?: Geology, v. 34, p. 697–700, doi: 10.1130/G22624.1.

Toscano, M.A., and Macintyre, I.G., 2003, Corrected western Atlantic sea-level curve for the last 11,000 years based on calibrated ^{14}C dates from *Acropora* palmate framework and intertidal mangrove peat: Coral Reefs, v. 22, p. 257–270, doi: 10.1007/s00338-003-0315-4.

Turnbull, J.C., Lehman, S.J., Miller, J.B., Sparks, R.J., Southon, J.R., and Tans, P.P., 2007, A new high precision $^{14}CO_2$ time series for North American continental air: Journal of Geophysical Research, v. 112, p. D11310, doi: 10.1029/2006JD008184.

White, W.A., Morton, R.A., and Holmes, C.W., 2002, A comparison of factors controlling sedimentation rates and wetland loss in fluvial-deltaic systems, Texas Gulf Coast: Geomorphology, v. 44, p. 47–66, doi: 10.1016/S0169-555X(01)00140-4.

Zervas, C., 2001, Sea-level variations in the United States, 1854–1999: Silver Spring, Maryland, National Oceanic and Atmospheric Administration (NOAA) Technical Report NOS CO-OPS 36, NOAA, 80 p.

Manuscript Accepted by the Society 10 March 2008

Printed in the USA

The Geological Society of America
Special Paper 443
2008

Response of Mobile Bay and eastern Mississippi Sound, Alabama, to changes in sediment accommodation and accumulation

Antonio B. Rodriguez[†]
Institute of Marine Sciences, University of North Carolina at Chapel Hill, 3431 Arendell Street, Morehead City, North Carolina 28557, USA

D. Lawrence Greene Jr.
ConocoPhillips, 600 North Dairy Ashford, Houston, Texas 77079, USA

John B. Anderson
Department of Earth Science, Rice University, 6100 Main Street, MS 126, Houston, Texas 77005, USA

Alexander R. Simms
Boone Pickens School of Geology, Oklahoma State University, 105 NRC, Stillwater, Oklahoma 74074, USA

ABSTRACT

The rate of creation of sediment accommodation and the rate of sediment accumulation are the primary variables that define the evolution of depositional environments, but they are seldom quantified in studies of coastal evolution. From a detailed map of antecedent topography, a sea-level curve, and measured and modeled sedimentation rates, we quantify these variables for Mobile Bay, Alabama, throughout the Holocene. The timing of recorded rapid changes in depositional environments is compared to the calculated changes in sediment accommodation and sediment accumulation to infer causality.

A comparison of cumulative changes in sediment accommodation to cumulative sediment volume changes shows that the estimated volume of sediment accumulation in the estuary kept pace with and slightly exceeded sediment accommodation until around 8.2 ka. Prior to 8.2 ka, Mobile Bay was an intertidal to supratidal marsh, likely part of the delta-plain environment. Remnants of this environment are preserved discontinuously across the bay directly above the exposure surface as peat. At 8.2 ka, parts of the study area were initially submerged, and the central-basin depositional environment was created. Between 8680 and 8200 cal yr B.P., the bay shoreline transgressed up the axis of the study area at a rate of ~100 m/yr, while the extensive delta plain and marsh were eroded and replaced by a central-basin environment. After 8200 cal yr B.P., the locations of the depositional environments of Mobile Bay were relatively static and aggraded throughout the remainder of the Holocene rise in sea level. This record suggests that a threshold was crossed during the early Holocene

[†]E-mail: abrodrig@email.unc.edu.

Rodriguez, A.B., Green, D.L., Jr., Anderson J.B., and Simms, A.R., 2008, Response of Mobile Bay and eastern Mississippi Sound, Alabama, to changes in sediment accommodation and accumulation, *in* Anderson, J.B., and Rodriguez, A.B., eds., Response of Upper Gulf Coast Estuaries to Holocene Climate Change and Sea-Level Rise: Geological Society of America Special Paper 443, p. 13–29, doi: 10.1130/2008.2443(02). For permission to copy, contact editing@geosociety.org.
©2008 The Geological Society of America. All rights reserved.

in which low-gradient antecedent topography was rapidly flooded by sea-level rise. This promoted sediment accretion landward of the study area and facilitated erosion in the study area through inundation and ravinement, which are reflected as a dramatic decrease in the rate of estuarine-basin sediment accumulation between 8680 and 8400 cal yr B.P. Some estuaries and coastal areas are inherently unstable and prone to threshold responses to forcing mechanisms, particularly those characterized by low-gradient and low-elevation topography.

Keywords: estuary, sea level, climate change, Holocene, bayhead delta.

INTRODUCTION

The most important mechanism in the Holocene evolution of estuaries, or any depositional environment, is the interplay between the rate of creation of sediment accommodation and the rate of sediment accumulation in the environment (Jervey, 1988; Reading, 2000). In an estuary, sediment accommodation is a function of changes in relative sea level, which is the position of the sea surface relative to a fixed datum near the seafloor (Posamentier et al., 1988; Posamentier and Allen, 1999), and the shape of the paleovalley. Sediment accumulation is a function of sediment flux and sequestration. These variables are difficult to quantify in three dimensions. Consequently, previous studies of Holocene estuarine evolution commonly explain abrupt and rapid spatial changes to depositional environments as being the result of variations in the rate of sea-level rise, sediment supply, antecedent topography, and/or hydrodynamics (Wilkinson and Byrne, 1977; Fletcher et al., 1990; Colman et al., 1992; Allen and Posamentier, 1993; Thomas and Anderson, 1994; Nichol et al., 1996; Lin et al., 2005; Rodriguez et al., 2005). Although these individual forcing mechanisms are components of sediment accommodation and sediment accumulation, translation of estuarine depositional environments is not solely a reflection of a change in any individual component. This severely limits the applicability of the spatial extent of a depositional environment as a proxy for an individual forcing mechanism.

Over millennial time scales and in tectonically stable areas, changes in sediment accommodation are strongly controlled by sea-level rise and the morphology of the land being inundated (antecedent topography). Variations in antecedent topography caused rapid fluctuations in the rate of creation of sediment accommodation and shoreline transgression during inundation of river mouths and the initial formation of Chesapeake Bay (Bratton et al., 2003) and Galveston Bay (Rodriguez et al., 2005). Figure 1 illustrates the control that variable antecedent topography has on the creation of sediment accommodation through continuous inundation of idealized fluvial end-member topographies. For each 30 time steps, new sediment accommodation was measured as the change in water volume between sea level and the antecedent topography. Flooding of terraced topography at a continuous rate results in a rapid increase in the rate of creation of sediment accommodation as sea level crosses the break in slope from the steep thalweg flank to the low-gradient terrace (Fig. 1A). Flooding of U-shaped topography at a continuous rate results in a rapid decrease in the rate of creation of sediment accommodation as sea level crosses the break in slope from the low-gradient and broad valley base to the steeper valley flank (Fig. 1B). A V-shaped valley has a regular shape; hence, the rate of creation of sediment accommodation is constant and similar to the rate of sea-level rise (Fig. 1C). In Figures 1A and 1B, the stratigraphic record would likely have preserved a flooding surface or back-stepping event during the high rate of creation of sediment accommodation, assuming a constant rate of sediment supply to the area. Without knowledge of antecedent topography in the area, the depositional change would likely be interpreted as an increase in the rate of sea-level rise.

Variations in sediment supply can be just as important as variations in the rate of creation of sediment accommodation for the evolution of an estuary. Ideally, observed changes to estuarine depositional environments should be placed in context along with a comprehensive sediment budget that quantifies the volume of various sources and sinks. Here, estuarine-basin sediment accumulation is defined as the volume of sediment that accumulates in the estuary through time. Using this definition, sediment accumulation in the estuary should always be less than the total sediment volume delivered to the estuary, which is the sum of fluvial, marine, shoreline erosion, biogenic, and runoff sources, because it does not include the volume of sediment eroded during storms, bypassed through inlets, and/or sequestered landward in the alluvial plain during floods.

Studies of Holocene estuarine evolution typically highlight profound changes to the depositional system that occurred over a short time period. For example, in Chesapeake Bay, initiation of estuarine sedimentation was essentially simultaneous over an upstream distance of at least 100 km (Bratton et al., 2003), in Galveston Bay, the Trinity River mouth and bayhead delta shifted landward at a rate of 6.5 km/century at the same time that the associated barrier shoreline was stranded on the inner continental shelf (Rodriguez et al., 2005), and in the Holocene Changjiang estuary of China, distributary-channel deposits were replaced by tidal-flat and estuarine deposits over a landward distance of >150 km in 1000 yr (Hori et al., 2002). It is important to place observed episodic changes in the location of the estuarine

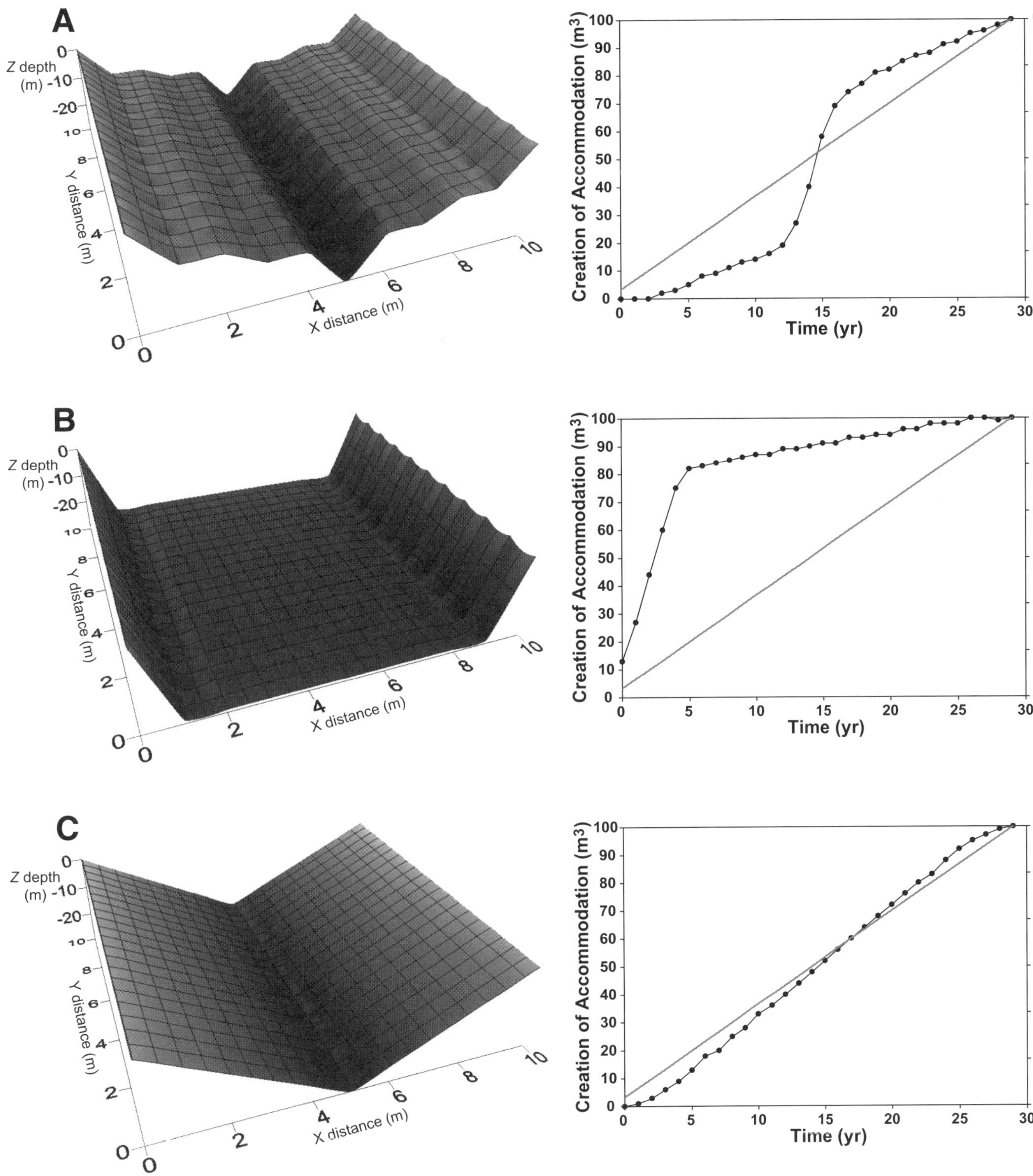

Figure 1. Graphs showing the creation of sediment accommodation for three idealized valley shapes inundated by a linear rise in sea level. The potential creation of sediment accommodation is calculated as the change in volume between sea level and the valley for each time step (30 time steps in total). Inundation of a terraced valley (A) results in an increase and decrease in the rate of creation of sediment accommodation as different gradients are inundated. Inundation of a U-shaped valley (B) results in a decrease in the rate of creation of sediment accommodation as steeper topography is flooded. Inundation of a V-shaped valley (C) results in a continuous rate of creation of sediment accommodation.

depositional environment in context with changes in sediment accommodation and sediment accumulation to quantify thresholds. The objective of this study is to quantify changes in the creation of sediment accommodation and estuarine-basin sediment accumulation within Mobile Bay, Alabama, over the last 10,000 yr. The Holocene evolution of the estuary, preserved in the stratigraphic record, is placed in context with these forcing mechanisms to help determine evolutionary thresholds.

STUDY AREA

The Mobile Bay estuary, which includes Bon Secour Bay in the southeast, is restricted from the Gulf of Mexico by Dauphin Island and Morgan Peninsula (Fig. 2). The estuary is shallow (average depth of 3–4 m) and microtidal, with a diurnal tidal range of 0.35–0.80 m (Knowles, 1989; Hummell, 1996). The piedmont Mobile-Tensaw River system, which has a drainage

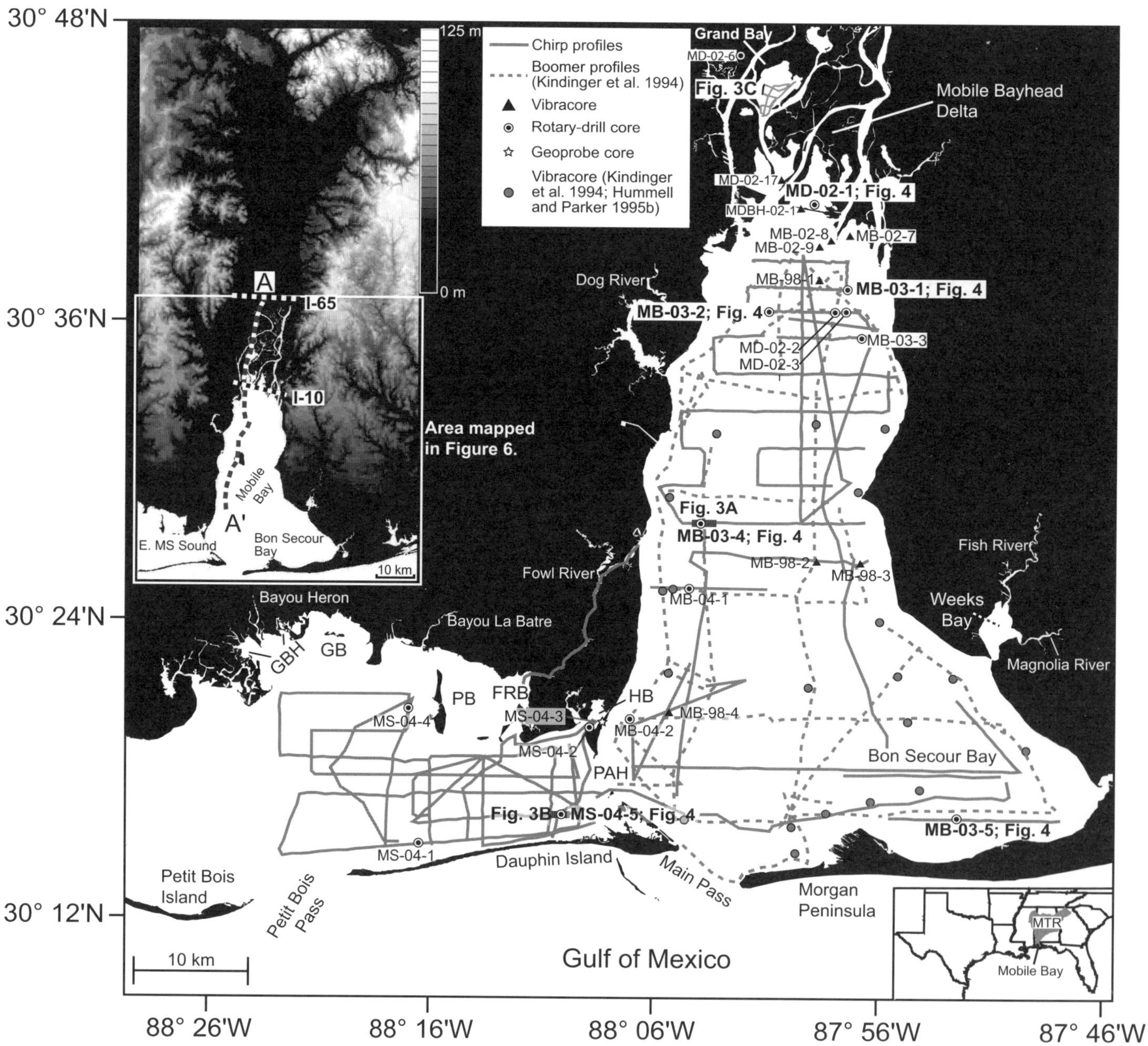

Figure 2. Study area map showing the locations of seismic lines and cores collected and incorporated into this study. Data shown in subsequent figures are indicated by dark-gray lines. The locations of Grand Batture Headlands (GBH), Grand Bay (GB), Portersville Bay (PB), Fowl River Bay (FRB), Heron Bay (HB), Pass aux Herons (PAH), and Mobile Tensaw River Basin (MTR) are also indicated. MS—Mississippi.

area of ~133,500 km^2, discharges into northern Mobile Bay and forms an extensive (465 km^2) bayhead delta (Fig. 2; Gastaldo, 1989). The delta is constrained by steep topography on the west and east (Fig. 2). The location of the lower Mobile-Tensaw River system and associated bayhead delta may be, in part, controlled by the interpreted Mobile graben (Smith, 1988).

Average water discharge for the Mobile-Tensaw River is ~1788 m^3/s (Hummell and Parker, 1995a). Approximately 85% of the Mobile-Tensaw River water discharge passes through the Main Pass tidal inlet into the Gulf of Mexico, while Pass aux Herons, which connects Mobile Bay to eastern Mississippi Sound, transmits the remainder into Mississippi Sound (Ryan, 1969). Discharge from coastal streams and rivers results in variations in surface salinity from 0‰ to 30‰ (McBride et al., 1991; Hummell and Parker, 1995a, 1995b; Hummell, 1996).

Eastern Mississippi Sound, including Heron, Fowl River, Portersville, and Grand Bays, is restricted from the Gulf of Mexico by Dauphin Island, Petit Bois Island, and Petit Bois Pass (Fig. 2). The northern shoreline of the estuary is marked by a series of small islands and tidal marshes and has low-gradient topography (0–5 m elevation 4–6 km inland). The Fowl and Bayou La Batre coastal-plain fluvial systems flow into eastern Mississippi Sound and have small drainage basins of 197 km^2 and 75 km^2, respectively. Average water discharge is 1 m^3/s for the Fowl River and 0.4 m^3/s for Bayou La Batre (Greene, 2006). In general, turbidity is greater in the northern part of eastern Mississippi Sound, which is largely attributed to sediment-laden water from Mobile Bay entering the sound through Pass aux Herons (McPherson, 1970).

Fluvial systems along the Alabama coast were incised into the coastal plain during the last lowstand in sea level (oxygen isotope stage 2; Kindinger et al., 1994; Bartek et al., 2004). The morphology that developed at that time (the antecedent topography) was inundated during the subsequent transgression to form Mobile Bay and eastern Mississippi Sound. Previous studies that have mapped the antecedent topography in Mobile Bay differ in regard to valley location and depth. Mars et al. (1992), Davies and Hummell (1994), and Hummell and Parker (1995a) mapped the valley as 13–15 m deep, extending from the modern bayhead delta to 15 km north of Morgan Peninsula. Here, it bifurcates into a western valley that continues through the modern tidal inlet and an eastern valley that extends below Morgan Peninsula (Mars et al., 1992; Davies and Hummell, 1994; Hummell and Parker, 1995a). Kindinger et al. (1994) mapped the valley as 25–45 m deep, extending along eastern Mobile Bay from south of the modern bayhead delta to 15 km north of Morgan Peninsula. At this location, it bifurcates into a western valley that extends through the modern tidal inlet and an eastern valley that extends below Morgan Peninsula (Kindinger et al., 1994). From newly collected seismic data and cores that penetrate through the exposure surface, Greene et al. (2007) demonstrated that the antecedent topography that developed during the last lowstand is somewhat a reflection of older, underlying sequence boundaries. Given the control that variations in antecedent topography have on the creation of sediment accommodation and estuarine evolution, it is essential for this surface to be mapped in detail.

METHODS

The antecedent topography and depositional history of Mobile Bay and eastern Mississippi Sound were constrained from 500 km of high-resolution seismic data, nine vibracores, 15 rotary-drill cores, and one Geoprobe core. These data were incorporated with ~250 km of high-resolution single-channel (200–400 J) boomer data published in Kindinger et al. (1994), 26 vibracores published in Hummell and Parker (1995a), and borings from bridges for Interstate 65 (75) and Interstate 10 (14) collected by the state of Alabama Highway Department (May, 1976). Chronostratigraphy was developed from 36 radiocarbon dates (Table 1).

Seismic data were collected using an EdgeTech SB-216S chirp system that emits a 2–15 kHz pulse at 0.25 s intervals (Fig. 2). A velocity of 1500 m/s was used to convert the two-way traveltime to depth. This velocity was verified by cores that sampled changes in lithology at the same depths as seismic facies transitions (Figs. 3 and 4). The cores were primarily used to ground truth seismic facies to lithology and to provide shell, plant, and wood material for environmental analysis and radiocarbon dating. Radiocarbon dating (using accelerator mass spectrometry [AMS]) of shell and organic material was performed by the Woods Hole Oceanographic Institution and Beta Analytic (Table 1), and dates were calibrated using CALIB 5.0.2 software (Stuiver and Reimer, 1993; Hughen et al., 2004; Reimer et al., 2004). To ensure that dated material was preserved in situ, whenever possible, individual wood fragments and articulated bivalves were chosen for dating instead of bulk peat or unpaired valves. Species from well-defined macrofossil assemblages indicative of the depositional environment were preferentially chosen for dating. No local reservoir corrections were applied to estuarine and marine carbonate during calibration.

RESULTS AND INTERPRETATIONS

Antecedent Topography

Greene et al. (2007) recognized four regional unconformity-bounded stacked seismic units in the study area. The shallowest unconformity truncates underlying reflectors (Fig. 3A) and, in places, is characterized by a high-amplitude reflection (Fig. 3C) and high relief (Fig. 3B). The unconformity was sampled in cores and is a sharp contact with dark greenish gray (Munsell color 5 GY 4/1) to medium gray (N5) clay and silt, sandy mud, or peat above a stiff, oxidized medium gray (N5), greenish gray (5GY 6/1) to yellowish orange–reddish orange clay (10 YR 6/6-10 R 6/6; Figs. 4 and 5). The stiff and oxidized nature of the clay below the contact is indicative of subaerial exposure. Clay rip-up clasts, which were frequently found above the contact, are indicative of erosion (Figs. 4 and 5). Wood and peat material from below the

TABLE 1. RADIOCARBON AGES AND TYPE OF MATERIAL ANALYZED

Lab code	Sample name (depth, cm)[†]	Material	Conventional ^{14}C age (yr B.P., 1σ)	Calibrated ^{14}C age (cal yr B.P., 1σ)[‡]
OS-38355	MD-02-1 (217)	Wood	720 ± 30	661–683
OS-37366	MD-02-1 (683)	Wood	5180 ± 30	5912–5942
OS-37365	MD-02-1 (851)	Wood	825 ± 30	693–743
OS-38356	MD-02-1 (851)	Wood	1080 ± 25	956–999
OS-37368	MD-02-1 (903)	Wood	6150 ± 30	6959–7159
OS-37364	MD-02-1 (942)	Wood	5070 ± 30	5752–5827
OS-38357	MD-02-1 (942)	Wood	5180 ± 30	5912–5942
OS-37574	MD-02-1 (1116)	Wood	6220 ± 50	7021–7121
OS-40279	MD-02-1 (1229)	Wood	7190 ± 35	7967–8017
OS-37367	MD-02-1 (1310)	Wood	8000 ± 55	8856–8997
OS-38354	MD-02-1 (1492)	Wood	8210 ± 40	9119–9259
OS-37363	MD-02-1 (1554)	Wood	8480 ± 35	9481–9523
OS-37370	MD-02-1 (1610)	Wood	8810 ± 35	9741–9909
OS-37369	MD-02-1 (1726)	Wood	8710 ± 35	9595–9697
OS-37362	MD-02-1 (1809)	Wood	8770 ± 50	9687–9892
OS-37382	MD-02-2 (856)	*Crassostrea virginica*	4090 ± 30	4078–4189
OS-37381	MD-02-2 (1066)	*Crassostrea virginica*	5180 ± 30	5507–5582
OS-44044	MB-03-5 (622)	*Semele proficua*	4630 ± 25	4812–4868
OS-44045	MB-03-5 (909)	*Nuculana concentrica*	5160 ± 30	5502–5568
OS-41368	MB-03-5 (1151)	*Nuculana concentrica*	6290 ± 40	6680–6794
OS-41176	MB-03-5 (1368)	*Corbula contracta*[§]	7680 ± 40	8069–8190
OS-44826	MB-03-5 (1412)	Wood	8060 ± 40	8976–9029
OS-42767	MB-03-2 (912)	Wood	5460 ± 30	6272–6295
OS-42943	MB-03-2 (1026)	Wood	7070 ± 50	7851–7906
OS-42769	MB-03-1 (723)	Wood	6220 ± 35	7025–7117
OS-42770	MB-03-1 (760)	Wood	5100 ± 35	5758–5822
OS-42949	MB-03-1 (828)	*Balanus* sp.	5370 ± 30	5669–5786
OS-42768	MB-03-1 (1095)	Wood	6980 ± 35	7782–7855
OS-42942	MB-03-4 (1514)	Wood (from peat)	8270 ± 60	9196–9318
OS-44717	MS-04-1 (776)	*Probythenella louisianae*	4030 ± 30	3981–4092
OS-44718	MS-04-1 (866)	Wood	>48,000	N.A.
OS-44719	MS-04-1 (1052)	Wood	>48,000	N.A.
OS-44720	MS-04-4 (591)	Wood (from peat)	5470 ± 50	6210–6308
OS-44722	MS-04-5 (1258)	Wood	38,400 ± 330	N.A.
AA35692	MB-98-4 (637)	*Crassostrea virginica*	4980 ± 55	5278–5416
AA35693	MB-98-4 (697)	Bulk peat	>47,000	N.A.

Note: Samples were processed at the National Science Foundation accelerator mass spectrometry (AMS) laboratory at the Woods Hole Oceanographic Institution and Beta Analytic.
[†]Sample depths (reported parenthetically) are relative to sea level.
[‡]Calibrated using Stuiver and Reimer (1993), Hughen et al. (2004), and Reimer et al. (2004).
[§]Articulated bivalve.

contact in three different cores was either radiocarbon dead (MS-04-1 and MB-98-4) or Pleistocene in age (MS-04-5; Table 1). All dates obtained above the contact were Holocene (Table 1).

The exposure surface was mapped regionally throughout the area from the Interstate 65 cross section in the north, to Dauphin Island and Morgan Peninsula in the south (Figs. 2 and 6). This surface is the antecedent topography that was flooded to form the modern estuary. The depth to the exposure surface below the modern delta is mostly based on interpolation between the Interstate 10 cross section at the delta front and the Interstate 65 cross section 30 km to the north (Greene et al., 2007; May, 1976), except for one rotary-drill core (MD-02-6) and a small grid of seismic data collected in Grand Bay (Figs. 2 and 3). The exposure surface defines multiple, broad (0.6–5.4 km wide) valleys that extend to the modern La Batre, Fowl, Dog, and Mobile-Tensaw River systems (Fig. 6). A narrow and shallow paleovalley extending from Weeks Bay into Bon Secour Bay was recognized on some seismic lines, but it was below mapping resolution. The main trunk valleys are generally in the same location as those mapped on older unconformities (Greene et al., 2007). Valley thalwegs are 10–25 m deep (measured from sea level). Broad and flat interfluvial areas separate the paleovalleys (Fig. 6).

Holocene Depositional Environments and Stacking Patterns

The acoustic character of the sediment package above the exposure surface in Mississippi Sound, Mobile Bay, and Grand Bay is generally transparent to planar (Fig. 3). The relative homogeneity of this seismic unit makes differentiating depositional environments difficult. Hence, depositional environments were primarily characterized from the lithologic data. Above the exposure surface at the modern Mobile delta front, core MD-02-1 sampled carbonaceous clay and silt with sand-filled burrows, sandy mud, and a fining-upward interval from sand at 15.38 m to clay and silt at 14.65 m (Fig. 4). This unit is interpreted

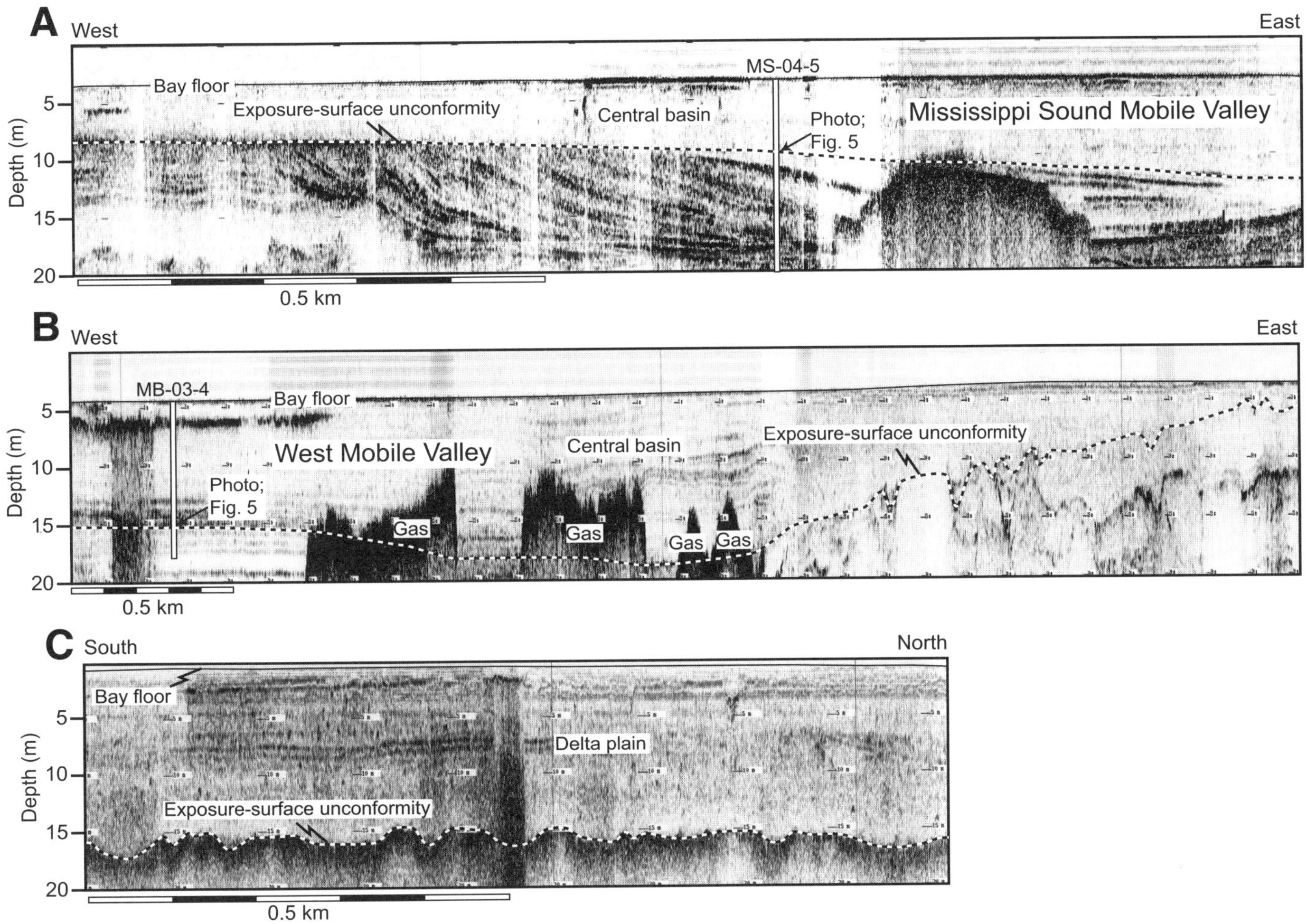

Figure 3. The shallowest regional unconformity in the study area is an exposure surface, which is recognized in seismic data by erosional truncation of underlying reflectors (A), high relief (B), and a high-amplitude reflection (C). Examples are from southeastern Mississippi sound (A), the middle of Mobile Bay (B), and Grand Bay, which is an interdistributary bay in the southern part of the Mobile bayhead delta (C). The specific profile locations are shown on Figure 1.

as a delta-plain deposit. The oldest shell material (*Rangia*) above the exposure surface in core MD-02-1 is at ~9.0 m. At 8.25 m, grain size coarsens upward from clay and silt to sand (2.5–1.0ϕ) and the abundance of shell material (*Rangia*) increases (Fig. 4). The sandy depositional environment above 8.25 m is interpreted as delta front. The delta plain is distinguished from the delta front by its finer texture, overall greater abundance of carbonaceous material, and absence of shell material.

A bioturbated, dark greenish gray (5 GY 4/1) to medium gray (N5), clay and sandy-clay unit containing *Crassostrea*, *Mulinia*, and *Rangia* valves was sampled above the exposure surface in most cores collected from the modern estuary (central basin) environment (Fig. 4). Sand-size particles are 3.0–4.0ϕ in size. This facies extends to the bay floor and is distinguished from bayhead delta facies by its greater abundance of shell material, presence of higher-salinity macrofauna, finer texture, and less carbonaceous material. It is interpreted as a central-basin deposit. The facies averages 10 m in thickness in the middle of the estuary. Overall, the amount of carbonaceous material between the sediment-water interface and the exposure surface decreases south of the bayhead delta, and most cores from the central basin contain the highest concentrations of carbonaceous material directly above the exposure surface. Peat was sampled directly above the exposure surface in cores MB-03-04and MS-04-4 (Figs. 4 and 5), and it is interpreted as a marsh deposit. The peat is <0.5 m thick and overlain by central-basin sediment.

The cores show an overall transgressive succession. Dalrymple et al. (1992) presented an idealized transgressive vertical succession of facies at a central-basin locale that included bayhead delta deposits above the exposure surface. At the modern central-basin locale in Mobile Bay, central-basin mud either exists directly above the exposure surface or is separated from

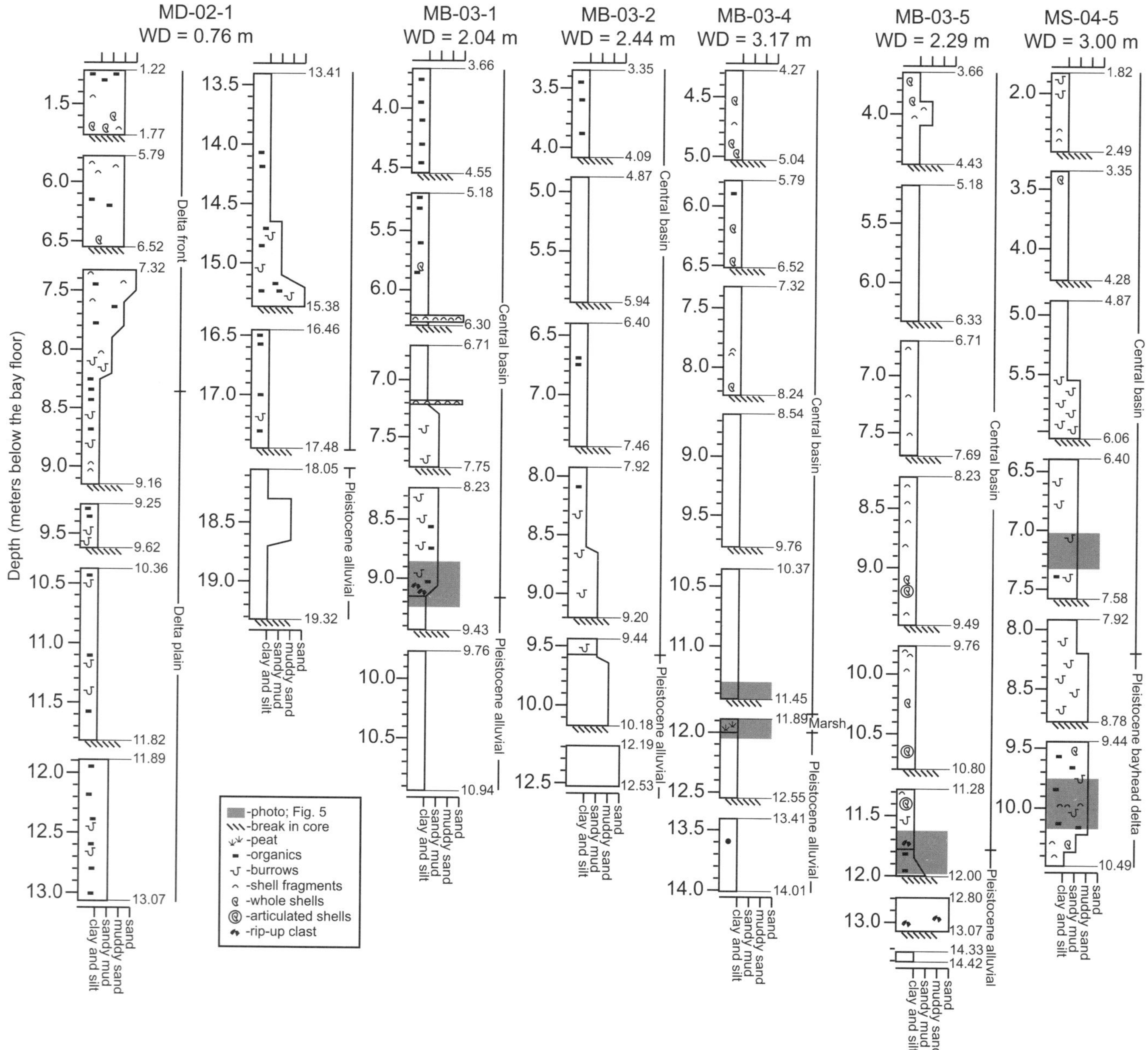

Figure 4. Rotary-drill cores collected from the modern delta front (MD-02-1), upper Mobile Bay (MB-03-1 and MB-03-2), middle Mobile Bay (MB-03-4), lower Mobile Bay (MB-03-5), and eastern Mississippi Sound (MS-04-5) illustrate the sedimentary facies and the exposure surface. With the exception of core MB-03-4, which sampled peat above the exposure surface, the modern depositional environment each core was collected from extends down to the exposure surface. The depth scale is relative to the bay floor, and water depth at the coring location (WD) is indicated below the core name. See Figure 2 for core locations.

Figure 5. Core photographs showing the lithologic character of the exposure surface and overlaying facies. Peat was preserved above the exposure surface in core MB-03-4; however, MB-03-1, MB-03-5, and MS-04-5 are from modern central-basin locations and sampled central-basin clay directly above the surface. The exposure surface is not easily identifiable in core MS-04-5; therefore, seismic data (Fig. 3) and a radiocarbon date of 38,400 yr B.P. at 12.58 m (Table 1) were used to confirm its placement. Depths are in m relative to sea level. See Figure 2 for core locations and Figure 4 for photographed intervals.

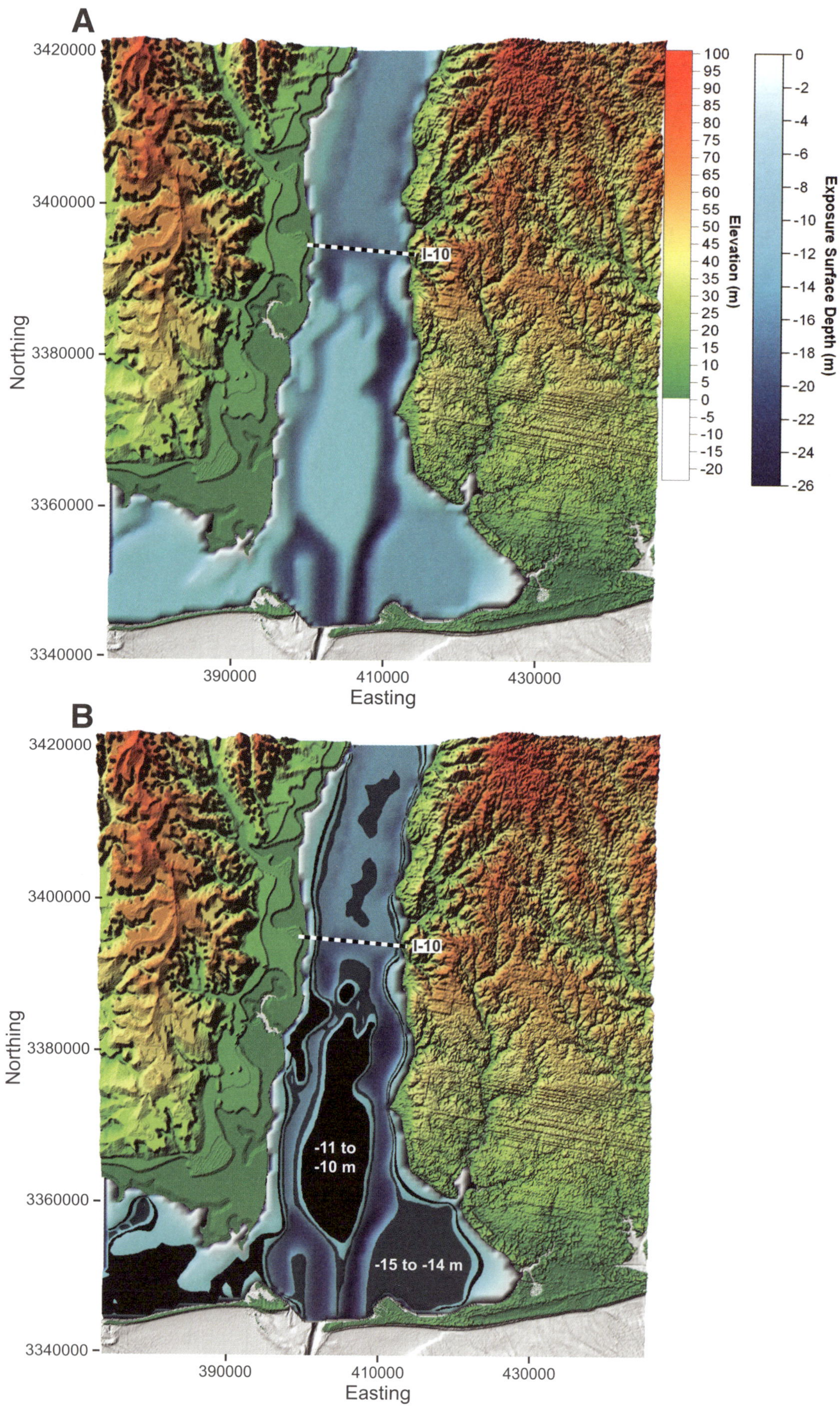

Figure 6. (A) Map showing topography of the exposure surface (blue color scale; relative to sea level) below the Mobile bayhead delta from the Interstate 65 crossing in the north to Dauphin Island and Morgan Peninsula in the south. Topography of surrounding land area, from the National Oceanic and Atmospheric Administration (NOAA) National Geophysical Data Center, is shown in red, yellow, and green. (B) The same map, but the low-gradient areas from −15 to −14 m and −11 to −10 m below sea level are colored gray and black, respectively.

the exposure surface by a thin marsh unit. The marsh unit may be part of a paleobayhead delta plain (Fig. 7). The discontinuous marsh and/or bayhead delta plain may have been more extensive in area than the two cores that sampled it would suggest. Initial inundation of the valley resulted in erosion of the valley floor, as evidenced in many cores by clay rip-up clasts recognized directly above the exposure surface in the central-basin unit. At the modern bayhead delta front, delta-plain sediment lies directly above the exposure surface and is overlain by delta-front sediment. Hummell and Parker (1995a) concluded that during the middle Holocene, the central basin extended an average of 3 km to the north prior to delta progradation. This conclusion was based on engineering boring descriptions (no material was evaluated by the authors; Hummell and Parker, 1995a) from the same location as core MD-02-1. Data presented here support delta aggradation and retrogradation throughout the Holocene (Fig. 7).

A prominent bayhead delta back-stepping event likely occurred during central-basin formation because bayhead delta deposits are thin or absent above the exposure surface in Mobile Bay and Mississippi Sound but were recognized in seismic data offshore on the Alabama shelf (Cabote, 1998; Bartek et al., 2004). It is likely that a thin bayhead delta was emplaced but was subsequently mostly eroded during inundation. Bayhead delta deposition was initiated at the modern location at 9790 ± 100 cal yr B.P. based on a radiocarbon date from a single piece of wood sampled at 18.09 m below sea level (mbsl) directly above the exposure surface in core MD-02-1. Marsh deposition on the exposure surface at modern central-basin locales was initiated soon thereafter, at 9260 ± 60 cal yr B.P. This is based on the radiocarbon date of a single piece of wood sampled at 15.14 mbsl in core MB-03-4 from the marsh unit directly above the exposure surface. Central-basin deposition was initiated between 8130 ± 60 and 9000 ± 30 cal yr B.P. and persists to the present. This is based on radiocarbon dates in core MB-03-5 of an articulated *Corbula contracta* shell sampled at 13.68 mbsl, ~0.4 m above the exposure surface (younger age), and a single piece of wood sampled at the exposure surface (older age). Given the short period of time (~800 yr) between the oldest bayhead delta deposits dated above the exposure surface at the base of the modern bayhead delta (core MD-02-1), and the oldest central-basin sediment behind Morgan Peninsula (core MB-03-5), 45 km to the south, the bayhead delta retreated up the valley very rapidly, ~550 m/century. This rate of delta-front retreat is a conservative estimate considering that the depth of the exposure surface below the modern delta front is at a lower stratigraphic level, 5.5 m deeper, than its depth behind Morgan Peninsula, and bay ravinement likely removed material during inundation. Modern bayhead delta and central-basin depositional environments formed close to their modern positions and aggraded throughout the Holocene (Fig. 7).

Creation of Sediment Accommodation

The creation of sediment accommodation is defined here as new space (volume) made available for potential sediment accumulation over a given time period. Because the study area is in a tectonically stable region, and the timeframe of interest is

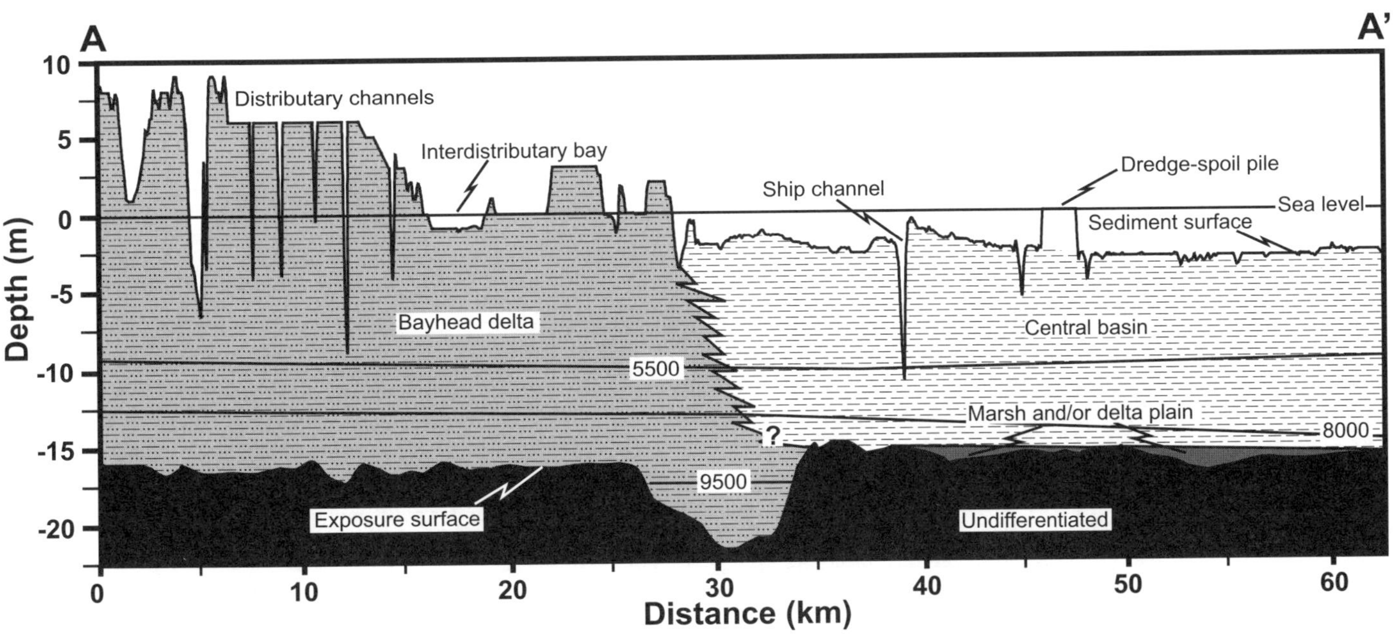

Figure 7. Cross section A–A′ through the western paleochannel from the Interstate 65 crossing of the bayhead delta in the north to Main Pass tidal inlet in the south based on seismic data and cores (see Fig. 2 for location). Isochrons are flat, indicating aggradation of bayhead delta and central-basin sediment since the early Holocene. Thin, discontinuous bayhead delta deposits are preserved above the exposure surface in the central basin.

relatively short (the Holocene), the creation of sediment accommodation is primarily a function of relative sea-level rise and the topography of the land being inundated. Sea-level change in the northern Gulf of Mexico during the Holocene is a topic of debate. Blum et al. (2002, 2003) argued that sea level rose to +2 m or more during the middle Holocene, based in part on the elevations and dates of beach ridges on Morgan Peninsula (Fig. 2). Rodriguez and Meyer (2006) presented stratigraphic data from the same beach ridges that support the Holocene sea-level curve of Törnqvist et al. (2004) and Otvos (2004), which show sea level rising continuously at a decreasing rate and at elevations below the present level (Fig. 8). Sea-level data points (basal peats) from Louisiana published in Törnqvist et al. (2004) do not extend past 8.5 ka but do correspond closely with sea-level data points (peat and *Acropora palmata* coral) from the western Atlantic published in Toscano and Macintyre (2003), which span the entire

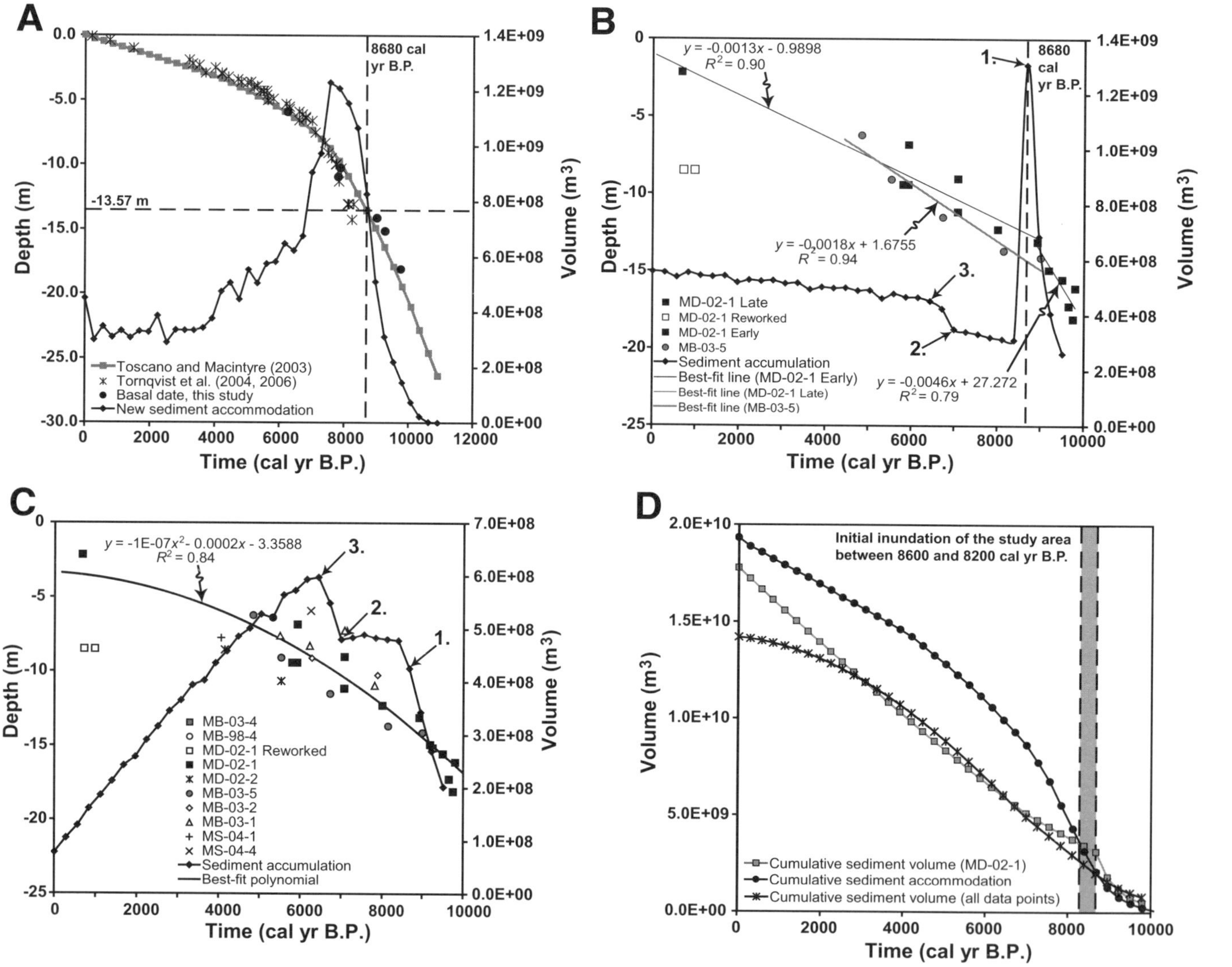

Figure 8. Sea-level data points for Louisiana and ages obtained directly above the exposure surface for this study fall on the sea-level curve of Toscano and Macintyre (2003; sampled every 280 yr), indicating the curve is appropriate for basing calculations of sediment accommodation (A). The most rapid rate of creation of sediment accommodation occurred at 8680 cal yr B.P., when sea level was −13.57 m, just above low-gradient antecedent topography in Bon Secour Bay (A). Early to middle Holocene sedimentation rates (mm/yr) for cores MD-02-1 and MB-03-5 are similar, and since MD-02-1 was dated in greater detail and extends over a longer time period, calculations of estuarine-basin sediment accumulation were based on this core (B). Based on core MD-02-1, a decrease in the rate of sedimentation is interpreted at 8925 ± 70 cal yr B.P. (B). Estuarine-basin sediment accumulation was also calculated using a less interpretative method for deriving sedimentation rates, where all Holocene radiocarbon ages obtained in the study area were included, and a best-fit polynomial was calculated (C). Cumulative sediment accommodation and cumulative sediment volume, using both methods, are similar up to 8200–8600 cal yr B.P., where the curves cross, indicating initial inundation of the study area (D).

Holocene (Fig. 8). Our dates (6) were obtained from basal peat and basal central-basin clay deposits, and they plot within 1 m of the Toscano and Macintyre (2003) sea-level curve, suggesting that this curve is applicable to the area (Fig. 8). A single piece of wood sampled at –5.9 m from a basal peat in core MS-04-4 yielded a radiocarbon date of 6260 ± 50 cal yr B.P. (Table 1). This point falls directly on the Toscano and Macintyre (2003) sea-level curve and provides further evidence against the existence of a middle Holocene sea-level highstand in the area.

Sediment accommodation was calculated using the Toscano and Macintyre (2003) sea-level curve and a contour map of the exposure surface. Sea level was calculated every 280 yr from 10.92 ka to present (39 iterations; Fig. 8). This time interval was chosen because sea level was close to the lowest depth of the exposure surface, about –25 m, around 10.92 ka (Toscano and Macintyre, 2003; Fig. 6). The volume between sea level and the contour map of the exposure surface (Fig. 6) was measured for each iteration, and the creation of sediment accommodation through time was calculated by subtracting successive volumes. For example, the creation of sediment accommodation at iteration 2 was calculated by subtracting the volume between sea level and the contour map measured at 10,640 cal yr B.P. from the volume measured at time 10,360 cal yr B.P. The contour map of the exposure surface is only an approximation of morphology before inundation because this surface was continually modified by ravinement processes as sea-level rose. Estuarine shoreline ravinement creates sediment accommodation and is neither spatially constant nor constant through time. Basing calculations of sediment accommodation on a static representation of the exposure surface will have the effect of placing the sediment accommodation created by ravinement processes earlier than when it was actually created. This is because ravinement processes principally modify a shoreline from sea level to some depth.

Overall, the rate of creation of sediment accommodation increases from 10,920 to 7560 cal yr B.P., decreases from 7560 to 3360 cal yr B.P., and is relatively constant from 3360 cal yr B.P. to the present. The rate of increase in the creation of sediment accommodation is highest between 8960 and 8680 cal yr B.P., which is around the period of time when the rate of sea-level rise began to decrease. This high rate of creation of sediment accommodation occurred when sea level rose from –14.9 to –13.6 m, and it is associated with inundation of low-gradient antecedent topography in Bon Secour Bay that is between –14 and –15 mbsl (Fig. 6B). Low-gradient antecedent topography also exists in the middle of Mobile Bay and Mississippi Sound between –11 and –10 mbsl, but when this area was inundated between 8120 and 7840 cal yr B.P., it was not associated with an increase in the rate of creation of sediment accommodation (Figs. 6B and 8). Furthermore, the area inundated when sea level rose from –11 to –10 m is comparable to the area inundated when sea level rose from –15 to –14 m. The reason there was not an associated increase in the rate of creation of sediment accommodation is because the rate of sea-level rise was less between 8120 and 7840 cal yr B.P. than between 8960 and 8680 cal yr B.P. Although this illustrates the importance of both the rate of sea-level rise and antecedent topography on the creation of sediment accommodation in the study area, whether an area is inundated or not is also highly dependent on sediment supply.

Estuarine-Basin Sediment Accumulation

Short-term (century-scale) estuarine sedimentation rates should vary spatially, especially around inlets and bayhead deltas. Based on a comparison of historical bathymetric charts, Ryan (1969) and Ryan and Goodell (1972) measured short-term sedimentation rates in Mobile Bay as 9 mm/yr in the north and south and <3 mm/yr in the center. These rates do not reflect the long-term Holocene average and are likely a result of human intervention (Nichols, 1989). Furthermore, Nichols (1989) concluded that short-term sedimentation rates and the minimum depth to which an estuarine basin can accrete are modulated through time by energy dissipation of waves and currents. This suggests that long-term Holocene sedimentation rates should show significantly less spatial variation than what is measured in the short term.

Estuarine-basin sediment accumulation is defined here as new sediment (volume) that fills the basin over a given time period. Estuarine-basin sediment accumulation during the Holocene was calculated using sedimentation rates (mm/yr) measured in cores and a contour map of the exposure surface. In calculating estuarine-basin sediment accumulation, the long-term sedimentation rates derived are assumed to reflect the entire estuary. Although there should be some spatial variation in long-term sedimentation rates, the observation that the post–9 ka sedimentation rate is similar in cores MD-02-1 and MB-03-5 (0.5 mm/yr difference), even though these intervals encompass delta-plain, delta-front, and central-basin depositional environments, lends some legitimacy to the assumption (Fig. 8B). Additionally, incorporation of spatial variability in long-term sedimentation rates into the calculations of sediment accumulation, as short-term measurements of sedimentation rates would suggest (Nichols, 1989), yields an unrealistic modern estuarine morphology, where large variations in bathymetry exist and areas that are subaqueous today become emergent.

Holocene sedimentation rates (mm/yr) were calculated using an interpretive method and radiocarbon ages from cores MD-02-1 and MB-03-5, obtained from the Mobile delta front and Bon Secour Bay, respectively (Figs. 2 and 8B; Table 1). These cores were selected because they are from opposite ends of the estuary, sample the entire estuarine succession, and were dated in detail. Core MD-02-1 was collected from an area where the exposure surface is deep, and this is the only core that spans the entire Holocene. Some dates in core MD-02-1 are inverted (Table 1), especially those above 9.2 m, which corresponds to the contact between delta plain below and delta front above. Inverted dates are likely the result of material being stored in the delta plain and later remobilized to the delta front (Stanley and Hait, 2000; Stanley, 2001). A decrease in sedimentation rate in core MD-02-1 is interpreted at 8925 ± 70 cal yr B.P. (–13.1 mbsl), and regression lines were derived for the data points before and after this date. With the exception of the two very young dates at 8.51 m, all of the data points were included in the regressions. Early sedimentation rates in core MD-02-1 are 4.6 mm/yr and decrease to 1.3 mm/yr at 8925 ± 70 cal yr B.P., >3 m below the contact between delta-front and delta-plain facies.

All dates from core MB-03-5 are in stratigraphic succession (Table 1), the record does not extend back prior to 9 ka, and a single regression line was derived for the data points (Fig. 8B). Given the high elevation of the exposure surface in Bon Secour Bay, the area was exposed during the early Holocene when sediment was accreting rapidly in the vicinity of core MD-02-1. The Holocene sedimentation rate in core MB-03-5 is 1.8 mm/yr, which is similar to the post–9 ka sedimentation rate in core MD-02-1 (1.3 mm/yr), even though these intervals encompass delta-plain, delta-front, and central-basin depositional environments; therefore, only the long-term sedimentation rates derived from core MD-02-1 were used in the calculations of estuarine-basin sediment accumulation.

A more conservative method of calculating sedimentation rates was also used, where all Holocene radiocarbon ages obtained in the study area were included, and a best-fit polynomial was calculated (Figs. 2 and 8C; Table 1). With the exception of the two very young dates at ~8.51 m, all of the data points presented on the graph were included in the regression. The long-term sedimentation rates defined by the polynomial are assumed to reflect the entire estuary. Figure 8C shows that by using this method, the sedimentation rate decreases gradually through time. This is unlike our interpretation of the data presented in Figure 8B, which shows an abrupt decrease in the rate of sedimentation at 8925 ± 70 cal yr B.P.

It is important to convert sedimentation rates (mm/yr) into estuarine-basin sediment accumulation because the decrease in sedimentation rates, observed using both methods, could be the result of a decrease in sediment sequestration in the estuary, a decrease in sediment supplied to the estuary, and/or an increase in sediment accommodation associated with flooding of low-gradient topography (Rodriguez et al., 2005). Compaction of sediment above the exposure surface should also play a role in the measured sediment-accumulation rates. Sediment compaction was not quantified for the area because compaction varies spatially with the thickness and composition of Holocene sediment and water depths, making it difficult to quantify in three dimensions through time. For clay, the total compaction at any depth is likely 5%–10% of the unit thickness above the exposure surface (Paul and Barras, 1998; Massey et al., 2006).

Calculations of estuarine-basin sediment accumulation (the volume of sediment that accumulates in the estuary over a given time period) take into account the direct control that variations in antecedent topography have on sedimentation rates measured at any given location. For example, given a constant rate of sediment supply to the estuarine basin, as sea-level rises and the area of the estuarine basin and the space available for deposition increases through time, sedimentation rates (mm/yr) measured at any single location will decrease, but estuarine-basin sediment accumulation (m^3/yr) will be constant. The elevation of the estuary bottom, relative to sea level, was calculated every 280 yr from 9.8 ka to the present (36 iterations) from the sediment-accretion regression lines of both methods (Fig. 8). The volume between the estuary floor and the contour map of the exposure surface (Fig. 6) was measured for each iteration, and estuarine-basin sediment accumulation through time was calculated by subtracting iterations. For example, estuarine-basin sediment accumulation at iteration 5 was calculated by subtracting the volume between the estuary floor and the contour map measured at 9800 cal yr B.P. from the volume measured at time 9520 cal yr B.P. The bathymetry of the paleoestuary used in the calculations has relatively little relief, as indicated by the flat central-basin seismic reflectors and the modern estuarine bathymetry. Comparison of both calculations (Figs. 8B and 8C) helped to determine if observed variations were simply the product of decisions made during construction of the sediment-accretion curves (e.g., the decision to make an abrupt decrease in sedimentation rate at 8680 cal yr B.P.; Fig. 8B).

Results for the interpreted sediment-accretion curve, which has an abrupt decrease in rate at 8925 ± 70 cal yr B.P., show that estuarine-basin sediment accumulation increases sharply from 9520 to 8680 cal yr B.P., decreases sharply from 8680 to 8400 cal yr B.P., and increases at a low rate from 8400 cal yr B.P. to present (Fig. 8B). The significant decrease in estuarine-basin sediment accumulation after 8680 cal yr B.P. shows that there was a decrease in estuarine-basin sediment sequestration or supply at this time. This decrease corresponds to an event when sea level rose from −14.90 to −13.57 m, and to the highest rate of creation of sediment accommodation in the study area, indicating that there could be an indirect link between estuarine-basin sediment accumulation and sea-level rise over the low-gradient antecedent topography in Bon Secour Bay. Furthermore, 8680 cal yr B.P. is close to the time when the rate of sea-level rise in the Gulf of Mexico region began to decrease dramatically (Toscano and Macintyre, 2003).

Results for the sediment-accretion curve based on all cores show that estuarine-basin sediment accumulation increases sharply from 9520 to 8400 cal yr B.P., is relatively constant from 8400 to 7000 cal yr B.P., increases sharply again from 7000 to 6440 cal yr B.P., and decreases at a near-constant rate from 6440 cal yr B.P. to present (Fig. 8C). The change from an increasing to a constant rate of estuarine-basin sediment accumulation at 8400 cal yr B.P. is close (one time step or 280 yr later) to when estuarine-basin sediment accumulation decreases suddenly in Figure 8B (denoted on the graphs by a 1). This is additional evidence that there could be an indirect link between estuarine-basin sediment accumulation and sea level reaching the elevation of low-gradient topography. The change from a constant to an increasing rate of estuarine-basin sediment accumulation at 7 ka in Figure 8C corresponds to a similar increase in the rate of estuarine-basin sediment accumulation shown in Figure 8B (denoted on the graphs by a 2), indicating an increase in estuarine-basin sediment sequestration or supply at this time. Estuarine sediment accumulation after 6440 cal yr B.P. increases in Figure 8B and decreases in Figure 8C (denoted on the graphs by a 3), which reflects differences in the shapes of the sediment-accretion curves, which increase at a constant rate for Figure 8B and increase at a decreasing rate for Figure 8C.

DISCUSSION

Sea level reaching the elevation of low-gradient antecedent topography either increases the volume of sediment exported out of an estuary (e.g., through inlets to the continental shelf) and/or increases the volume of sediment sequestered in landward alluvial environments. Comparisons of cumulative changes in sediment accommodation with cumulative sediment volume changes, derived from two different sediment-accretion models (Figs. 8B and 8C), show that the volume of sediment in the estuary kept pace with and slightly exceeded sediment accommodation up to around 8.6–8.2 ka, when the curves cross in Figure 8C. Prior to this interval of time, Mobile Bay was an intertidal to supratidal marsh, likely part of the delta-plain environment (Fig. 9A). Remnants of this environment are preserved directly above the exposure surface in the middle of the study area as peat, sampled in core MB-03-4 between −15.15 mbsl and <−14.55 mbsl and dated at 9196–9318 cal yr B.P. The calculations shown in Figure 8C indicate that between 8.6 and 8.2 ka, parts of the study area were initially submerged, and the central-basin depositional environment was created. Sedimentary facies and dates in the cores verify these calculations and the assumptions on which they are based; no central-basin sediment predates 8.2 ka. The oldest central-basin sediment date, 8069–8190 cal yr B.P., is from an articulated *Corbula contracta* at −13.68 mbsl in core MB-03-5. The pre–8.2 ka date of 8976–9029 cal yr B.P. at −14.12 mbsl was obtained from a piece of wood sampled at the exposure surface, which was likely remnant marsh that existed in Bon Secour Bay prior to inundation and associated erosion.

Even though between 8960 and 8680 cal yr B.P. the rate of creation of sediment accommodation in the study area was the highest, because sea-level rise was rapid and reached the elevation of the low-gradient antecedent topography in Bon Secour Bay, estuarine-basin sediment accumulation kept pace with accommodation, which prevented inundation. Between 8680 and 8400 cal yr B.P., estuarine-basin sediment accumulation decreased sharply (Fig. 8B). Low-gradient antecedent topography also exists in the area of the Mobile Delta, north of Interstate 10 (Figs. 2 and 6). As sea level approached the elevation of low-gradient antecedent topography, sediment accommodation was also being created in the floodplain, inducing sediment accretion updip and causing a decrease in estuarine-basin sediment accumulation. Although the more conservative approach to modeling estuarine-basin sediment accumulation (Fig. 8C) does not show a decrease, it does show a change from an increasing to constant rate of estuarine-basin sediment accumulation, which also suggests a landward shift in deposition. By 8200 cal yr B.P., the central-basin depositional environment formed (Fig. 9B), and, immediately prior to this time, erosion associated with bay ravinement removed material along the bay shorelines. The eroded delta-plain sediment was likely transported through tidal channels into the Gulf of Mexico, contributing to the decrease in estuarine-basin sediment accumulation. This explains the discontinuous preservation of early Holocene delta-plain sediment in the middle of Mobile Bay. Both Figures 8B and 8C show an increase in modeled estuarine-basin sediment accumulation around 7000 cal yr B.P., a time when the creation of sediment accommodation was decreasing at a rapid rate as a result of sea-level rising at a low rate and inundating relatively high-gradient antecedent topography (Fig. 8A). This decrease in creation of accommodation also influenced the floodplain and delta plain and may have increased sediment bypass updip, causing the observed increase in estuarine-basin sediment accumulation.

CONCLUSIONS

Based on comparisons of changes in modeled estuarine-basin sediment accumulation with cumulative changes in sediment

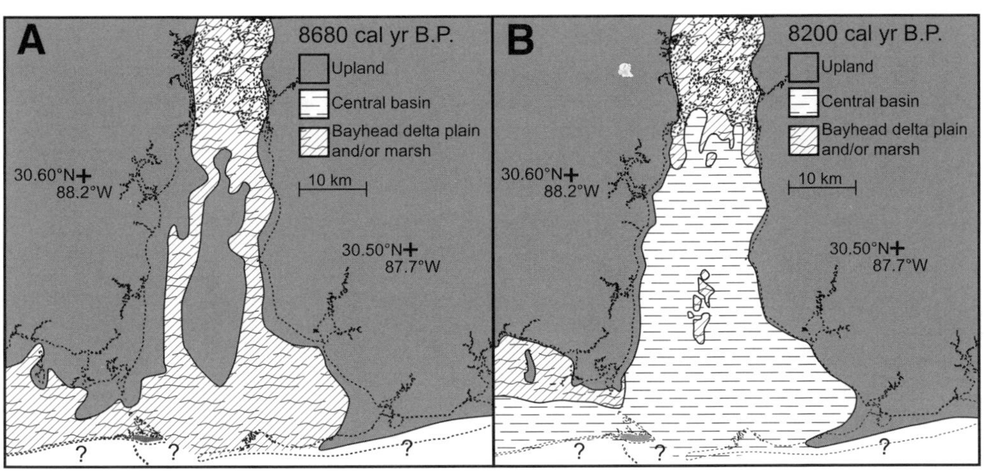

Figure 9. Change in extent of depositional environments between 8680 cal yr B.P. (A) and 8200 cal yr B.P. (B). During this time period, the bayhead delta-plain depositional environment is replaced with central-basin and Mobile Bay forms.

accommodation and sediment volume, between 8680 and 8200 cal yr B.P., the Mobile Bay shoreline transgressed up the axis of the valley at a minimum rate of 100 m/yr. During this transgression, a widespread delta plain and marsh was eroded and replaced by a central-basin environment (Fig. 9). The time period is based on the sharp change in modeled estuarine-basin sediment accumulation (8400 and 8680 cal yr B.P.) and the time when the two curves of cumulative changes in sediment accommodation cross the sediment volume curve (8200 and 8600 cal yr B.P.). This is a much more accurate approximation of inundation rate than simply basing the calculation on the distance and age of the oldest central-basin sediment in Bon Secour Bay (downdip) and the oldest bayhead delta sediments at the modern bayhead delta (updip), because the modeling results incorporate the whole study area, as opposed to just two points. After 8200 cal yr B.P., the bayhead delta and central-basin depositional environments of the study area were relatively static. This suggests that a threshold was crossed during the early Holocene that created a short period of rapid environmental change. The largest post–8200 cal yr B.P. change was accretion of Morgan Peninsula and formation of Main Pass tidal inlet around 6.7 ka (Blum et al., 2003; Rodriguez and Meyer, 2006), which corresponds to the period when creation of sediment accommodation was decreasing and estuarine-basin sediment accumulation was increasing.

Between 8960 and 8680 cal yr B.P., the rate of creation of sediment accommodation and estuarine-basin sediment accumulation was highest. Between 8680 and 8400 cal yr B.P., the rate of estuarine-basin sediment accumulation decreased dramatically, while the rate of creation of sediment accommodation was still increasing rapidly (Figs. 8A and 8B). This change in sediment sequestration and/or supply was likely a result of the high rate of creation of sediment accommodation that shifted deposition landward and facilitated erosion in the vicinity of Mobile Bay through inundation and ravinement. At 7000 cal yr B.P., the rate of estuarine-basin sediment accumulation increased, while the rate of creation of sediment accommodation was decreasing rapidly (Fig. 8), suggesting a shift in deposition seaward (toward the estuary); however, this change was not significant enough to shift the location of depositional environments.

Constraints on sediment accommodation space during the evolution of an estuary are essential for linking observed depositional changes with their associated forcing mechanisms. Some estuaries and coastal areas are inherently unstable, characterized by low-gradient and low-elevation topography. Without the low-gradient antecedent topography between −14 and −15 mbsl, depositional environments would likely have responded to sea-level rise between 9 and 8 ka in a much more gradual manner. The modern Mobile Delta is relatively flat and close to sea level. As the creation of sediment accommodation continues to increase as a result of accelerated rates of sea-level rise and dams continue to sequester sediment away from the bayhead delta, the area may be close to reaching another threshold. The Highway 90 causeway parallels Interstate 10 at the delta front. This structure acts as a buffer between the delta and central-basin depositional environments by restricting the exchange of water and sediment. If the causeway is removed, which is a common topic of local debate, it will certainly alter the delicate balance between sediment sequestration and the creation of sediment accommodation that exists in the area today. The likely effect would be retreat of the Mobile bayhead delta up the valley.

ACKNOWLEDGMENTS

This research was supported by the National Science Foundation grant EAR-0107650 and the American Chemical Society Petroleum Research Fund grant ACS-PRF 36694-G8. Valuable field and laboratory assistance was provided by Craig Meyer, Robin Mattheus, Joe Lambert, and Tom Creech. Christopher Sommerfield and John Jaeger are thanked for providing insightful and constructive reviews that improved the manuscript.

REFERENCES CITED

Allen, G.P., and Posamentier, H.W., 1993, Sequence stratigraphy and facies model of an incised valley fill; the Gironde Estuary, France: Journal of Sedimentary Research, v. 63, p. 378–391.

Bartek, L.R., Cabote, B.S., Young, T., and Schroeder, W., 2004, Sequence stratigraphy of a continental margin subjected to low-energy and low-sediment-supply environmental boundary conditions: Late Pleistocene–Holocene deposition offshore Alabama, USA, *in* Anderson, A.B., and Fillon, R.H., eds., Late Quaternary Stratigraphic Evolution of the Northern Gulf of Mexico Margin: Society for Sedimentary Geology Special Publication 79, p. 85–109.

Blum, M.D., Carter, A.E., Zayac, T., and Goble, R., 2002, Middle Holocene sea-level and evolution of the Gulf of Mexico: Journal of Coastal Research, v. 36, Special Issue, p. 65–80.

Blum, M.D., Sivers, A.E., Zayac, T., and Goble, R., 2003, Middle Holocene sea-level and the evolution of the Gulf of Mexico coast: Transactions of the Gulf Coast Association of Geological Societies, v. 53, p. 64–77.

Bratton, J.F., Colman, S.M., Thieler, E.R., and Seal, R.R.I., 2003, Birth of the modern Chesapeake Bay estuary between 7.4 and 8.2 ka and implications for global sea-level rise: Geo-Marine Letters, v. 22, p. 188–197.

Cabote, B.S., 1998, Seismic Stratigraphy and Stochastic Models of Late Quaternary Lowstand and Transgressive Deposition, Mobile Incised Valley System, Offshore Alabama [M.S. thesis]: Tuscaloosa, University of Alabama, 311 p.

Colman, S.M., Halka, J.P., and Hobbs, C.H., III, 1992, Patterns and rates of sediment accumulation in the Chesapeake Bay during the Holocene rise in sea level, *in* Fletcher, C., and Wehmiller, J.F., eds., Quaternary coasts of the United States: Marine and Lacustrine Systems: Society for Sedimentary Geology Special Publication 48, p. 101–111.

Dalrymple, R.W., Zaitlin, B.A., and Boyd, R., 1992, Estuarine facies models: Conceptual basis and stratigraphic implications: Journal of Sedimentary Petrology, v. 62, p. 1130–1146.

Davies, D.J., and Hummell, R.L., 1994, Lithofacies evolution from transgressive to highstand systems tracts, Holocene of the Alabama coastal zone: Gulf Coast Association of Geological Societies Transactions, v. 44, p. 145–153.

Fletcher, C.H., Knebel, H.J., and Kraft, J.C., 1990, Holocene evolution of an estuarine coast and tidal wetlands: Geological Society of America Bulletin, v. 102, p. 283–297, doi: 10.1130/0016-7606(1990)102<0283:HEOAEC>2.3.CO;2.

Gastaldo, R.A., 1989, Preliminary observations on phytotaphonomic assemblages in a subtropical/temperate Holocene bayhead delta: Mobile Delta, Gulf Coastal Plain, Alabama: Review of Palaeobotany and Palynology, v. 58, p. 61–83, doi: 10.1016/0034-6667(89)90057-2.

Greene, D.L., 2006, Seismic Stratigraphy and Late Quaternary Evolution of Mobile Bay and Mississippi Sound, Alabama—A Record of Large- and Small-Scale Fluvial Systems through Multiple Sea-level Cycles [Ph.D. thesis]: Tuscaloosa, University of Alabama, 234 p.

Greene, D.L., Rodriguez, A.B., and Anderson, J.B., 2007, Seaward-branching coastal-plain and piedmont incised-valley systems through multiple sea-level cycles: Late Quaternary examples from Mobile Bay and Mississippi Sound, USA: Journal of Sedimentary Research, v. 77, p. 139–158, doi: 10.2110/jsr.2007.016.

Hori, K., Saito, Y., Zhao, Q., and Wang, P., 2002, Evolution of the coastal depositional systems of the Changjiang (Yangtze) River in response to late Pleistocene–Holocene sea-level changes: Journal of Sedimentary Research, v. 72, p. 884–897, doi: 10.1306/052002720884.

Hughen, K.A., Baillie, M.G.L., Bard, E., Bayliss, A., Beck, J.W., Bertrand, C.J.H., Blackwell, P.G., Buck, C.E., Burr, G.S., Cutler, K.B., Damon, P.E., Edwards, R.L., Fairbanks, R.G., Friedrich, M., Guilderson, T.P., Kromer, B., McCormac, F.G., Manning, S.W., Bronk Ramsey, C., Reimer, P.J., Reimer, R.W., Remmele, S., Southon, J.R., Stuiver, M., Talamo, S., Taylor, F.W., van der Plicht, J., and Weyhenmeyer, C.E., 2004, Marine04: Marine radiocarbon age calibration 26–0 ka BP: Radiocarbon, v. 46, p. 1059–1086.

Hummell, R.L., 1996, Holocene Geologic History of the West Alabama Inner Continental Shelf, Alabama: Geological Survey of Alabama Circular 189, p. 131.

Hummell, R.L., and Parker, S.J., 1995a, Holocene Geologic History of Mobile Bay, Alabama: Geological Survey of Alabama Circular 186, 97 p.

Hummell, R.L., and Parker, S.J., 1995b, Holocene Geologic History of Mississippi Sound, Alabama: Geological Survey of Alabama Circular 185, 91 p.

Jervey, M.T., 1988, Quantitative geological modeling of siliciclastic rock sequences and their seismic expression, *in* Wilgus, C.K., et al., eds., Sea-Level Changes: An Integrated Approach: Society for Sedimentary Geology Special Publication 42, p. 47–69.

Kindinger, J.L., Balson, P.S., and Flocks, J.G., 1994, Stratigraphy of the Mississippi-Alabama shelf and the Mobile River incised-valley system, *in* Dalrymple, R.W., Boyd, R., and Zaitlin, B.A., eds., Incised-Valley Systems: Origin and Sedimentary Sequences: Society for Sedimentary Geology Special Publication 51, p. 83–95.

Knowles, S.C., 1989, Analysis of barrier island dynamics for ship channel planning at Ship Island, Mississippi, *in* Stauble, D.K., and Magoon, O.T., eds., Barrier Islands: Process and Management: Charleston, American Society of Civil Engineers, p. 238–252.

Lin, C.-M., Zhuo, H.-C., and Gao, S., 2005, Sedimentary facies and evolution in the Qiantang River incised valley, eastern China: Marine Geology, v. 219, p. 235–259, doi: 10.1016/j.margeo.2005.06.009.

Mars, J.C., Shultz, A.W., and Schroeder, W.W., 1992, Stratigraphy and Holocene evolution of Mobile Bay in southwestern Alabama: Gulf Coast Association of Geological Societies Transactions, v. 42, p. 529–542.

Massey, A.C., Paul, M.A., Gehrels, W.R., and Charman, D.J., 2006, Autocompaction in Holocene coastal back-barrier sediments from south Devon, southwest England, UK: Marine Geology, v. 226, p. 225–241, doi: 10.1016/j.margeo.2005.11.003.

May, E.B., 1976, Holocene Sediments of Mobile Bay, Alabama: Alabama Marine Resources Bulletin 11, 25 p.

McBride, R.A., Byrnes, M.R., Penland, S., Pope, D.L., and Kindinger, J.L., 1991, Geomorphic history, geologic framework, and hard mineral resources of the Petit Bois Pass area, Mississippi-Alabama: Houston, Gulf Coast Section, Society for Sedimentary Geology (SEPM) Foundation, 12th Annual Research Conference, Program and Abstracts, p. 116–127.

McPherson, R.M., 1970, The hydrography of Mobile Bay and Mississippi Sound, Alabama: Alabama Marine Science Institute: Marine Science Journal, v. 1, p. 1–83.

Nichol, S.L., Boyd, R., and Penland, S., 1996, Sequence stratigraphy of a coastal-plain incised valley estuary: Lake Calcasieu, Louisiana: Journal of Sedimentary Research, v. 66, p. 847–857.

Nichols, M.M., 1989, Sediment accumulation rates and relative sea-level rise in lagoons: Marine Geology, v. 88, p. 201–219, doi: 10.1016/0025-3227(89)90098-4.

Otvos, E.G., 2004, Holocene gulf levels: Recognition issues and an updated sea-level curve: Journal of Coastal Research, v. 20, p. 680–699, doi: 10.2112/1551-5036(2004)20[680:HGLRIA]2.0.CO;2.

Paul, M.A., and Barras, B.F., 1998, A geotechnical correction for post-depositional sediment compression: Examples from the Forth valley, Scotland: Journal of Quaternary Science, v. 13, p. 171–176, doi: 10.1002/(SICI)1099-1417(199803/04)13:2<171::AID-JQS369>3.0.CO;2-Z.

Posamentier, H.W., and Allen, G.P., 1999, Siliciclastic Sequence Stratigraphy—Concepts and Applications: Society for Sedimentary Geology, Concepts in Sedimentology and Paleontology, no. 7, 210 p.

Posamentier, H.W., Jervey, M.T., and Vail, P.R., 1988, Eustatic controls on clastic deposition. I: Conceptual framework, *in* Wilgus, C.K., et al., eds., Sea-Level Changes: An Integrated Approach: Society for Sedimentary Geology Special Publication 42, p. 109–124.

Reading, H.G., 2000, Sedimentary Environments: Processes, Facies and Stratigraphy: Oxford, Blackwell Science, 688 p.

Reimer, P.J., Baillie, M.G.L., Bard, E., Bayliss, A., Beck, J.W., Bertrand, C.J.H., Blackwell, P.G., Buck, C.E., Burr, G.S., Cutler, K.B., Damon, P.E., Edwards, R.L., Fairbanks, R.G., Friedrich, M., Guilderson, T.P., Hogg, A.G., Hughen, K.A., Kromer, B., McCormac, F.G., Manning, S.W., Ramsey, C.B., Reimer, R.W., Remmele, S., Southon, J.R., Stuiver, M., Talamo, S., Taylor, F.W., van der Plicht, J., and Weyhenmeyer, C.E., 2004, IntCal04: Terrestrial radiocarbon age calibration 26–0 ka BP: Radiocarbon, v. 46, p. 1029–1058.

Rodriguez, A.B., and Meyer, C., 2006, Sea-level variation during the Holocene deduced from the morphologic and stratigraphic evolution of Morgan Peninsula, Alabama, USA: Journal of Sedimentary Research, v. 76, p. 257–269, doi: 10.2110/jsr.2006.018.

Rodriguez, A.B., Anderson, J.B., and Simms, A.R., 2005, Terrace inundation as an autocyclic mechanism for parasequence formation: Galveston estuary, Texas, USA: Journal of Sedimentary Research, v. 75, p. 608–620, doi: 10.2110/jsr.2005.050.

Ryan, J.J., 1969, A Sedimentological Study of Mobile Bay, Alabama [Ph.D. thesis]: Tallahassee, Florida State University, 110 p.

Ryan, J.J., and Goodell, H.G., 1972, Marine geology and estuarine history of Mobile Bay, Alabama, *in* Nelson, B.W., ed., Environmental Framework of Coastal Plain Estuaries: Geological Society of America Memoir 133, p. 517–554.

Smith, W.E., 1988, Geomorphology of the Mobile Delta: Geological Survey of Alabama Circular 132, 131 p.

Stanley, D.J., 2001, Dating modern deltas: Progress, problems and prognostics: Annual Review of Earth and Planetary Sciences, v. 29, p. 257–294, doi: 10.1146/annurev.earth.29.1.257.

Stanley, D.J., and Hait, A.K., 2000, Deltas, radiocarbon dating, and measurements of sediment storage and subsidence: Geology, v. 28, p. 295–298, doi: 10.1130/0091-7613(2000)28<295:DRDAMO>2.0.CO;2.

Stuiver, M., and Reimer, P.J., 1993, Extended ^{14}C database and revised CALIB radiocarbon calibration program: Radiocarbon, v. 35, p. 215–230.

Thomas, M.A., and Anderson, J.B., 1994, Sea-level controls on the facies architecture of the Trinity/Sabine incised-valley system, Texas continental shelf, *in* Dalrymple, R.W., Boyd, R., and Zaitlin, B.A., eds., Incised-Valley Systems: Origin and Sedimentary Sequences: Society for Sedimentary Geology Special Publication 51, p. 63–82.

Törnqvist, T.E., Gonzalez, J.L., Newsom, L., van der Borg, K., de Jong, A.F.M., and Kurnik, C.W., 2004, Deciphering Holocene sea-level history on the U.S. Gulf Coast: A high-resolution record from the Mississippi Delta: Geological Society of America Bulletin, v. 116, p. 1026–1039, doi: 10.1130/B2525478.1.

Törnqvist, T.E., Bick, S.J., van der Borg, K., and de Jong, A.F.M., 2006, How stable is the Mississippi Delta?: Geology, v. 34, p. 697–700, doi: 10.1130/G22624.1.

Toscano, M.A., and Macintyre, I.G., 2003, Corrected western Atlantic sea-level curve for the last 11,000 years based on calibrated ^{14}C dates from *Acropora palmata* framework and intertidal mangrove peat: Coral Reefs, v. 22, p. 257–270, doi: 10.1007/s00338-003-0315-4.

Wilkinson, B.H., and Byrne, J.R., 1977, Lavaca Bay—Transgressive deltaic sedimentation in a central Texas estuary: American Association of Petroleum Geologists Bulletin, v. 61, p. 527–545.

Manuscript Accepted by the Society 10 March 2008

The Geological Society of America
Special Paper 443
2008

Sediment accommodation control on estuarine evolution: An example from Weeks Bay, Alabama, USA

Antonio B. Rodriguez[†]
Institute of Marine Sciences, University of North Carolina at Chapel Hill, 3431 Arendell Street, Morehead City, North Carolina 28557, USA

Diana M. Duran
ExxonMobil Exploration Company, Houston, Texas 77210, USA

Christopher R. Mattheus
Institute of Marine Sciences, University of North Carolina at Chapel Hill, 3431 Arendell Street, Morehead City, North Carolina 28557, USA

John B. Anderson
Department of Earth Science, Rice University, 6100 Main Street, MS 126, Houston, Texas 77005, USA

ABSTRACT

The evolution of depositional environments is strongly controlled by the rate at which sediment accommodation is created; however, this factor is seldom quantified in studies of estuarine evolution. From a detailed map of the exposure surface, or bay line, and a precise sea-level curve, we calculated sediment accommodation over the last 8.4 k.y. in Weeks Bay, Alabama, and compared sediment accommodation with the late Quaternary evolution of the bay derived from seismic and lithologic data. The stratigraphy of the Fish and Magnolia paleovalleys, the two fluvial systems that discharge into the bay, is composed of multiple sequences. Due to data limitations, only the upper sequence boundary was mapped regionally. This surface formed in response to the oxygen isotope stage 2 lowstand, and it is overlain by alluvial and estuarine deposits that are separated by an exposure surface. Estuarine sediments were first deposited at 7200 cal yr B.P., and no bayhead delta sediments are recognized above the bay line, indicating that the Fish and Magnolia bayhead deltas back-stepped across the bay. Bayhead delta back-stepping occurred when the rate of sea-level rise was decreasing. A map of the exposure surface shows it to be broad and relatively flat. When this antecedent topography was flooded, the rate at which sediment accommodation was created increased rapidly, forcing the bayhead delta to step back. The morphology of the land being inundated is the most significant control on changes in sediment accommodation and, therefore, the most important factor to quantify when predicting the future response of coastal systems to sea-level rise.

Keywords: estuary, bayhead delta, sea level, Holocene, incised valley.

[†]E-mail: abrodrig@email.unc.edu.

Rodriguez, A.B., Duran, D.M., Mattheus, C.R., and Anderson, J.B., 2008, Sediment accommodation control on estuarine evolution: An example from Weeks Bay, Alabama, USA, *in* Anderson, J.B., and Rodriguez, A.B., eds., Response of Upper Gulf Coast Estuaries to Holocene Climate Change and Sea-Level Rise: Geological Society of America Special Paper 443, p. 31–42, doi: 10.1130/2008.2443(03). For permission to copy, contact editing@geosociety.org. ©2008 The Geological Society of America. All rights reserved.

INTRODUCTION

Estuaries are the seaward portion of drowned valley systems (Dalrymple et al., 1992). The impact that morphologic variations in the underlying exposure surface (antecedent topography) have on estuarine evolution in response to sea-level rise should be profound. This is because the rate of relative sea-level rise and the morphology of the land being inundated dictate the rate at which sediment accommodation is created. The evolution of depositional environments is predominantly controlled by both the rate at which sediment accommodation is created and the rate of sediment accumulation (Posamentier et al., 1988; Van Wagoner et al., 1990).

To fully understand the forcing mechanisms behind past nonlinear changes in estuarine depositional environments, it is essential to constrain the antecedent topography. Heap and Nichol (1997), Bratton et al. (2003), and Rodriguez et al. (2005) recognized that the evolution of estuaries is strongly controlled by variations in antecedent morphology. The absence of bayhead delta facies in the Weiti River (New Zealand) incised valley-fill sequence was partially attributed to the presence of a fluvial terrace that affected the creation of sediment accommodation during sea-level rise (Heap and Nichol, 1997). Initiation of estuarine sedimentation in Chesapeake Bay around 7.4–8.2 cal yr B.P. was essentially simultaneous (within analytical error) over an upstream distance of at least 100 km (Bratton et al., 2003). This rapid marine flooding event was attributed to either an increase in the rate of sea-level rise or sea level rising at a constant rate and crossing a topographic break in slope from high gradient to low gradient (Bratton et al., 2003). Rodriguez et al. (2005) recognized that terrace inundation in the Galveston Estuary, Texas, caused a sudden change in the slope of inundation, which resulted in reorganization of the estuarine environments and modification of the estuarine processes. Specifically, when sea level inundated a −10 m fluvial terrace, the Trinity River mouth and bayhead delta shifted landward at a rate of 6.5 km/century, and the associated barrier shoreline was stranded on the inner continental shelf, forming Heald Bank (Rodriguez et al., 2005). Although these previous studies have demonstrated that sea-level rise, and the resulting inundation of irregular coastal topography, forces depositional environments to change at nonlinear rates, they fall short of truly examining changes in sediment accommodation as a forcing mechanism.

Modern coastal topography will strongly influence how estuaries respond to future accelerated rates of sea-level rise (Church et al., 2001). Quantification of Holocene changes in sediment accommodation and examination of estuarine responses will improve predictive coastal-change models. The objective of this study is to examine the impact of interactions between sea-level rise and antecedent topography on the Holocene evolution of Weeks Bay, Alabama.

STUDY AREA

Weeks Bay, a small subsidiary bay located in the northeastern Gulf of Mexico along the eastern shore of Mobile Bay, Alabama (Fig. 1), is well suited for this study. The bay is ~8.0 km^2 and has an average water depth of 1.25 m and volume of 12,000 m^3 (Crance, 1971). Its small size facilitates generation of a detailed map of the antecedent topography (the subaerial exposure surface). Holocene sea level is well constrained for this part of the Gulf of Mexico from basal peat deposits in coastal Louisiana (Törnqvist et al., 2004) and basal peat and shell material collected along the northern Gulf of Mexico coast (Otvos, 2004). These data show that sea level was rising continuously at a decreasing rate from ca. 8.5 ka to present (Törnqvist et al., 2004; Otvos, 2004). The main fluvial systems that drain into Weeks Bay are the Fish River, from the north, and the Magnolia River, from the east (Fig. 1). The Fish and Magnolia Rivers are small, draining 427 km^2 and 111 km^2, respectively, and they are constrained to the coastal plain. Weeks Bay receives saltwater from the Gulf of Mexico through Main Pass tidal inlet via Mobile Bay. Average salinities range from 0 to 20 psu and are strongly related to the salinity of the Mobile Bay estuary, which is controlled by seasonal discharge of the Mobile River system and water discharge of the Fish and Magnolia Rivers (Schroeder et al., 1990, 1992). The bay has a diurnal microtidal regime, and tidal currents range in amplitude from 0.15 to 0.3 m/s (Schroeder et al., 1992).

METHODS

To evaluate the control of interactions between sea level and antecedent topography on the evolution of Weeks Bay, it is necessary to create a contour map of the exposure surface, determine the type of depositional environment(s) that formed above the exposure surface and the timing of inundation, and place these data into context with changing sea level. This was accomplished by seismic and core studies and radiocarbon dating.

Seismic Data

An EdgeTech SB-216S seismic system was used to collect 35 km of two-dimensional seismic data from Weeks Bay (Fig. 1). The system was configured to emit a 20 ms linear sweep from 2 to 15 kHz at a rate of 4 Hz and was towed at 3–4 knots. Time to depth conversion was calculated using a velocity of 1500 m/s. This velocity was verified by cores that sampled changes in lithology at the same depths as seismic facies transitions. Hard copies of the seismic data were used for data interpretation. Seismic surfaces identified were digitized using the EdgeTech MIDAS data acquisition system and imported into Surfer® 8.0 to generate contour maps.

Cores

Vibracores (9) and rotary-drill cores (2) were obtained from Weeks Bay, and cores oriented in an east-west transect across the Fish River bayhead delta (6) were collected using a truck-mounted GeoProbe™ coring system (Fig. 1). Cores collected

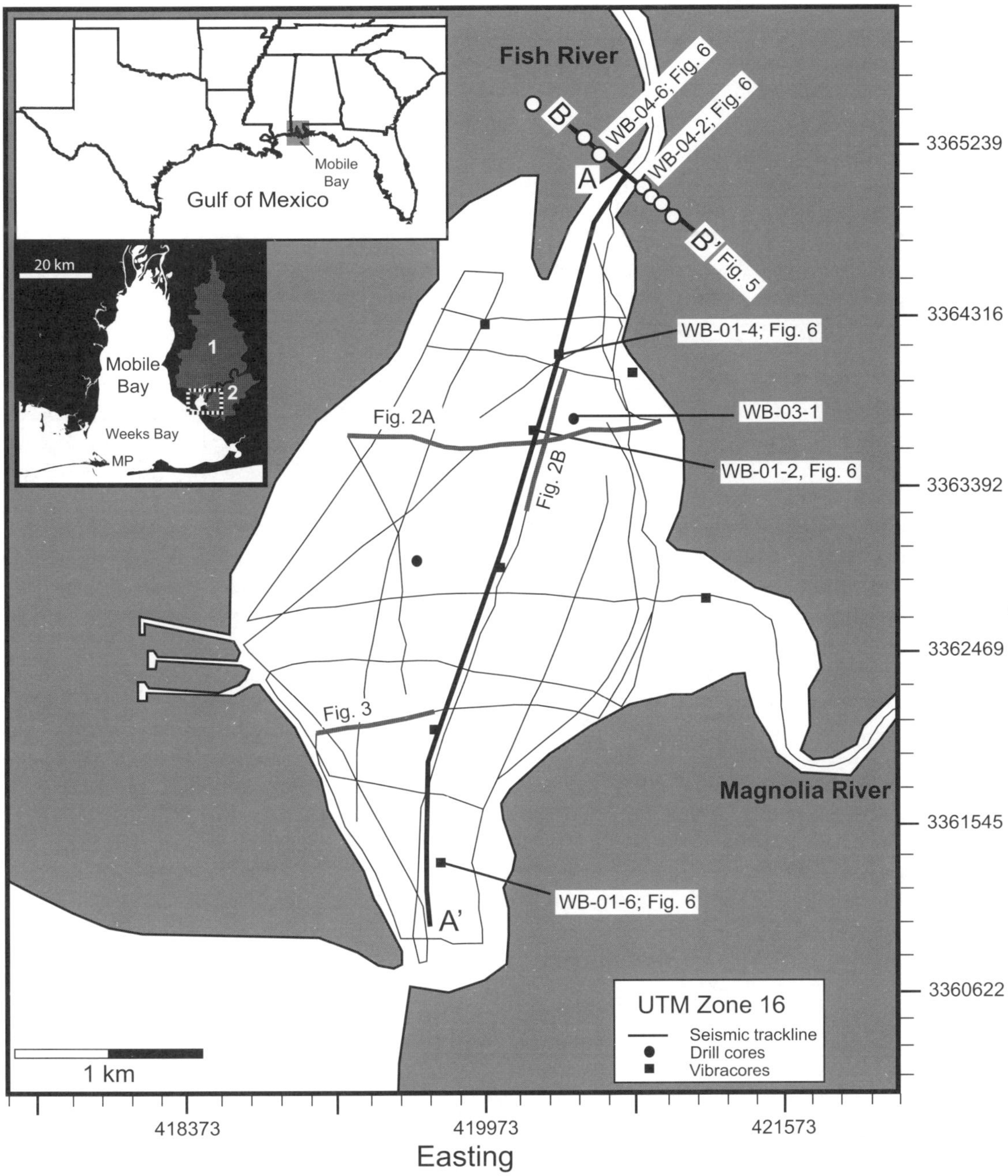

Figure 1. Study area map showing the locations of seismic lines and cores collected for this study. Inset map shows the extent of the Fish River (1) and the Magnolia River (2) drainage basins. MP—Main pass tidal inlet.

across the Fish River bayhead delta were supplemented with platform-boring descriptions (21) from the Alabama Department of Transportation's Bureau of Materials and Tests. These cores were collected in an east-west transect ~30 m to the south of the GeoProbe[tm] cores. Cores were used to characterize depositional environments, ground truth seismic facies to lithology, and to provide shell and wood material for radiocarbon dating (Fig. 1). Grain-size analysis, using laser diffraction (Malvern Mastersizer 2000) for the vibracores and sieves for the Geoprobe cores, was conducted to aid in differentiating depositional environments. The Mastersizer 2000 measures particle sizes from 0.02 µm to 2000 µm, and the selected sieves separated particles into four classes, <63 µm, 0.25–0.5 mm, 0.5–1.0 mm, and 1.0–2.0 mm.

Dating

Radiocarbon dating (accelerator mass spectrometry [AMS]) of shell and organic material was performed by the Woods Hole Oceanographic Institution and Beta Analytic (Table 1). The dates were calibrated using CALIB 5.0.2 software (Stuiver and Reimer, 1993). To ensure that dated material was preserved in situ, whenever possible, individual wood fragments and articulated bivalves were chosen for dating instead of bulk peat or unpaired valves. Species from well-defined macrofossil assemblages, indicative of the depositional environment, were preferentially chosen for dating.

DEPOSITIONAL UNITS AND BOUNDING SURFACES

Regional Seismic-Stratigraphic Surfaces

In Weeks Bay, two regional seismic surfaces (A and B) were recognized (Figs. 2 and 3). Both surfaces show high-relief, erosional truncation of seismic reflectors below and onlap above. A deeper seismic surface with similar characteristics as surfaces A and B was also recognized (Figs. 2 and 3; dashed line); however, this surface could only be traced for short distances and in many places did not rise in elevation along the bay shoreline. Seismic surfaces A and B are deepest in the center of the bay; the lower surface (A) downcuts to a maximum of –26 m, and the upper surface (B) downcuts to a maximum of –15 m. Along the bay shoreline, surfaces A and B rise in elevation to –11 m and –1 m, respectively. The surfaces extend into Mobile Bay, where Greene et al. (2007) interpreted them as sequence boundaries associated with oxygen isotope stages (OIS) 6 and 2, respectively. Due to seismic attenuation, surface A was not mapped throughout the entire bay. A structure-contour map of surface B shows that east-west–oriented cross sections are generally V-shaped (Fig. 4A), interpreted to represent the Fish and Magnolia paleovalleys. Just south of the Magnolia River mouth, in the center of Weeks Bay, the depth of surface B increases where the paleovalleys merge.

Seismic-Stratigraphic Units

Above surface A, four main seismic units (SU-1 to SU-4) were recognized (Figs. 2 and 3). All units thicken toward the center of the bay, where surfaces A and B are deepest. Cores only sampled SU-3 and SU-4. SU-1 and SU-2 are located between surfaces A and B. SU-1, the deepest unit, is characterized by acoustically wavy to parallel-planar reflections, and it ranges between 5 and 10 m thick. The contact between SU-1 and the overlying SU-2 unit is erosional and is characterized by small-scale channel scouring (4–8 m deep and 5–10 m wide; Fig. 2A). SU-2 is characterized by inclined-parallel reflections and is 3–6 m thick. This unit was only identified in the middle of the bay, above the area where surface A is deepest. Surface B overlies SU-2 in the center of the bay and SU-1 along the bay margin. Interpretation of SU-1 and SU-2 is difficult without lithologic data, but, based on seismic character and bounding surfaces, SU-1 could be an alluvial plain, and SU-2 may be a bayhead delta. The stratigraphy of Greene et al. (2007) and radiocarbon "dead" samples obtained at the Fish bayhead delta below surface B (Fig. 5; Table 1) suggest that SU-1 and SU-2 were deposited sometime between oxygen isotope stages 6 and 3, but the units could have been deposited earlier.

SU-3 and SU-4 are between surface B and the bay floor (Figs. 2 and 3). SU-3 is characterized by inclined-parallel reflections to a chaotic reflection pattern and is 3–5 m thick. Cores sampled SU-3 as a light brown (5 YR 6/4) to light olive gray (5 Y 6/1)

TABLE 1. RADIOCARBON AGES AND TYPE OF MATERIAL ANALYZED

Lab. code	Sample name	Depth (m)†	Material	Conventional ^{14}C age (yr B.P.; 1σ)	Calibrated ^{14}C age (cal yr B.P.; 1σ)‡
OS-32702	WB-01-2	2.29	Wood	2710 ± 45	2810 ± 40
Beta-207539	WB-01-2	2.81	Wood	4100 ± 40	4580 ± 50
OS-32703	WB-01-2	3.49	Wood	5240 ± 35	5965 ± 35
Beta-207541	WB-01-2	4.84	Wood	6010 ± 40	6845 ± 50
OS-32704	WB-01-2	4.98	Peat	6310 ± 40	7200 ± 20
OS-32699	WB-01-1	2.90	*Ischadium recurvum*	2840 ± 40	2960 ± 40
OS-32701	WB-01-1	4.68	Wood	7080 ± 35	7865 ± 15
OS-32705	WB-01-9	2.08	Wood	3400 ± 45	3585 ± 55
Beta-207540	WB-03-1	5.39	Wood	5710 ± 40	6495 ± 50
OS-46460	WB-03-1	6.39	Wood	6320 ± 35	7260 ± 20
OS-46678	WB-04-3	6.25	Wood	>48,000	N.A.
OS-46679	WB-04-7	6.74	Wood	4620 ± 25	5430 ± 15
OS-46680	WB-04-7	9.82	Wood	12,150 ± 45	14,010 ± 65
OS-47774	WB-04-2	14.27	Wood	20,500 ± 100	24,535 ± 175
OS-46677	WB-04-2	16.45	Wood	>48,000	N.A.

Notes: Samples were processed at the National Science Foundation (NSF) accelerator mass spectrometry (AMS) laboratory at the University of Arizona and Beta Analytic.

†Sample depths (reported parenthetically) are relative to sea level.

‡Calibrated using Stuiver and Reimer (1993), Hughen et al. (2004), and Reimer et al. (2004).

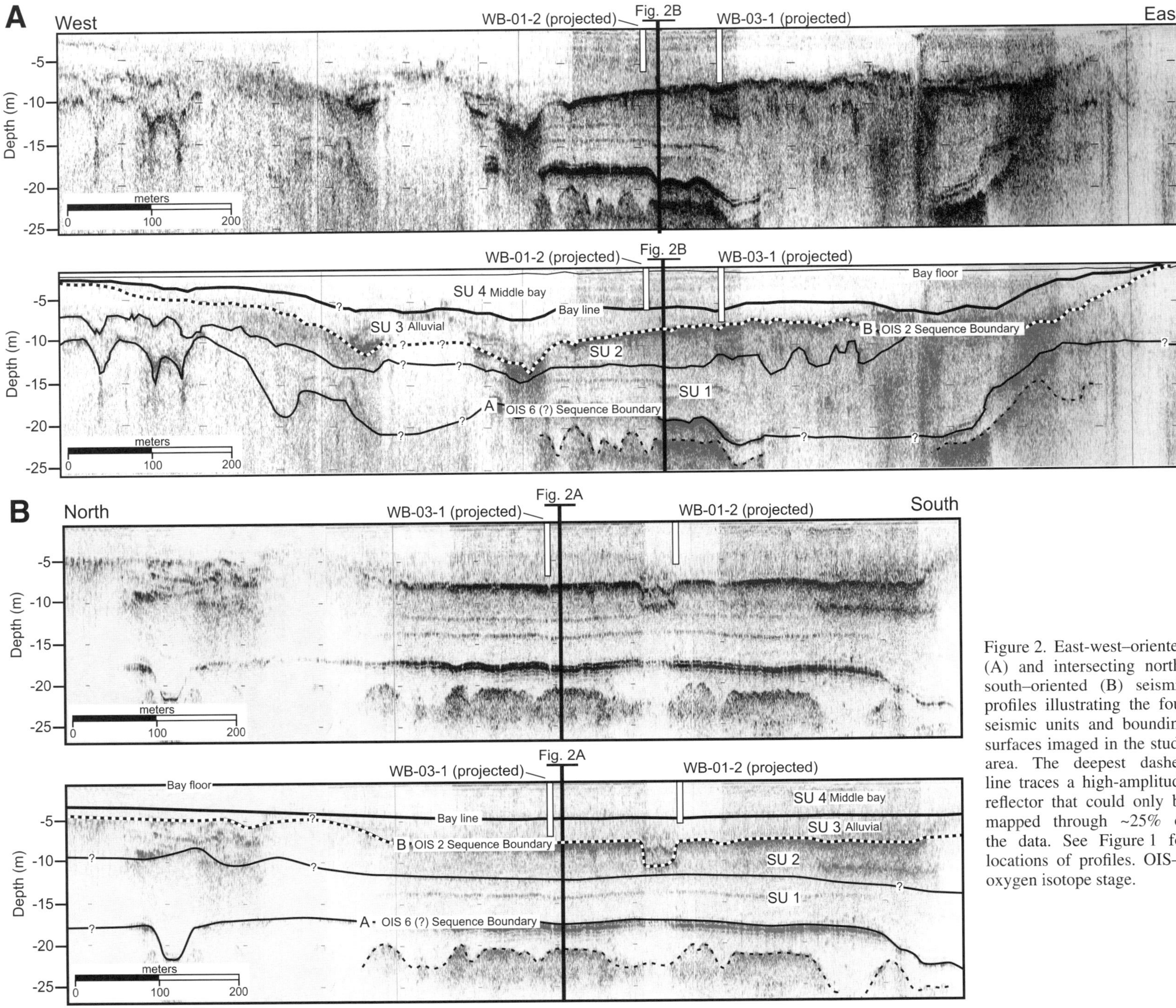

Figure 2. East-west–oriented (A) and intersecting north-south–oriented (B) seismic profiles illustrating the four seismic units and bounding surfaces imaged in the study area. The deepest dashed line traces a high-amplitude reflector that could only be mapped through ~25% of the data. See Figure 1 for locations of profiles. OIS—oxygen isotope stage.

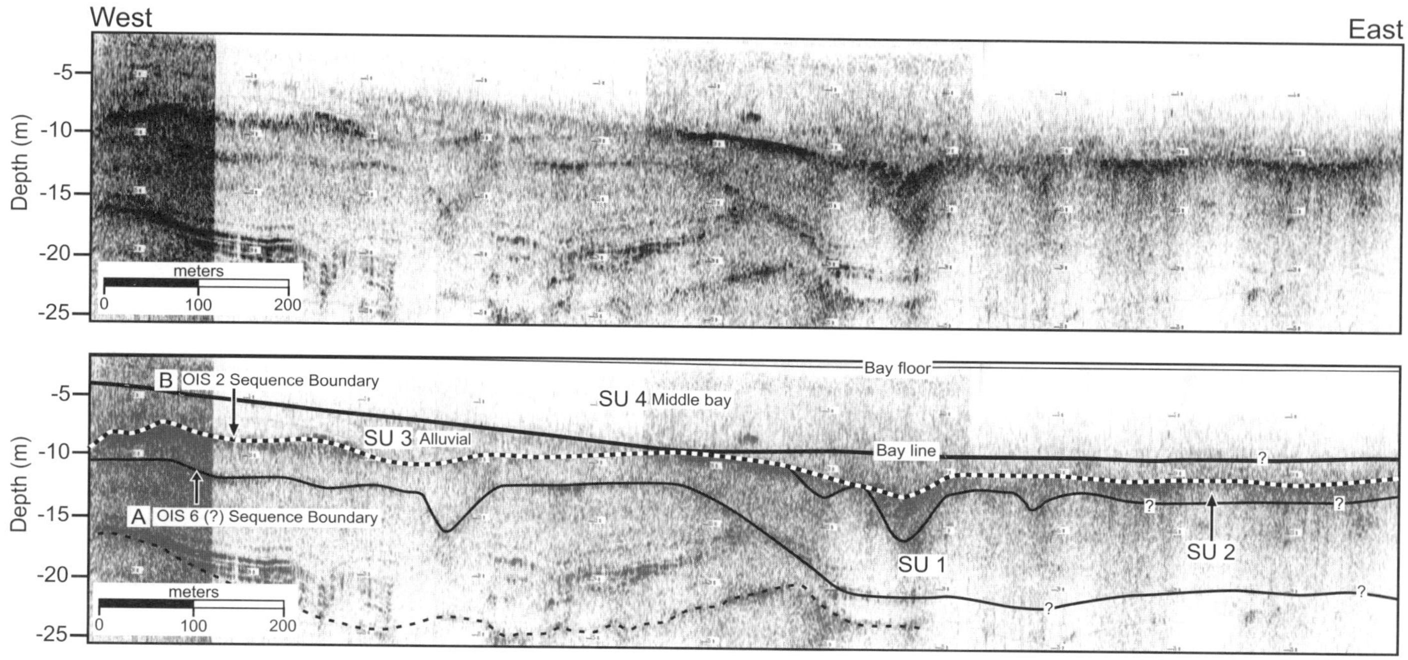

Figure 3. Seismic profile showing the western flank of the Fish and Magnolia paleovalleys. Throughout Weeks Bay, only middle-bay seismic facies were imaged above the bay line. See Figure 1 for location of profile.

sandy clay to an indurated oxidized clay. Based on the lithology and stratigraphic position, directly above the bounding surface of the paleovalleys, SU-3 is interpreted as alluvial (core WB-01-2; Fig. 6). The contact between SU-3 and the overlying SU-4 is recognized as a low-amplitude reflection that amalgamates with surface B toward the bay margin (Fig. 2). Cores show the contact to be sharp, and it is commonly associated with clay rip-up clasts and root penetration (core WB-01-2; Fig. 6). SU-4 is characterized by an acoustically laminated to transparent facies that is 3–9 m thick and extends to the bay floor. At the base of SU-4, cores sampled either a 15–25-cm-thick peat layer interpreted as marsh, or an olive-gray (5 Y 4/1) to dark greenish gray (5 GY 4/1) homogeneous clayey silt (negatively skewed grain-size distribution with a mode of 6.7ϕ) interpreted as middle bay (cores WB-01-6 and WB-01-2; Figs. 6 and 7). The surface separating SU-3 and SU-4 is interpreted as an exposure surface, or bay line, across which the first estuarine deposits accumulated. An individual piece of wood sampled from peat preserved directly above the bay line in core WB-01-2 was radiocarbon dated as 7200 ± 20 cal yr B.P. (Fig. 6; Table 1). The oldest material radiocarbon dated from middle-bay facies directly above the bay line was sampled in core WB-03-1 at −6.39 m (a single piece of wood) and has an age of 7260 ± 20 cal yr B.P. (Table 1). A structure-contour map of the bay line, based on cores and seismic data, shows that east-west–oriented cross sections are generally U-shaped (Fig. 4B). In the lower parts of Weeks Bay, SU-4 contains abundant shell material and is olive-gray sandy clay (5 Y 4/1; core WB-01-6; Fig. 6). Grain-size analyses show a bimodal distribution with modes at 1.9ϕ and 6.8ϕ (Fig. 7). In the upper parts of Weeks Bay, SU-4 was sampled as olive-gray (5 Y 4/1) bioturbated sandy clay with organics and abundant wood fragments (core WB-01-4; Fig. 6). Typical grain-size curves for this upper-bay facies are positively skewed with a mode of 2.3ϕ (Fig. 7).

Bayhead Delta

Core WB-01-4, collected just south of the Fish River mouth, sampled a pale yellowish brown (10 YR 6/2) sand (mode of 1.5ϕ), interpreted as bayhead delta front (channel-mouth bar) overlying upper-bay facies (Figs. 6 and 7). Cores collected on the west side of the Fish River, just landward of the river mouth, sampled <3 m of modern alluvial plain and marsh deposits above an overall coarsening-upward 9–13-m-thick unit (Figs. 5 and 6). The coarsening-upward unit is medium gray (N5) bioturbated sand with wood and clay rip-up clasts at the base and medium dark gray to dark greenish-gray bioturbated sandy clay with wood fragments and clay rip-up clasts at the top; it is interpreted as a prograding bayhead delta (Figs. 5 and 6). Cores collected on the east side of the Fish River, just landward of the river mouth, sampled modern alluvial deposits above up to 9.0 m of homogeneous grayish blue-green (5 BG 5/2) clay interpreted as middle-bay facies (Fig. 5). Below bayhead delta and middle-bay facies, a fine- to coarse-grained, subangular to subrounded, well-sorted, oxidized, dark yellowish-orange to grayish-orange sand and clayey sand was sampled and interpreted as alluvial (Figs. 5 and 6). The contact between the lower alluvial unit and

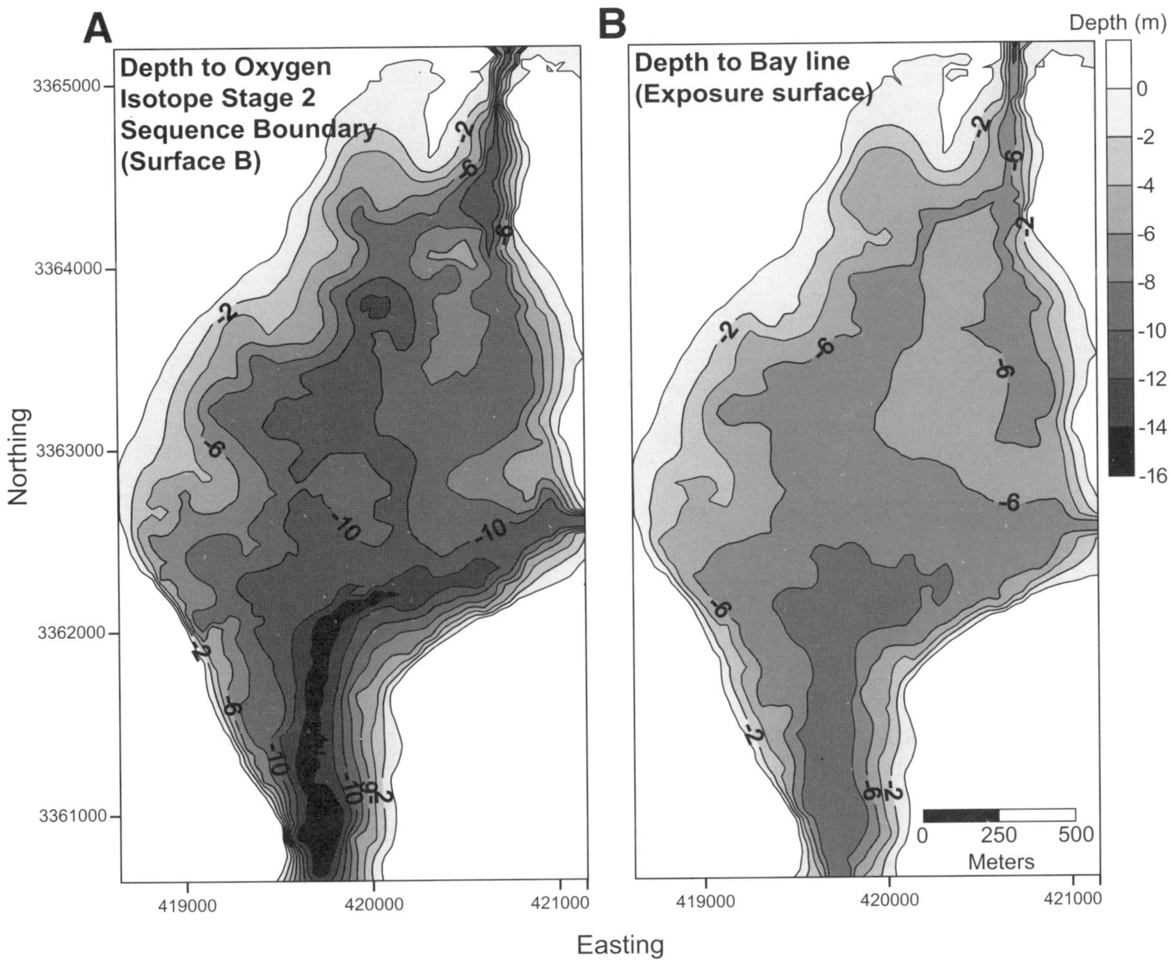

Figure 4. Structure maps of the oxygen isotope stage 2 sequence boundary (A) and the bay line (B). These surfaces amalgamate toward the shoreline. The bay line is relatively flat in the middle of Weeks Bay.

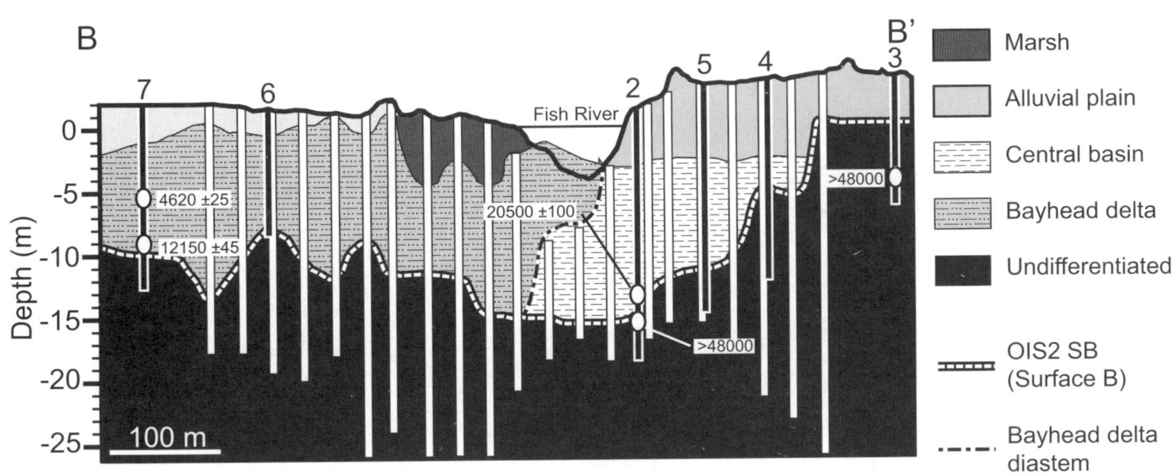

Figure 5. Cross section B–B′ based on platform-boring descriptions obtained from the Alabama Department of Transportation's Bureau of Materials and Tests (white vertical lines) and GeoProbe[tm] cores (black vertical lines). In this location, the bay line is amalgamated with the oxygen isotope stage 2 sequence boundary (OIS2 SB). See Figure 1 for location of cross section and Table 1 for additional information on dates (cal yr B.P.).

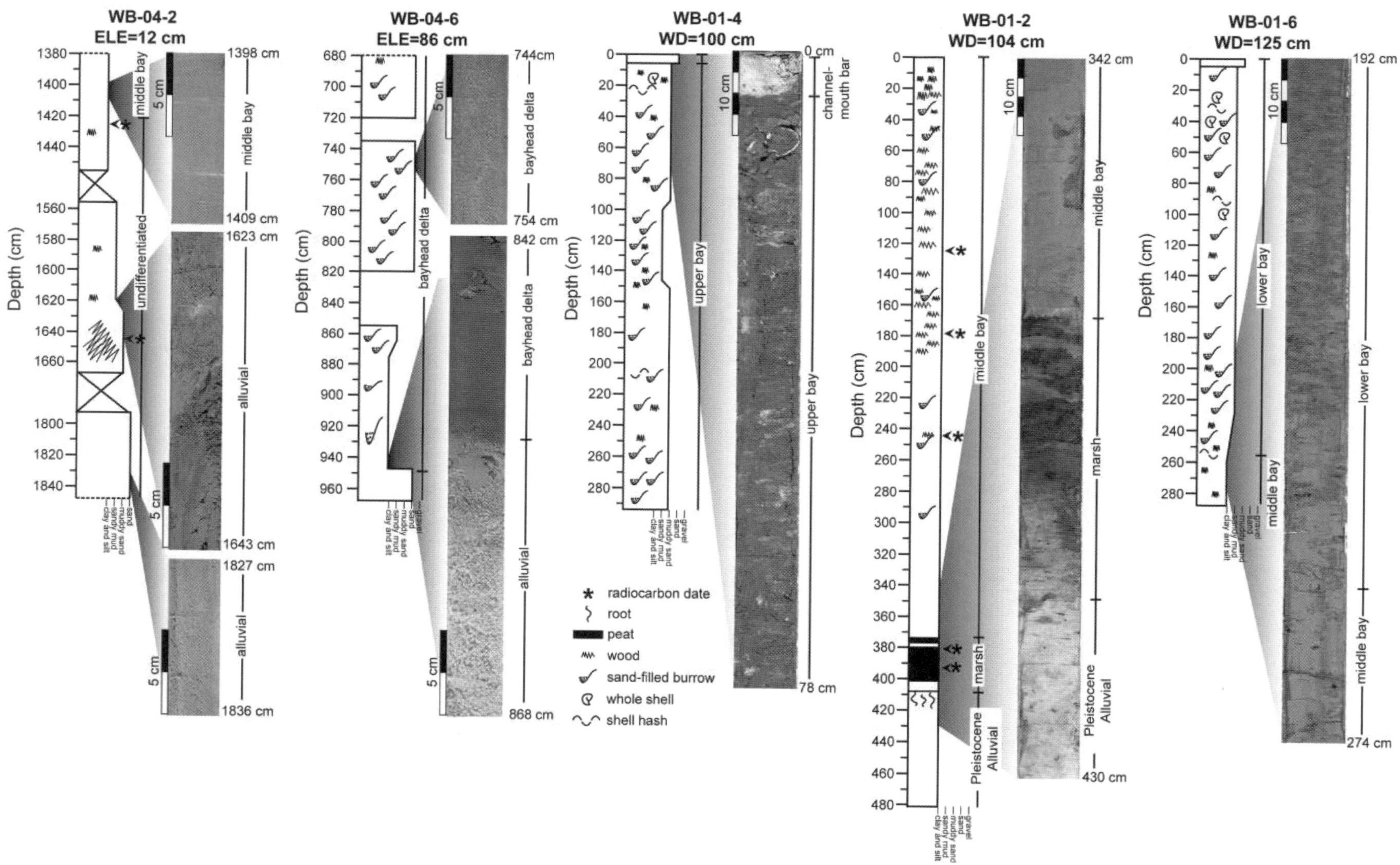

Figure 6. Core photographs and logs showing upper-bay (core WB-01-4), middle-bay (cores WB-04-2, WB-01-2, and WB-01-6), lower-bay (core WB-01-6), alluvial (WB-01-2), bayhead delta (cores WB-04-6 and WB-01-4), and marsh (core WB-01-2) facies. The bay line is shown in core WB-01-2 at the contact between alluvial and marsh deposits. The oxygen isotope stage 2 sequence boundary is shown in cores WB-04-2 and WB-04-6 as the contact between alluvial and middle-bay and bayhead delta deposits, respectively. Note missing sections in core WB-04-2. See Figure 1 for core locations and Table 1 for additional information on dates. ELE—elevation; WD—water depth.

overlying bayhead delta and middle bay units is sharp. Large individual pieces of wood sampled at and below the contact in core WB-04-2 were radiocarbon dated at 24,535 ± 175 cal yr B.P. and >48,000 yr B.P. (radiocarbon dead), respectively. Individual pieces of wood sampled at and 3.08 m above the contact in core WB-04-7 were radiocarbon dated at 14,010 ± 65 and 5430 ± 15 cal yr B.P., respectively. The contact is interpreted as the oxygen isotope stage 2 sequence boundary (surface B), which, at this updip location, is amalgamated with the bay line.

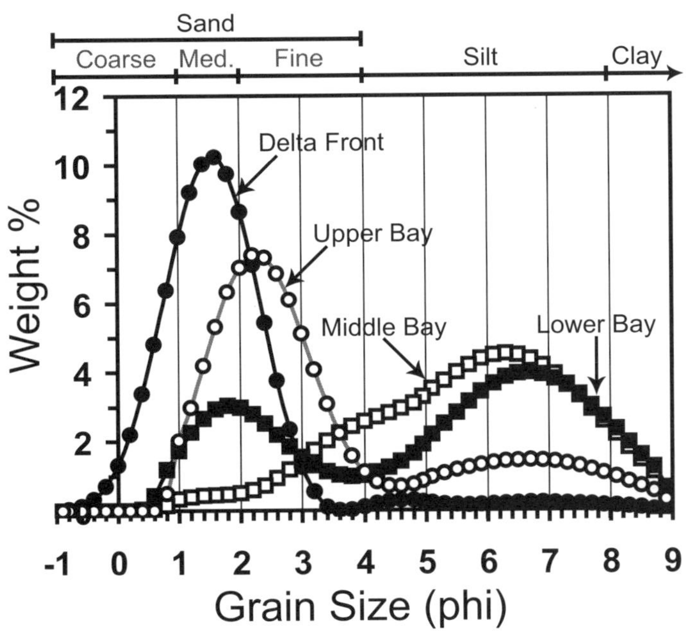

Figure 7. Representative grain-size curves for delta front, upper-bay, middle-bay, and lower-bay depositional environments.

Holocene Stratigraphy

Weeks Bay shows a general transgressive to regressive stratigraphic succession above surface B, the oxygen isotope stage 2 sequence boundary. In the middle of Weeks Bay, alluvial sediments are overlain by discontinuous marsh deposits and up to 9 m of estuarine facies (Fig. 8). In the center of Weeks Bay, middle-bay clay accumulated at a rate of 0.4 mm/yr from ca. 6000 to 2800 cal yr B.P. based on core WB-01-2 (Table 1; Figs. 6 and 9). This value corresponds to middle-bay sediment accumulation rates reported in Haywick et al. (2003) for Weeks Bay, and the *y*-intercept of the linear regression line ($R^2 = 0.97$) is −1.2 m, which is close to the water depth at which this core was collected in (Fig. 6).

Alluvial sediments that postdate formation of the stage 2 sequence boundary pinch out landward of the modern bayhead delta (Fig. 5). The modern bayhead delta prograded over estuarine facies at the vicinity of core WB-01-04 (Fig. 8). Landward of the river mouth, on the west side of cross section B–B′ (Fig. 5), no estuarine sediments were sampled below bayhead delta facies. There is also more relief on the stage 2 sequence boundary on the west side of B–B′ than on the east side, where estuarine sediments exist directly above the sequence boundary. Estuarine facies were likely removed on the west side during delta progradation by the bayhead delta diastem. The diastem also may have removed the alluvial sediments, which are recognized above the sequence boundary in Weeks Bay at this location. The diastem is amalgamated with the sequence boundary and the bay line on the west side of cross-section B–B′. This amalgamation is likely the reason that the surface has greater relief here than on the east side. Below the modern Fish River at cross-section B–B′, the bayhead delta diastem marks the contact between bayhead delta and estuarine sediments.

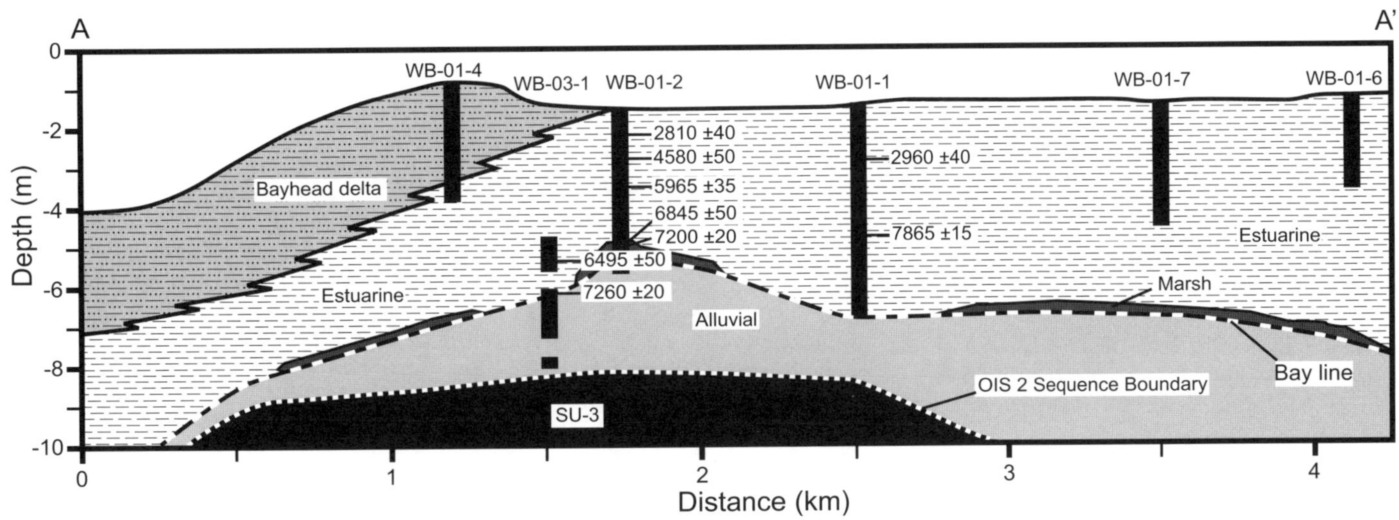

Figure 8. Cross section A–A′ based on vibracores, rotary-drill cores, and seismic data illustrating the stratigraphy above the oxygen isotope stage 2 sequence boundary in Weeks Bay. No bayhead delta deposits were recognized in the middle of the bay. See Figure 1 for location of cross section and Table 1 for additional information on dates (cal yr B.P.).

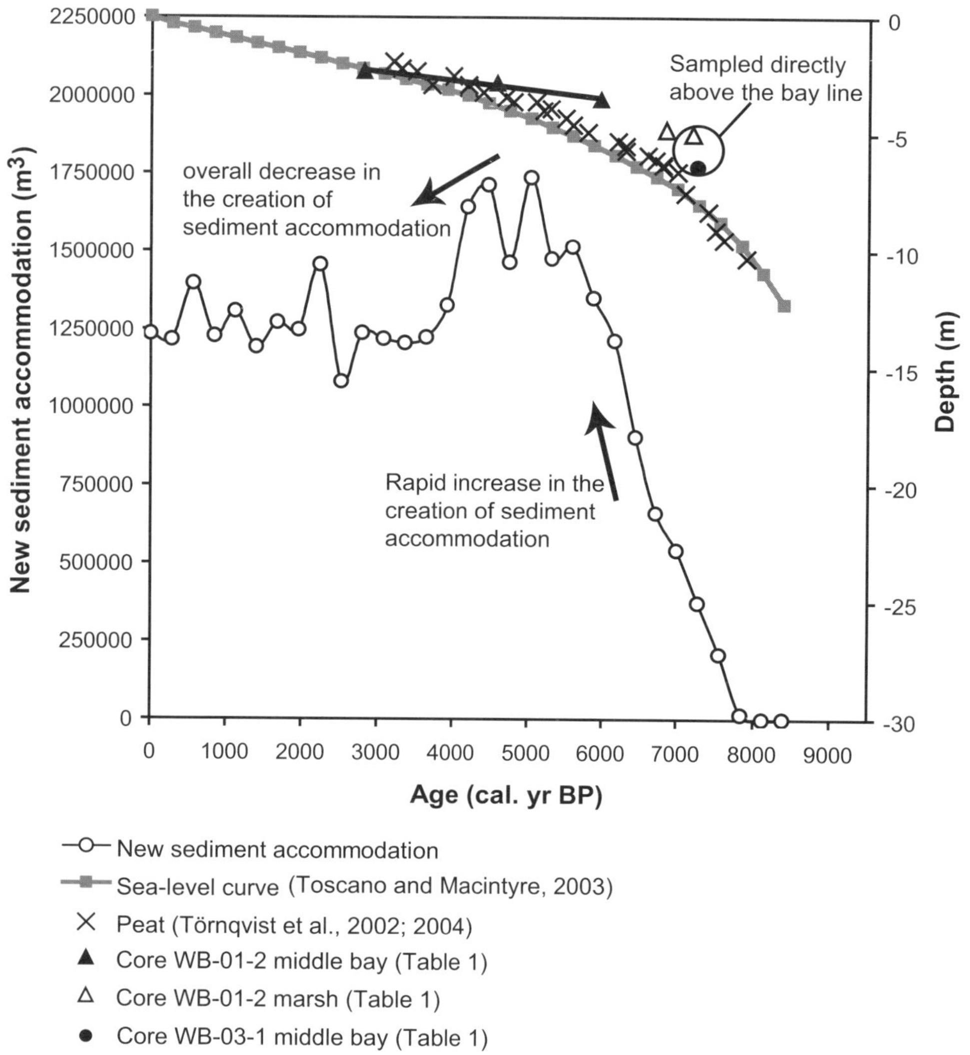

Figure 9. Graph showing the creation of sediment accommodation space (new sediment accommodation), sea level, and sediment accumulation through time in Weeks Bay. Core WB-01-2 is from the center of the paleovalley and did not sample bayhead delta deposits above the bay line (Figs. 1 and 6). The bay line flooding surface formed during a period of time when the rate of sea-level rise was decreasing; however, the rate of creation of sediment accommodation was at a maximum. See Tables 1 and 2 for additional information.

CHANGES IN SEDIMENT ACCOMMODATION THROUGH TIME

In response to Holocene sea-level rise, a continuous succession of depositional environments should have accumulated above the bay line in Weeks Bay; however, no bayhead delta deposits were imaged or sampled below estuarine facies. Bayhead delta deposits were only sampled at and landward of the modern Fish River, and presumably the Magnolia River (no data were collected at the Magnolia River mouth). Only discontinuous marsh (peat) and middle-bay (clayey silt) environments were recognized above the bay line. Bayhead delta deposits either back-stepped across Weeks Bay or were deposited and subsequently removed. Deposition and removal of bayhead delta deposits is deemed unlikely because the depth and age of material sampled directly above the bay line, from marsh and middle-bay deposits, plot close to the sea-level curve of Törnqvist et al. (2004), leaving little time for bayhead delta deposits to have accumulated and been removed (Fig. 9). Furthermore, the contact between marsh and middle-bay deposits is neither erosional nor do radiocarbon dates above and below resolve a hiatus.

The bay line does not extend below –9.0 m in the middle of the bay, and when sea level was at this depth, the rate of rise was decreasing (Otvos, 2004; Törnqvist et al., 2004; Toscano and Macintyre, 2003). This indicates that an increase in the rate of sea-level rise is an unlikely mechanism for bayhead delta back-stepping. Given that the rate of sea-level rise and the morphology of the land being inundated principally control the creation of sediment accommodation, a decreasing rate of sea-level rise does not exclude an increasing rate of sediment accommodation as the mechanism for bayhead delta back-stepping. To test this, the creation of sediment accommodation was calculated for Weeks Bay over the last 8.4 k.y.

Creation of sediment accommodation through time (new sediment accommodation) was calculated from Holocene sea-level data points and the contour map of the bay line using the software Surfer® 8.0. Sea-level data points from Louisiana published in Törnqvist et al. (2004; basal peats) correspond well with the sea-level curve of Toscano and Macintyre (2003), which is based on peat and coral (*Acropora palmata*) from the western Atlantic (Fig. 9). The oldest material dated from middle-bay clay in core WB-01-2 plots 1.5 m above the sea-level curve (Fig. 9). This is within acceptable errors attributed to depth measurements, dating, and reworking and does not indicate deposition above sea level. Sea level was calculated every 280 yr from 8.4 ka to present (31 iterations; Table 2) from the Toscano and Macintyre (2003) sea-level curve. The volume between sea level and the contour map of the bay line (Fig. 4B) was measured for each iteration. Creation of sediment accommodation through time was calculated by subtracting iterations, for example, iteration 4 – iteration 3 = volume change at 7280 cal yr B.P. (Table 2). New sediment accommodation plotted against time shows that initial flooding of the estuary occurred between 7.8 and 7.5 cal yr B.P. This corresponds well to the age of material sampled directly above the bay line (ca. 7.2 ka; Fig. 9; Table 1) and helps validate applicability of the Toscano and Macintyre (2003) sea-level curve to this area. Because the rate of sea-level rise was decreasing between 8400 and 5800 cal yr B.P., the creation of sediment accommodation increased because low-gradient antecedent topography was being inundated. Rapid shoreline transgression, especially in a dip orientation, during initial inundation is what forced the bayhead delta to back-step across Weeks Bay. Given that the lowest gradient is in a dip orientation, the shoreline at the bayhead delta should have experienced the most rapid retreat rates. After around 5600 cal yr B.P., the change in sediment accommodation through time is more variable but overall shows a decreasing trend (Fig. 9). This was likely the time when the bayhead delta began to prograde.

TABLE 2. ACCOMMODATION SPACE CHANGE ABOVE THE BAY LINE

Iteration	Age (cal yr B.P.; 280 yr time steps)	Sea level (m)	Volume between sea level and bay line (m^3)	Volume change (m^3)
0	8400	–12.23	0	0
1	8120	–10.90	0	0
2	7840	–9.72	14,482	14,482
3	7560	–8.74	222,721	208,240
4	7280	–7.99	596,693	373,971
5	7000	–7.31	1,138,420	541,728
6	6720	–6.81	1,798,506	660,085
7	6440	–6.36	2,703,972	905,466
8	6160	–5.89	3,918,677	1,214,704
9	5880	–5.47	5,268,704	1,350,027
10	5600	–5.06	6,783,980	1,515,276
11	5320	–4.71	8,260,479	1,476,500
12	5040	–4.32	9,997,363	1,736,884
13	4760	–4.02	11,461,444	1,464,081
14	4480	–3.67	13,175,155	1,713,711
15	4200	–3.35	14,818,372	1,643,217
16	3920	–3.09	16,145,451	1,327,079
17	3640	–2.86	17,369,028	1,223,577
18	3360	–2.64	18,575,015	1,205,987
19	3080	–2.41	19,794,592	1,219,578
20	2800	–2.19	21,032,685	1,238,092
21	2520	–2.00	22,113,558	1,080,873
22	2240	–1.74	23,570,538	1,456,980
23	1960	–1.53	24,817,834	1,247,296
24	1680	–1.31	26,088,458	1,270,624
25	1400	–1.11	27,281,291	1,192,834
26	1120	–0.90	28,587,764	1,306,473
27	840	–0.70	29,815,124	1,227,360
28	560	–0.47	31,211,292	1,396,168
29	280	–0.28	32,427,766	1,216,474
30	0	–0.00	33,661,369	1,233,603

CONCLUSIONS

The Fish and Magnolia paleovalleys show a compound valley-fill stratigraphy, and two sequences were recognized in this study. Radiocarbon dates and correlation to other data sets (Kindinger et al., 1994; Greene et al., 2007) indicate that the upper sequence boundary formed in response to the oxygen isotope stage 2 lowstand. This unconformity is overlain by alluvial and estuarine deposits that are separated by an exposure surface or bay line. The first estuarine sediments were deposited in Weeks Bay at 7200 cal yr B.P. No bayhead delta deposits are recognized above the bay line. The Fish and Magnolia bayhead deltas back-stepped across Weeks Bay during a period of time when the rate of sea-level rise was decreasing (Törnqvist et al., 2004) and the rate of creation of sediment accommodation was increasing.

The slope of the bay line, in a dip orientation, is ~0.6 m/km. During initial inundation of Weeks Bay, sea level was rising at ~3.5 mm/yr (Törnqvist et al., 2004). This equates to a shoreline retreat rate of ~600 m/century, which is not extreme. However, the bay line not only has a low gradient in a dip orientation, but it is broad (~1 km) and flat in the middle of Weeks Bay. Flooding of this antecedent topography would make it difficult for a bayhead delta, with a constant sediment supply, to persist or gradually retreat landward in response to sea-level rise. The rapid increase in sediment accommodation in the middle of the bay forced the river mouth, the point where river flow expansion and associated deposition occur, to move landward almost instantaneously. Given these conditions, the bayhead delta back-stepped across Weeks Bay, with what must have been catastrophic ecologic consequences. The sea-level rise rate during bayhead delta back-stepping was similar to future predictions, which proves the importance of calculating future changes in sediment accommodation, as opposed to simply calculating rates of shoreline inundation, to predict coastal response to sea-level rise.

ACKNOWLEDGMENTS

Funding for this project provided in part by the Coastal Zone Management Act of 1972, amended, administered by the Office of Ocean and Coastal Resource Management, National Oceanic and Atmospheric Administration, the National Science Foundation grant EAR-0107650, and the Alabama Department of Conservation and Natural Resources, State Lands Division. The Alabama Department of Transportation's Bureau of Materials and Tests generously provided platform-boring descriptions, and Joe Lambert, Craig Meyer, and Larry Greene are thanked for their assistance in the field. Constructive reviews by Steven Goodbred and Ervin Otvos helped improve the paper.

REFERENCES CITED

Bratton, J.F., Colman, S.M., Thieler, E.R., and Seal, R.R.I., 2003, Birth of the modern Chesapeake Bay estuary between 7.4 and 8.2 ka and implications for global sea-level rise: Geo-Marine Letters, v. 22, p. 188–197.

Church, J.A., Gregory, J.M., Huybrechts, P., Kuhn, M., Lambeck, K., Nhuan, M.T., Qin, D., and Woodworth, P.L., 2001, Changes in sea level, *in* Houghton, J.T., Ding, Y., Griggs, D.J., Noguer, M., van der Linden, P.J., Dai, X., Maskell, K., and Johnson, C.A., eds., Climate Change 2001: The Scientific Basis: New York, Cambridge University Press, p. 639–694.

Crance, J.H., 1971, Description of Alabama estuarine areas—Cooperative Gulf of Mexico estuarine inventory: Alabama Marine Resources Bulletin, v. 6, p. 1–85.

Dalrymple, R.W., Zaitlin, B.A., and Boyd, R., 1992, Estuarine facies models: Conceptual basis and stratigraphic implications: Journal of Sedimentary Petrology, v. 62, p. 1130–1146.

Greene, D.L., Rodriguez, A.B., and Anderson, J.B., 2007, Seaward-branching coastal-plain and piedmont incised-valley systems through multiple sea-level cycles: Late Quaternary examples from Mobile Bay and Mississippi Sound, U.S.A.: Journal of Sedimentary Research, v. 77, p. 139–158, doi: 10.2110/jsr.2007.016.

Haywick, D.W., Fearn, M.L., and Sanders, J.M., 2003, Sedimentation rates in marginal marine environments in southern Alabama: Geological Society of America Abstracts with Programs, v. 35, no. 6, p. 282.

Heap, A.D., and Nichol, S.L., 1997, The influence of limited accommodation space on the stratigraphy of an incised-valley succession: Weiti River estuary, New Zealand: Marine Geology, v. 144, p. 229–252, doi: 10.1016/S0025-3227(97)00107-2.

Hughen, K.A., Baillie, M.G.L., Bard, E., Bayliss, A., Beck, J.W., Bertrand, C.J.H., Blackwell, P.G., Buck, C.E., Burr, G.S., Cutler, K.B., Damon, P.E., Edwards, R.L., Fairbanks, R.G., Friedrich, M., Guilderson, T.P., Kromer, B., McCormac, F.G., Manning, S.W., Bronk Ramsey, C., Reimer, P.J., Reimer, R.W., Remmele, S., Southon, J.R., Stuiver, M., Talamo, S., Taylor, F.W., van der Plicht, J., and Weyhenmeyer, C.E., 2004, Marine04: Marine radiocarbon age calibration 26–0 ka BP: Radiocarbon, v. 46, p. 1059–1086.

Kindinger, J.L., Balson, P.S., and Flocks, J.G., 1994, Stratigraphy of the Mississippi-Alabama shelf and the Mobile River incised-valley system, *in* Dalrymple, R.W., Boyd, R., and Zaitlin, B.A., eds., Incised-Valley Systems: Origin and Sedimentary Sequences: Society for Economic Paleontologists and Mineralogists (SEPM) Special Publication 51, p. 83–95.

Otvos, E.G., 2004, Holocene Gulf levels: Recognition issues and an updated sea-level curve: Journal of Coastal Research, v. 20, p. 680–699, doi: 10.2112/1551-5036(2004)20[680:HGLRIA]2.0.CO;2.

Posamentier, H.W., Jervey, M.T., and Vail, P.R., 1988, Eustatic controls on clastic deposition I—Conceptual framework, *in* Wilgus, C.K., et al., eds., Sea-Level Changes: An Integrated Approach: Society for Economic Paleontologists and Mineralogists (SEPM) Special Publication 42, p. 109–124.

Reimer, P.J., Baillie, M.G.L., Bard, E., Bayliss, A., Beck, J.W., Bertrand, C.J.H., Blackwell, P.G., Buck, C.E., Burr, G.S., Cutler, K.B., Damon, P.E., Edwards, R.L., Fairbanks, R.G., Friedrich, M., Guilderson, T.P., Hogg, A.G., Hughen, K.A., Kromer, B., McCormac, F.G., Manning, S.W., Ramsey, C.B., Reimer, R.W., Remmele, S., Southon, J.R., Stuiver, M., Talamo, S., Taylor, F.W., van der Plicht, J., and Weyhenmeyer, C.E., 2004, IntCal04: Terrestrial radiocarbon age calibration 26–0 ka BP: Radiocarbon, v. 46, p. 1029–1058.

Rodriguez, A.B., Anderson, J.B., and Simms, A.R., 2005, Terrace inundation as an autocyclic mechanism for parasequence formation: Galveston estuary, Texas, U.S.A.: Journal of Sedimentary Research, v. 75, p. 608–620, doi: 10.2110/jsr.2005.050.

Schroeder, W.W., Dinnel, S.P., and Wiseman, W.J.J., 1990, Salinity stratification in a river-dominated estuary: Estuaries, v. 13, p. 145–154, doi: 10.2307/1351583.

Schroeder, W.W., Dinnel, S.P., and Wiseman, W.J.J., 1992, Salinity structure of a shallow, tributary estuary, *in* Prandle, D., ed., Dynamics and Exchanges in the Coastal Zone: Washington, D.C., American Geophysical Union, p. 155–171.

Stuiver, M., and Reimer, P.J., 1993, Extended ^{14}C database and revised CALIB radiocarbon calibration program: Radiocarbon, v. 35, p. 215–230.

Törnqvist, T.E., Gonzalez, J., Newsom, L., van der Borg, K., and de Jong, A.F.M., 2002, Reconstructing "background" rates of sea-level rise as a tool for forecasting coastal wetland loss, Mississippi delta: Eos (Transactions, American Geophysical Union), v. 83, p. 525, 530–531.

Törnqvist, T.E., Gonzalez, J.L., Newsom, L., van der Borg, K., de Jong, A.F.M., and Kurnik, C.W., 2004, Deciphering Holocene sea-level history on the U.S. Gulf Coast: A high-resolution record from the Mississippi Delta: Geological Society of America Bulletin, v. 116, p. 1026–1039, doi: 10.1130/B2525478.1.

Toscano, M.A., and Macintyre, I.G., 2003, Corrected western Atlantic sea-level curve for the last 11,000 years based on calibrated ^{14}C dates from *Acropora palmata* framework and intertidal mangrove peat: Coral Reefs, v. 22, p. 257–270, doi: 10.1007/s00338-003-0315-4.

Van Wagoner, J.C., Mitchum, R.M., Campion, K.M., and Rahmanian, V.D., 1990, Siliciclastic Sequence Stratigraphy in Well Logs, Cores, and Outcrops: Concepts for High-Resolution Correlation of Time and Facies: American Association of Petroleum Geologists Methods in Exploration Series 7, v. 7, 55 p.

Manuscript Accepted by the Society 10 March 2008

Printed in the USA

The Geological Society of America
Special Paper 443
2008

Record of dramatic Holocene environmental changes linked to eustasy and climate change in Calcasieu Lake, Louisiana, USA

K.T. Milliken
John B. Anderson
Department of Earth Science, Rice University, 6100 Main Street, MS 126, Houston, Texas 77005, USA

Antonio B. Rodriguez
Institute of Marine Sciences, University of North Carolina at Chapel Hill, 3431 Arendell Street, Morehead City, North Carolina 28557, USA

ABSTRACT

Calcasieu Lake records a Holocene history of dramatic environmental change. The changes resulted in up to 20–30 km of landward translation of deltaic and estuarine depositional environments, and intervening periods of delta progradation, from 9600 cal B.P. to 1600 cal yr B.P. The sediments preserved beneath the lake record a series of back-stepping events. Seven events that affected the entire estuary are recognized: at 8900–8500 cal yr B.P., 8300–8000 cal yr B.P., 8000–7900 cal yr B.P., 7200 cal yr B.P., 5800–5600 cal yr B.P., 2000–2800 cal yr B.P., and 2500–1800 cal yr B.P. Only one of these events (8200 cal yr B.P.) can be linked to a known eustatic event that is also associated with climate change.

The Calcasieu system maintained a bay-head delta within the present-day lake during a period of relatively rapid eustatic sea-level rise (10,000–6000 cal yr B.P.). It was only after 3000 cal yr B.P. that a bay-head delta was not present in the modern Calcasieu Lake. This strongly suggests a climate–sediment supply control. Lower estuary tidal delta and inlet facies indicate restricted estuarine circulation since ca. 2800 cal yr B.P.

Keywords: incised valley, Quaternary, antecedent topography, estuary, transgression.

INTRODUCTION

Louisiana contains 60% of the Gulf of Mexico coastal wetlands and 40% of the U.S. coastal wetlands (Coast 2050, 1998; Mitsch and Gosselink, 2000). Canal and ship-channel dredging, pollutant introduction, levee construction, and petroleum and groundwater withdrawal have rapidly transformed much of the brackish marshes to open water over the past half century (Boesch et al., 1994; Turner, 1997; Day et al., 2000; Morton et al., 2006). Evidence for accelerated subsidence and intensification of Atlantic hurricanes has generated concern about the way(s) in which the western Louisiana coastal environment can counter in response to these trends (Michener et al., 1997; Goldenberg et al., 2001; Shinkle and Dokka, 2004; Pielke et al., 2005).

Coastal environments record their response to past environmental forcing in the sedimentary record. The sedimentary

Milliken, K.T., Anderson, J.B., and Rodriguez, A.B., 2008, Record of dramatic Holocene environmental changes linked to eustasy and climate change in Calcasieu Lake, Louisiana, USA, *in* Anderson, J.B., and Rodriguez, A.B., eds., Response of Upper Gulf Coast Estuaries to Holocene Climate Change and Sea-Level Rise: Geological Society of America Special Paper 443, p. 43–63, doi: 10.1130/2008.2443(04). For permission to copy, contact editing@geosociety.org. ©2008 The Geological Society of America. All rights reserved.

deposits in estuaries record the history of Holocene marine transgression and modern highstand conditions. Rising sea level, antecedent topography, and/or sediment supply variations have regulated the evolution of estuaries during the Holocene. The nature of the rate of environmental change as recorded in the sedimentary deposits needs to be first documented and then assigned to an appropriate controlling mechanism. The Holocene evolution of the Calcasieu river-estuary system provides an analog for future system changes. In the case of Lake Calcasieu, human activity has accelerated rates of relative sea-level rise and modified sediment supply through agricultural practices to a degree similar to the scale that occurred during the Holocene. Thus, examination of how these changes impacted the estuary in the past will provide insight into changes that are presently occurring.

SETTING

The Calcasieu system drains ~10,000 km^2 of Pliocene-Pleistocene sedimentary deposits. Drainage basin relief is less than 50 m, and vegetation type is southern mixed forest (Bailey, 1980). Present-day climatic conditions are subtropical and humid; temperatures average 10 °C (50 °F), and precipitation averages 1350 mm/yr (53 in/yr; www.cdc.noaa.gov). Low-gradient, small coastal-plain rivers, such as the Calcasieu River of west Louisiana, commonly terminate in a brackish estuary rather than an open-marine setting (Lee et al., 1990). The Calcasieu River discharges an average of 3.2×10^6 m^3/d of freshwater into the northern portion of the estuary, whereas semidiurnal to diurnal tides cycle a mean tidal prism of 39.1×10^6 m^3 of marine water via the tidal inlet at the southern portion of the estuary (Forbes, 1988; Nichol et al., 1992). Calcasieu Lake is classified as microtidal, with a spring-tide range of 0.6 m and upstream tide penetration of 100 km (Forbes, 1988; Nichol et al., 1992). Winds are predominantly from the south-southeast but switch to the north following a cold-front passage (Forbes, 1988; Nichol et al., 1992). The combination of relatively low freshwater inflow, shallow water depth, and wind regime produces a barotropic water mass with little to no stratification (Lee et al., 1990).

ANTECEDENT TOPOGRAPHY

During the most recent sea-level fall (120 ka to 20 ka), the rivers of east Texas and west Louisiana cut relatively deeply incised valleys that extended across the continental shelf (Fisk, 1948; LeBlanc, 1949; Nelson and Bray, 1970; Thomas and Anderson, 1994; Anderson et al., 2004). Identification of the most recently incised valley and sediment fill relies on a combination of core and seismic data; these are commonly called the Holocene-Pleistocene surface and Holocene sedimentary sequence, respectively.

Fisk (1948) and LeBlanc (1949) utilized core and boring data to map a Holocene-Pleistocene boundary that dips basinward from surface exposure near the northern end of the study area to approximately −10 m at the present-day coastline in the interfluvial areas. Beneath Calcasieu Lake, the Holocene-Pleistocene surface ranges from −10 m to −30 m below sea level (bsl) (Fig. 1). The surface exhibits moderate (meter-scale) relief regionally and significant (decameters) relief locally. Furthermore, the significant topographic relief on this surface is not randomly oriented but rather exhibits a dendritic pattern with convergence basinward, reminiscent of a fluvially dominated landscape (Fig. 1).

Criteria for determining the Holocene-Pleistocene surface in core data include a sharp increase in shear strength from overlying soft gray sediments to underlying hard, orange and tan mottled mud and sand. The sharp facies transition is interpreted to represent subaqueous deposition of the soft gray sediments following subaerial exposure and pedogenic modification of the underlying sediments. In west Louisiana, previous investigators (Fisk, 1948; LeBlanc, 1949; Bernard, 1950; Coleman and Smith, 1964; Kane, 1959; Gould and McFarlan, 1959; Coleman, 1966; Nelson and Bray, 1970; Winker, 1979) attributed sediment characteristics to eustatic fluctuations such that soft sediments were deposited during the Holocene to recent transgression and highstand, whereas the paleosol developed during the most recent Pleistocene eustatic lowstand, ca. 20,000 B.P., which correlates with oxygen isotope stage 2 (OIS2). The paleosol also marks the top of the Prairie Formation (Bernard, 1950; Doering, 1956).

Previous investigators have mapped the Calcasieu incised valley surface, determined general lithofacies associations, and chronostratigraphically constrained Holocene sedimentation in interfluvial areas (Fisk, 1948; LeBlanc, 1949; Gould and McFarlan, 1959; Byrne et al., 1959; McFarlan, 1961; Coleman, 1966; Nichol et al., 1996). However, only one study (Frazier, 1974) has provided any age constraints (one radiocarbon date) within the Calcasieu incised valley. Whereas the oldest sediments on the interfluvial areas are mid-Holocene (ca. 5000 yr B.P.; Gould and McFarlan, 1959; McFarlan, 1961), the Calcasieu incised valley contains a complete Holocene section.

The map of the Holocene-Pleistocene surface (antecedent topography) used in this study (Fig. 1) incorporates published results with new data and places the Pleistocene surface at 20–30 mbsl beneath Calcasieu Lake. This deviates slightly from interpretation of Nichol et al. (1996), who placed the base of the valley at the southern end of Calcasieu Lake at –60 m. This is nearly twice the depth of the other incised valleys of the Gulf Coast (Kane, 1959; Wright, 1980; Thomas and Anderson, 1994). Our interpretation of the available data is that the OIS 2 incised valley (–20 to –30 m; ca. 20 ka) cuts into an older, deeper valley (–60 m; ca. 140 ka), probably the OIS 6 valley.

METHODS

In 2003, 45 km of high-resolution seismic data were acquired aboard the research vessel (R/V) *Jenny* using an Edgetech Chirp 14 kHz subbottom profiler system; 19 km of data were acquired in 2004 using the R/V *Trinity* and a 4 kHz Boomer seismic system; and 30 km of 1995 Boomer data were provided by the

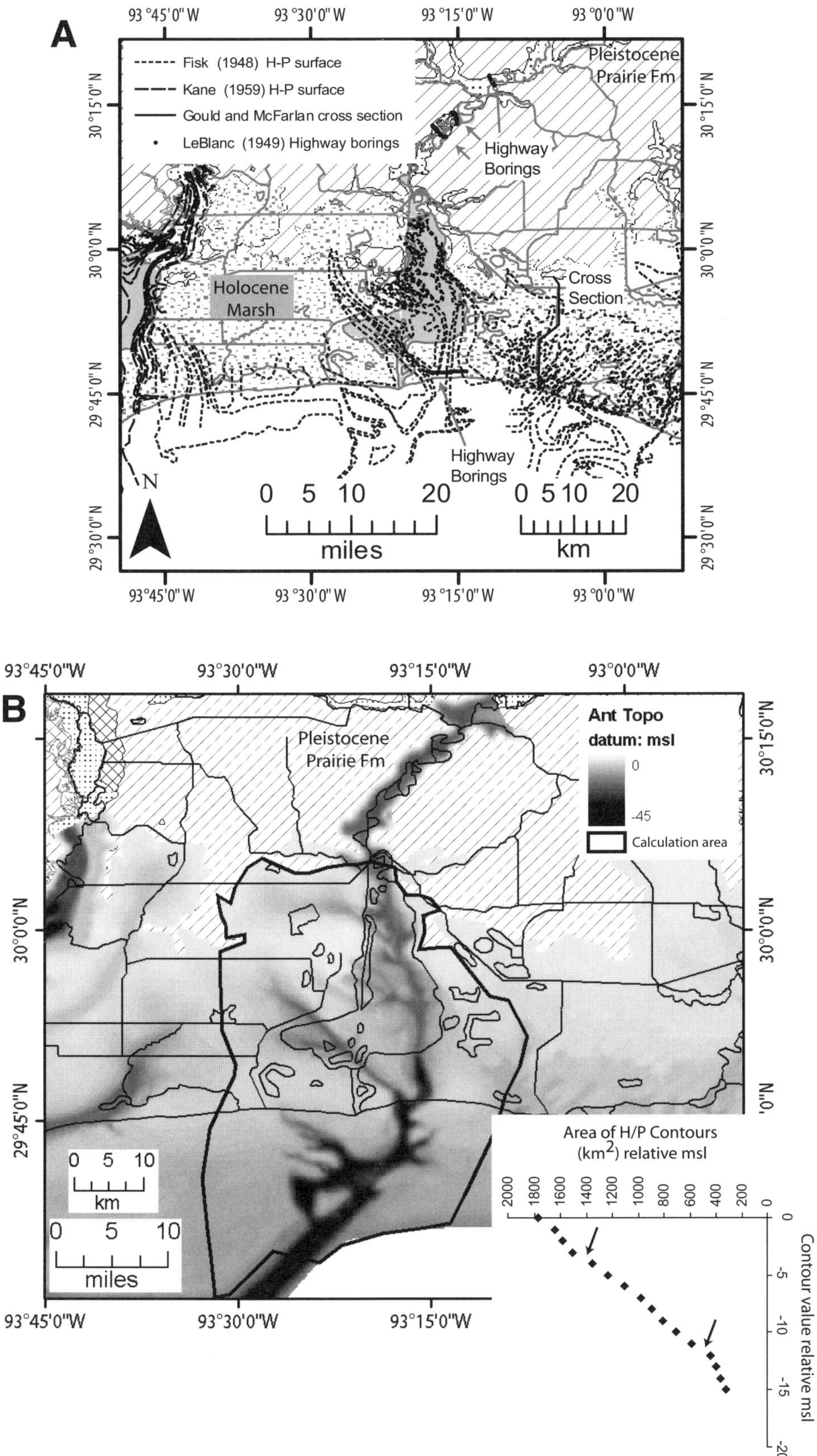

Figure 1. (A) Map showing preexisting data that were used to construct the Holocene-Pleistocene (H-P) surface. (B) The Holocene-Pleistocene surface as a variable-color, white to black image with a graph illustrating the area inundated by successive sea-level rise. The two arrows in the inset mark periods of 15%–30% increase in estuarine area. The shape of the curve indicates the lack of fluvial terraces. If antecedent topography exerts a control on estuarine evolution, as sea level rises to these levels, the system should respond with a dramatic estuarine reorganization and step back at ~11–12 mbsl and 5–6 mbsl.

U.S. Geological Survey (USGS) (Fig. 2). Seismic reflections are resolvable at the meter scale, and penetration depths are up to 100 ms or 75 m (1500 m/s assumed velocity). The core locations were strategically chosen in order to correlate with seismic data or cross sections derived from highway borings (LeBlanc, 1949; Nichol et al., 1996). Core data obtained in the Calcasieu Lake and Prien Lake include four rotary-drill cores (04cl05, 04cl06, 04cl07) from 6 to 20 mbsl and three vibracores (04cl08, 04cl09, 04cl10) up to 6 mbsl. Recovery of the top 2–3 m of water-saturated sediment was not possible using the push/rotary-drill rig, but vibracores were collected to augment the recovery.

LeBlanc (1949) utilized highway borings from the construction of State Highway 42 to draft a cross section that illustrates two possible incised valleys underlying the chenier plain south of Calcasieu Lake. Analysis and mapping of the planform geometry of the surface that corresponds to the incised valley suggest that the western valley is a tributary to the eastern trunk valley. Three cores (04cl01, 04cl03, 04cl04; Fig. 1) were collected on land along the preexisting LeBlanc (1949) cross section using a truck-mounted Geoprobe system with greater than 20 m coring capability. Drill sites 04cl01 and 04cl03 were sited to sample the western valley, and core 04cl04 sampled the eastern valley (Fig. 2). Sites 04cl01 and 04cl04 had 71% and 33% core recovery, respectively.

Macroscopic physical descriptions of all cores were conducted in the laboratory and included color, visual determination of lithology and grain size, macrofauna, presence or absence of sedimentary structures, and nature of bed and lamination contacts. Quantitative grain-size analysis was performed using a Malvern laser particle-size analyzer on 2–10 cm increments every 10–50 cm (McCave et al., 1986; Sperazza et al., 2004). Core facies and boundaries were integrated with seismic facies and seismic surfaces to map a three-dimensional subsurface stratigraphy. Depositional environments determined from core and seismic facies followed the nomenclature of Zaitlin et al. (1994).

Key surfaces, observed in core and seismic data, were chronologically constrained by radiocarbon dating of organic or carbonate material. Humic material was washed from organic-rich muds and muddy peat deposits, whereas carbonate material included mollusk shells and shallow-marine mixed assemblages of foraminifera. The samples were cleaned and sent to Woods Hole National Ocean and Sciences Accelerator Mass Spectrometry (NOSAMS) laboratory for accelerator mass spectrometry analyses. The reported ages were calibrated using Calib 2005 (Stuiver et al., 2005). The Northern Hemisphere calibration curve with no assumed reservoir was used for ages obtained from terrestrial organic matter (wood and plant remains); the marine calibration curve with a negative 350 yr reservoir was utilized for *Rangia* spp., *Nuculana* spp., and *Crassostrea* spp.; and the marine calibration curve with no additional reservoir was used for ages derived from foraminifera. The choice to treat ages derived from these three species was based on the assumption that these organisms acquire carbon from brackish water dissolved inorganic carbon (DIC). The Calcasieu River does not drain carbonate rocks

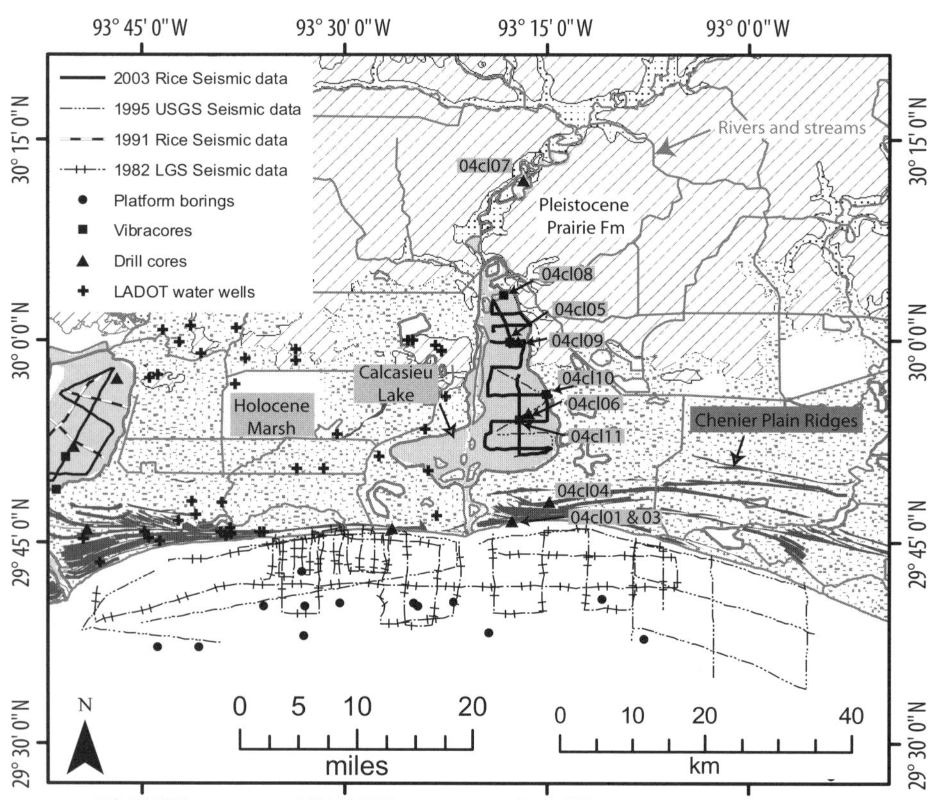

Figure 2. Map of study area showing new data sets utilized in this study. LADOT—Louisiana Department of Transportation; LGS—Louisiana Geological Society; USGS—U.S. Geological Survey.

or soils with abundant carbonate and should therefore have an atmospheric DIC pool, but we chose to remain consistent with research in other estuaries and utilize the marine calibration curve for mollusks (Anderson et al., this volume; Simms et al., this volume).

The ages of individual surfaces were assigned a range of a few hundred years rather than a single value. The range was determined using the 2σ age range of dated material above and/or below the surface and realistic limitations on sampling frequency of each surface with the limited number of core points.

Public and donated data sets utilized for stratigraphic information include Louisiana and Texas Department of Transportation water-well and highway boring descriptions, USGS and Louisiana Geological Survey seismic and core data, and Fugro-McClellan offshore platform borings. Previously published data utilized in this study to help constrain the Pleistocene surface include data from Fisk (1948) and LeBlanc (1949). Data published by Nichol et al. (1996) and Fearn (1995) allowed us to extend the sequence boundary and other chronostratigraphic surfaces updip to Prien Lake. The chenier plain stratigraphy and formation chronology was taken from Gould and McFarlan (1959), McFarlan (1961), Coleman (1966), and Frazier (1974).

RESULTS

Seismic Facies

Within the Calcasieu incised valley, a predictable vertical seismic facies succession is recognized, and it reveals sharp boundaries between seismic units (Fig. 3). For much of the seismic data at ~9–15 mbsl, a high-amplitude continuous subhorizontal reflector that exhibits a few meters of relief masks underlying data. A few seismic lines do image below this surface and show low-amplitude, chaotic reflectors in a unit between ~15–20 mbsl. This unit is assumed to exist at the base of the incised valley everywhere. A prominent surface with a few meters of relief separates seismic facies A (fluvial) from overlying facies B (deltaic). This surface is a boundary between chaotic seismic facies below and a more laminated seismic facies above, and it is interpreted as the bayline. Above the bayline surface, seismic facies B is characterized by discontinuous, subhorizontal, laminated reflectors that are truncated and onlapped by discontinuous laterally accreting high-amplitude reflectors (Fig. 3B). Facies B is sharply truncated and overlain by low-amplitude, continuous, laminated seismic facies C (central basin; Fig. 3C).

Seismic facies D (tidal) consists of inclined subparallel reflections and a laterally accreted geometry. The youngest seismic facies (facies E; shallow marine) is composed of subparallel, high-amplitude, continuous, laminated reflectors (Fig. 3D).

Lithofacies

Macroscopic description of the cores indicates six groups and ten subgroups of lithofacies based on sedimentologic characteristics and fauna (Fig. 4). In general, the lithofacies display a fining-upward trend followed by a coarsening-upward trend.

The A1 lithofacies consists predominantly of moderately sorted, very fine to fine sand with cross-bedding and some thin mud beds and organic matter. Lithofacies A1 consists of a unimodal size distribution centered at 125–250 µm and a fine tail (Fig. 5A). Although the base of the A1 lithofacies was not sampled, the thickness of this facies is constrained to approximately 10 m by seismic data (this study; Nichol et al., 1996) as well as highway borings (Le Blanc, 1949). Lithofacies A2 consists of organic-rich mud and abundant plant and woody fragments (Fig. 5B). The contact between lithofacies A1 and lithofacies A2 is sharp, and the thickness of lithofacies A2 is on the order of a few meters.

Lithofacies B1 is composed of dark gray sandy mud and muddy sand with a poorly sorted bimodal grain-size distribution (Fig. 5C). Lithofacies B2 consists of dark gray poorly sorted very fine to fine sandy mud and subvertical and subhorizontal sand lenses, occasional scattered broken *Rangia* spp., shells, and disseminated sand-sized wood and plant fragments. Grain-size data show bimodal populations with peaks at 5–10 µm and 100–110 µm, similar to lithofacies C (Fig. 5E). Facies B3 consists of centimeter-scale layers of tan, brown, and gray mud with silt and very fine sand lenses. This facies has a coarse skewed unimodal grain-size distribution (Fig. 5D). The C lithofacies consists of dark gray very fine to fine sandy mud and occasional layers of *Rangia* spp. shells and shell fragments. Grain-size data show bimodal populations with peaks at 5–10 µm and 100–110 µm (Fig. 5F). Lithofacies D1 is composed of gray-brown sandy mud and muddy sand, few shell fragments, scattered shell halves, and abundant subvertical brown sand lenses interpreted as burrows (Fig. 5G). The contact between lithofacies D1 and underlying facies is sharp, whereas the contact between underlying lithofacies D1 and overlying lithofacies D2 is gradational. Lithofacies D2 consists of decimeter-scale interbeds of brown shelly sand and gray muddy sand with broken shell fragments and gray sandy mud with some shell fragments (Fig. 5H). The grain-size data distinguish this alternation between sand and muddy sand in core 04cl01 from 4 to 6 mbsl (Fig. 6). Facies D1 and D2 are the only facies to contain and/or preserve significant macrofauna and microfauna. The faunal component includes macrofossils such as Ostracods, *Mulinia* sp., *Anadara* sp., *Abra* sp., *Polinices* sp., and *Retusa* sp., and microfossils such as *Quincoloculina* sp., *Elphidium* sp., and *Ammonia* sp. Facies E1 is composed of dark gray silty mud with diffuse organic material and iron-stained nodules (Fig. 5I). The F facies is an interbedded brown-gray shelly sand and sandy shell hash with decimeter-scale bedding (Fig. 5J).

Interpretation of Seismic Facies and Lithofacies

Figure 6 shows detailed lithologic logs for sediment cores and stratigraphic relationships of units described in this section. Cores collected from chenier ridges were projected onto a seismic line collected just offshore to show the tie between the

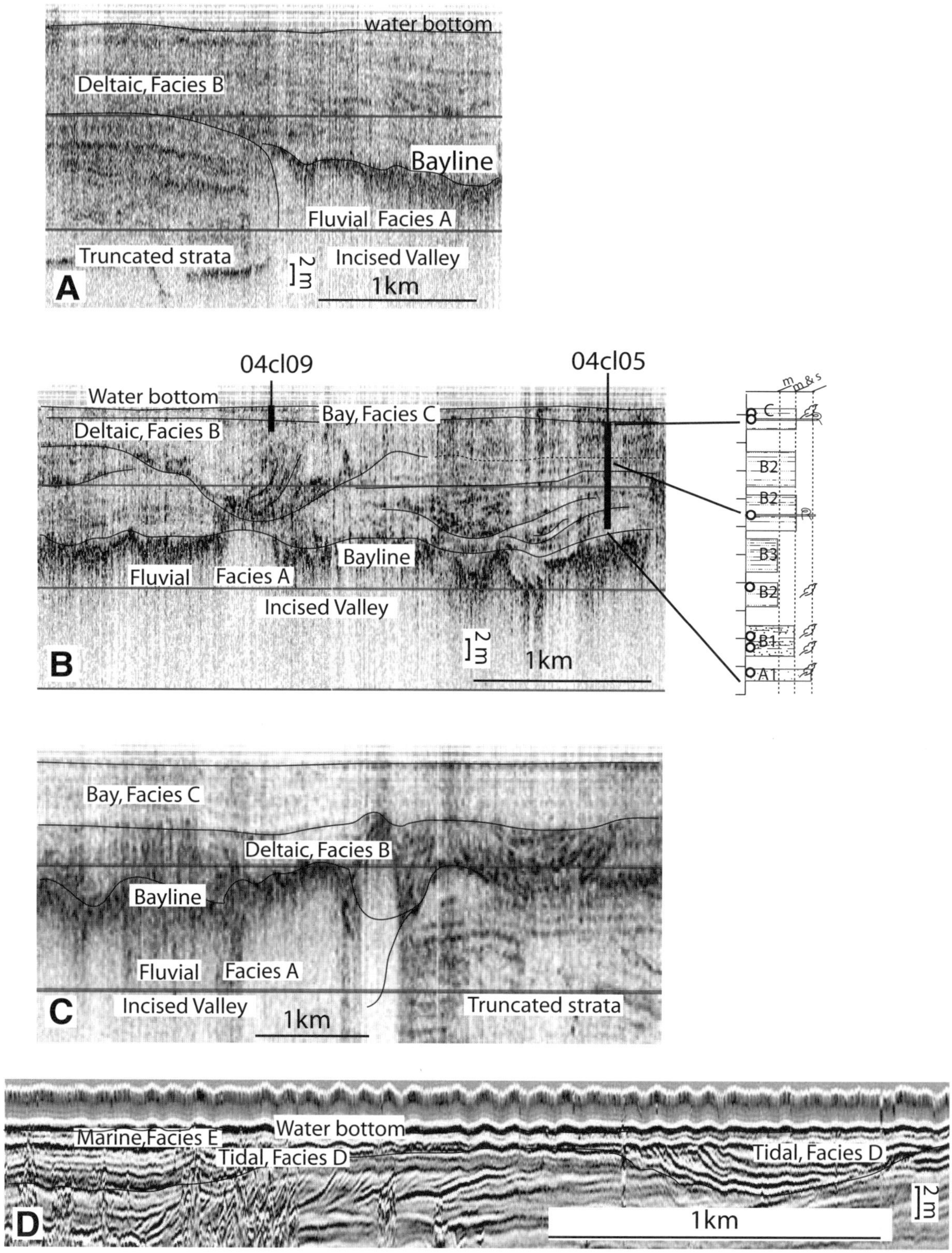

Figure 3. (A) Seismic line showing the bayline surface overlain by deltaic and very thin estuarine deposits. (B) Seismic line showing incised valley seismic facies linked to core facies, and fluvial deposits (not sampled here) underlying the bayline surface. The deltaic seismic facies (facies B) correlates to core lithofacies B1, B2, and B3. The estuarine seismic facies (facies C) is linked to core lithofacies C. Symbols as in Figure 6. (C) Seismic line illustrating the sharp boundary between deltaic (facies B) and estuarine (facies C) seismic facies. (D) Offshore seismic line showing tidal (facies D) and shallow-marine (facies E) seismic facies. Note horizontal and vertical scale bars.

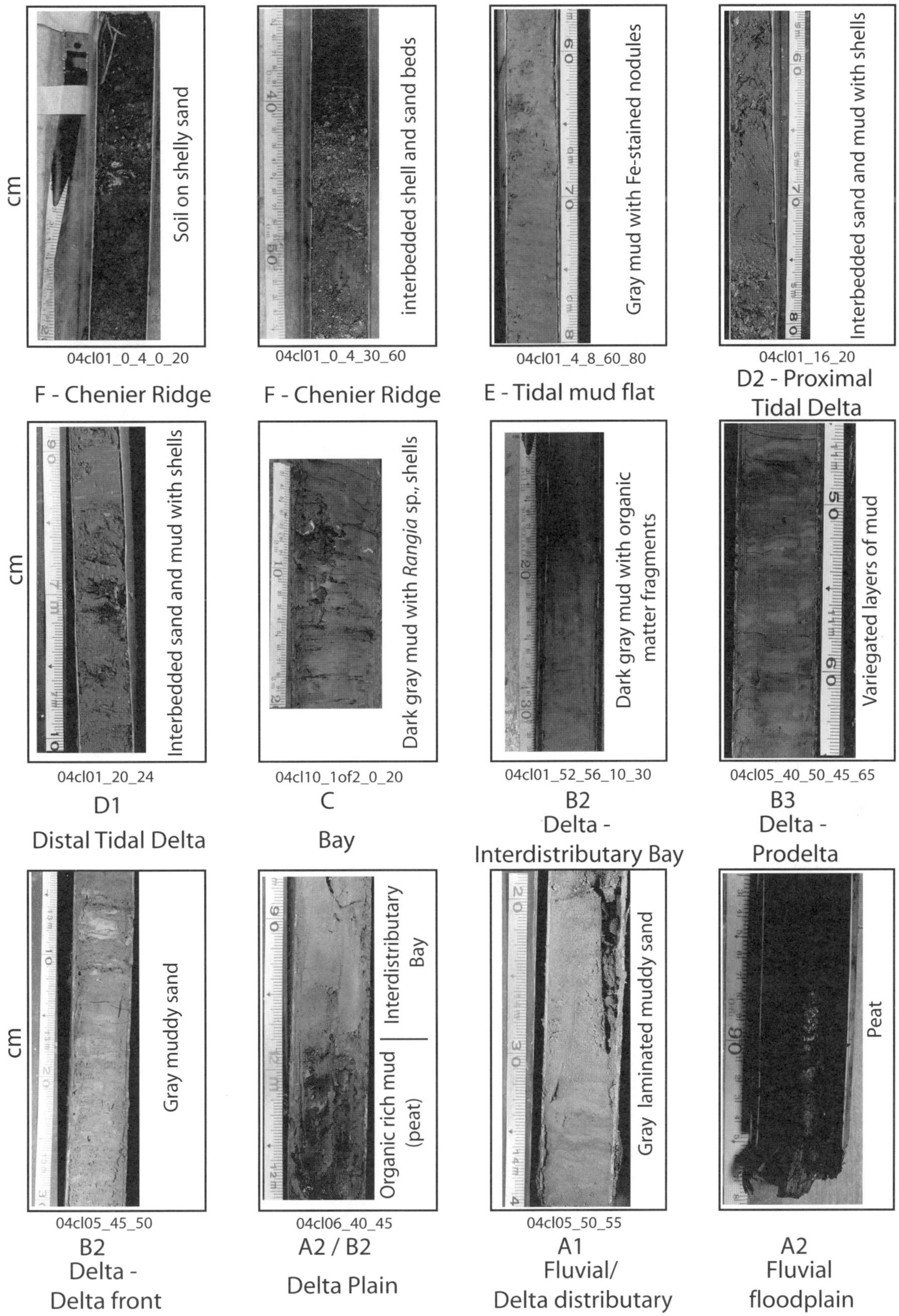

Figure 4. Core photos illustrating lithologic facies groups indicative of distinct depositional environments.

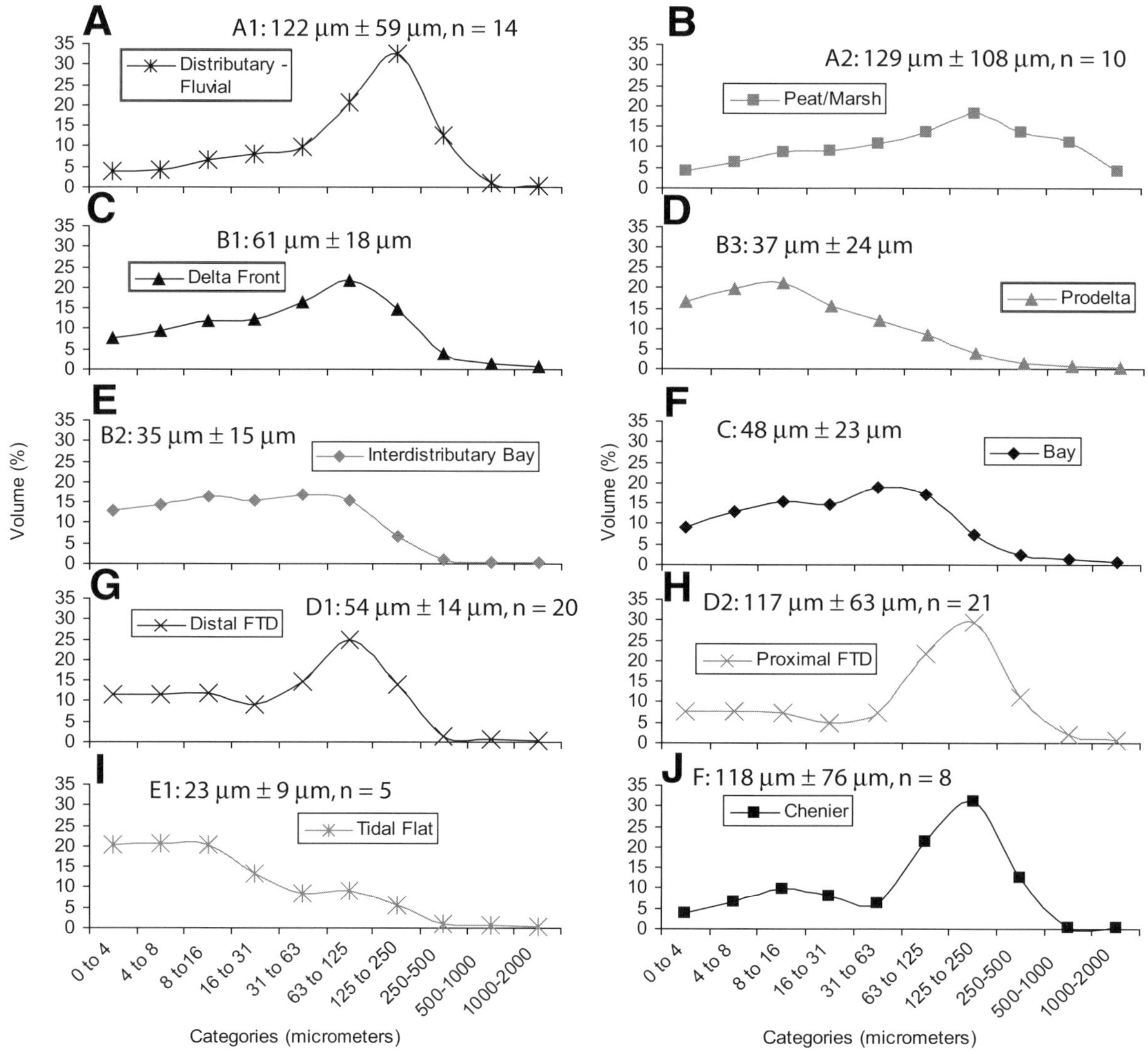

Figure 5. Average grain-size frequency curves of the 10 distinct lithofacies described in this study. FTD—flood tidal delta.

lithofacies and seismic facies (Fig. 7A). The facies succession is consistent with landward-migrating fluvial–estuary–tidal inlet to delta–shallow-marine depositional systems. Seismic facies A is correlated to sand and muddy sand within the cores (facies A1) and is interpreted to represent sandy fluvial channel, point bar, and overbank deposits or possibly deltaic distributary channel sand. The sand component is well-sorted fine to very fine sand. Most cores sampled only the top of seismic facies A.

Seismic facies B is interpreted to represent three or more deltaic subenvironments. The cut-and-fill character and lateral accretion (B1) observed in seismic profiles indicate that formation occurred in a deltaic distributary complex, where the subhorizontal laminated reflectors represent prodelta (B3), interdistributary bay (B2), or delta plain (A2) settings (Fig. 7B). The peat facies (A2) is ascribed to delta plain or fluvial floodplain environments. Macro- and microscopic inspection shows that wood and plant fragments make up the sand-size fraction, and mud forms the matrix.

Cores that sampled seismic facies B sampled peat overlain by variegated tan, gray, and brown mud with soft-sediment-deformation character (B3) and sandy mud and gray mud with few foraminifera and common sand-size plant and wood fragments (B2) (Fig. 4). The B3 facies ranges from clay to coarse silt and is the finest subgroup, with up to 15%–20% clay-sized material. This facies is interpreted as a prodelta deposit. The peat is interpreted as delta plain, and the gray mud is interpreted as an interdistributary bay deposit (B2).

The surface that marks the top of the fluvial and the base of the deltaic deposits is the bayline. It is commonly marked by peat, which provides excellent chronological control for the first definitive marine incursion into the valley. The deltaic subenvironments do not necessarily display a predictable vertical

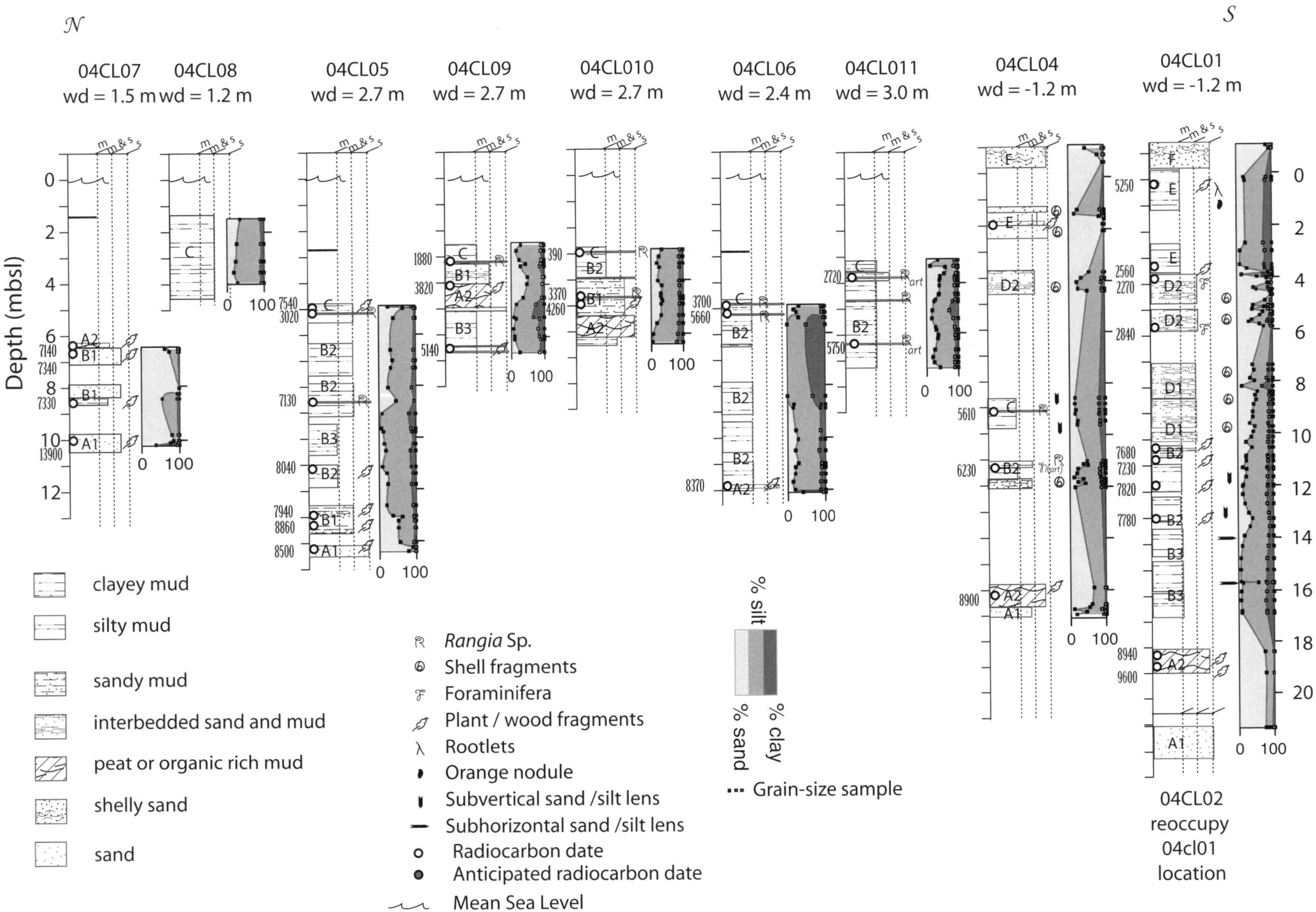

Figure 6. Lithologic logs of cores and grain-size data. Radiocarbon ages and sample locations are shown to the left of each curve. See Figure 1 for core locations. Cores are vertically aligned to mean sea level (msl) according to water depth (wd).

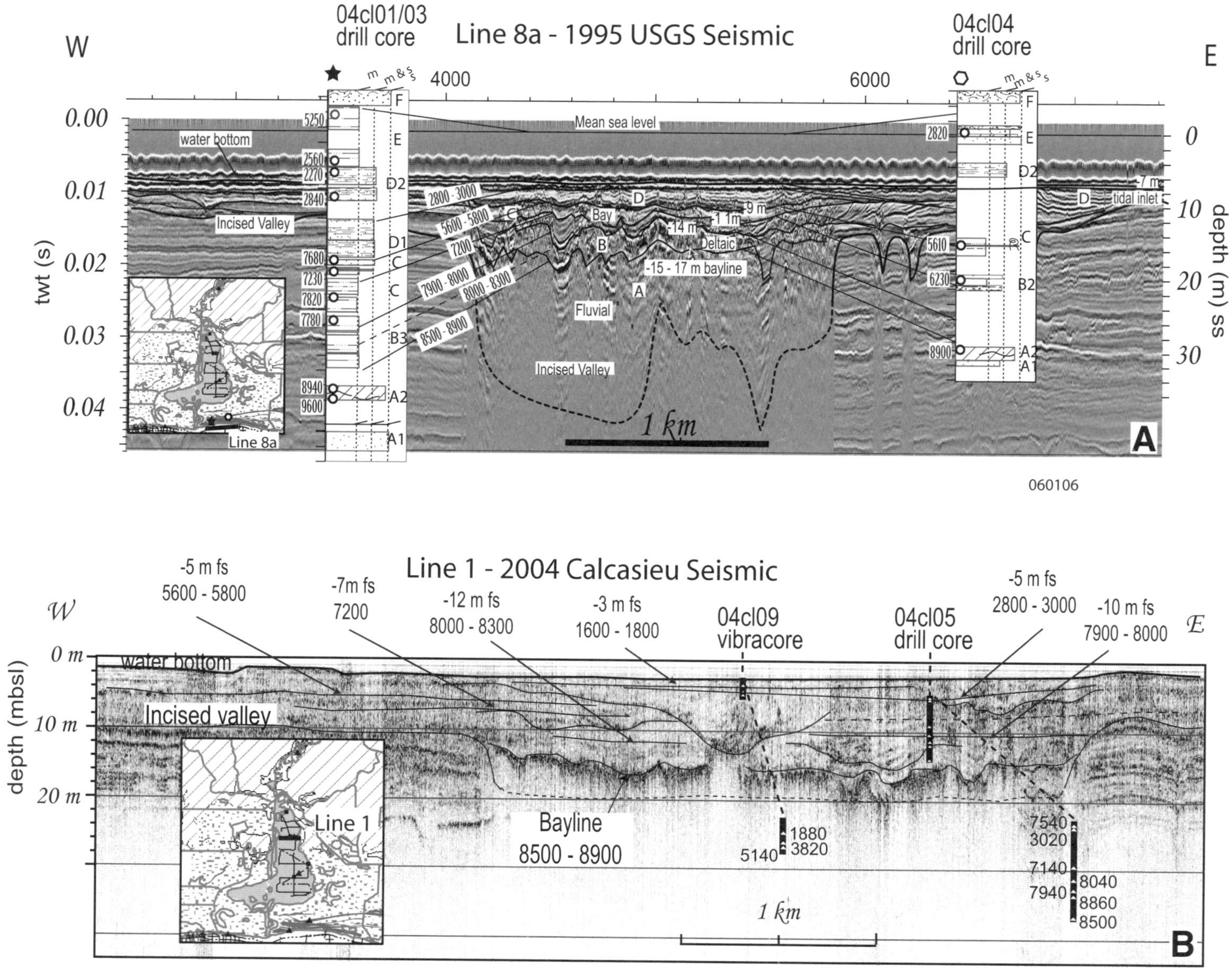

Figure 7. (A) Seismic line 8a is a strike-oriented seismic line showing projected cores 04cl01 and 04cl04. The interpreted incised valley with sediment fill and five surfaces at 15–17 m, 14 m, 11 m, 9 m, and 7 m are indicated. The 15–17 m surface marks the bayline and separates fluvial deposits below from bay-head delta deposits above. Two cores, lithology, and radiocarbon chronology are also shown. twt—two-way traveltime. (B) Seismic line 1 is a middle to upper estuarine seismic line showing the 15 m bayline, and 12 mbsl, 10 mbsl, 8 mbsl, 5 mbsl, and 3 mbsl disconformity (flooding) surfaces (fs). Two cores and their radiocarbon chronology are also shown. Dark black line demarcates sampled interval on each core. Incised valley surface demarcates Pleistocene below and Holocene above. The line location is shown on inset map.

succession (Fig. 6). Depending upon whether the delta is prograding or back-stepping, prodelta mud can underlie or overlie delta-front sand or delta-plain peat, respectively.

The deltaic facies (facies B) is sharply overlain by a soft dark gray mud with *Rangia cuneata* shell layers (Fig. 4; facies C). Facies C is interpreted as a central-basin mud deposit, where the shell layers represent wave ravinement surfaces. The central-basin mud ranges in grain size from medium to coarse silt and represents suspension deposits derived from the fluvial or marine systems and reworked by waves. A statistical comparison of the central-basin deposits (facies C) to interdistributary bay deposits (facies B2) suggests that the difference in the mean grain size of the two populations is significant at a 95% confidence limit. The central-basin mud is slightly coarser-grained relative to the interdistributary bay deposits. This indicates that waves and/or tidal currents may have winnowed the mud, or that silt was derived from offshore.

In the lower estuary and chenier plain, lithofacies D1 and D2 represent tidal inlet, tidal delta, and/or back-barrier estuarine environments. Distal tidal delta and/or back-barrier estuaries (D1) are characterized by millimeter- to centimeter-scale interbedded sand and mud laminations and layers with some shell fragment and foraminifera (lithofacies D1). Proximal tidal delta and tidal inlet deposits (D2) are composed of dark gray mud with subvertical lenses of sand and abundant fauna indicative of shallow inlet–influenced estuaries (Parker, 1960; Phleger, 1960; Andrews, 1981; Kaczorowski and Gernant, 1980). This includes a diverse group of species, including, *Elphidium* sp., *Quinqueloculina* sp., *Ammonia* sp., ostracods, *Abra* sp., *Anadara* sp., and *Mulinia* sp. Grain-size data show significant coarsening in core 04cl01 (Fig. 6) corresponding to a flooding surface that separates deltaic central-basin and lower-estuary deposits.

Facies E is only found in cores from the onshore chenier plain and potentially represents fine-grained mud that was deposited in the shoreface during the passage of cold fronts and hurricanes (Howe et al., 1935; Kemp, 1987; Huh et al., 1991). As the mud flat builds and becomes subaerial, the lithology changes to facies F, which represents the interbedded sand and shell layers of the chenier ridges. Alternatively, cores obtained from inter-ridge areas display mud flat overlain by shallow-estuary or marsh deposits.

Ternary diagrams display the sand, silt, and clay composition of the lithofacies and illustrate the relative separation and/or overlap of samples (Fig. 8). The prodelta (B3) facies is generally finer-grained than the central-basin (C) and interdistributary estuarine (B2) facies (Fig. 8A). The central-basin facies (C) is finer than the distal flood–tidal delta facies (D1), which is finer than the proximal flood–tidal delta facies (D2) (Fig. 8B). Figure 8C shows a comparison of the coarsest facies.

Flooding Surfaces and Chronostratigraphy

Seven surfaces coincident with seismic, litho-, and biofacies changes identified in seismic records and core data have been identified and dated. The samples of shells and organic material used for the ages are listed in Table 1, located in core descriptions in Figure 6, and located where appropriate in seismic data sections (Figs. 3 and 7). The surfaces shallow landward and are interpreted as flooding surfaces, since they place basinward depositional environments onto landward depositional environments (Fig. 9). In the lower portion of the estuary, the lowest surface, flooding surface 1, occurs at ~15–17 mbsl—this is the bayline surface between underlying fluvial and overlying upper-estuary–bay-head delta deposits in the lower portion of the estuary. The surface can be traced in seismic and core data over the southern two-thirds of the estuary. A maximum date on woody peat of 8900 ± 110 cal yr B.P. and minimum date on plant fragments of 8500 ± 80 cal yr B.P. constrain the age of the surface. A date of 8860 ± 140 cal yr B.P. from sand-sized plant fragments (core 04cl05) extracted from dark gray mud above the sample dated as 8500 ± 80 cal yr B.P. is assumed to be locally reworked material.

The 12–15 mbsl flooding surface, flooding surface 2, results in translation of the bay-head delta depositional system landward, and the age is constrained by three ages, younger than 8320 ± 100 cal yr B.P., 8500 ± 80 cal yr B.P., and older than 7940 ± 80 cal yr B.P. The surface is assigned an age of 8000–8300 cal yr B.P. This surface is not obvious in the lower portion of the modern estuary (04cl01, line 8a), but it is prominent in the upper part of the estuary (core 04cl05, Fig. 3B; core 04cl06, Fig. 3C) and constitutes the bayline further updip.

The initial incursion of estuary deposition into the study area occurs above the –10 to –14 mbsl flooding surface 3. An age of 7900–8000 cal yr B.P. is assigned to the –10 to –14 mbsl surface based on ages older than 7820 ± 120 cal yr B.P., 7780 ± 360 cal yr B.P., 7140 ± 140 cal yr B.P., 7300 ± 120 cal yr B.P., and younger than 8320 ± 100 cal yr B.P., 7940 ± 80 cal yr B.P., and 13,900 ± 140 cal yr B.P.

In core 04cl01, a lithologic change from sandy mud with abundant plant/wood fragments to muddy sand with shell fragments occurs at 10.6 mbsl (Fig. 6). Plant/wood fragments from 10.6 and 11 mbsl yielded ages of 7680 ± 100 cal yr B.P. and 7220 ± 280 cal yr B.P., respectively. The 7–11 mbsl surface, flooding surface 4, is therefore younger than 7220 ± 280 cal yr B.P., 7140 ± 140 cal yr B.P., and 7300 ± 120 cal yr B.P., but older than 7140 ± 120 cal yr B.P. This surface is assigned an age of 7200 cal yr B.P.

Three cores display facies changes that represent the 5–9 mbsl flooding surface 5. At 5.6 mbsl in core 04cl06, dark gray mud with a *Rangia* sp. shell layer overlies peat. A sample from the top of the peat yielded an age of 5660 ± 60 cal yr B.P. and a *Rangia* sp. shell layer at 9.3 mbsl in core 04cl04 provided a maximum age of 5610 ± 40 cal yr B.P. for the surface. In core 04cl11, an articulated *Rangia* sp. shell gave a maximum age of 5750 ± 90 cal yr B.P. (Fig. 7). This surface is therefore assigned an age range of 5600–5800 cal yr B.P. The first tidally influenced deposits overlie the 5–9 mbsl flooding surface.

The 5–7 mbsl surface, flooding surface 6, emplaces central-basin deposits onto bay-head delta in the majority of the study area and signifies a significant environmental change. Five cores

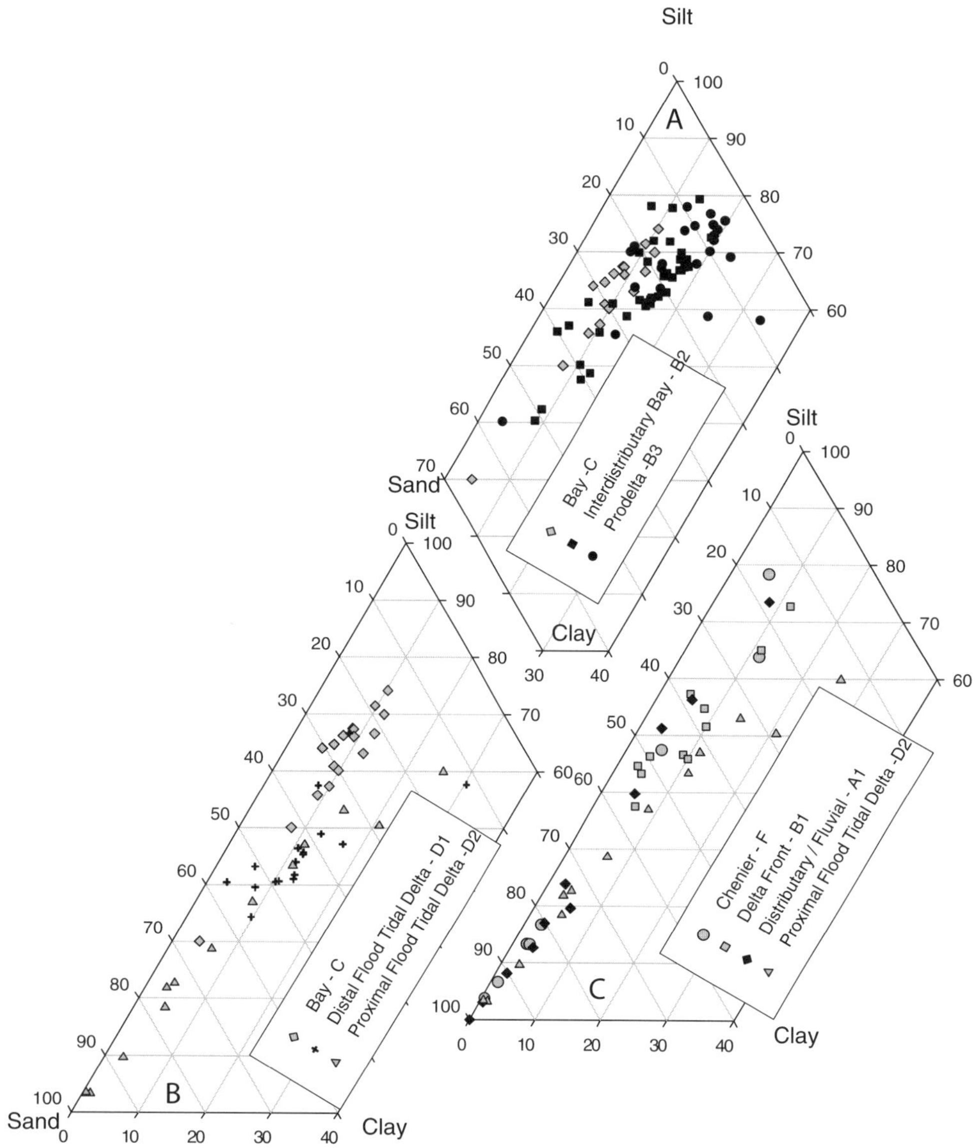

Figure 8. Three ternary grain-size plots (A, B, and C) showing sand, silt, and clay percentages of different facies groups, which represent the average grain size of various facies groups.

provided suitable datable material to date the 5–7 flooding surface. The age of the surface is constrained as older than 2720 ± 40 cal yr B.P. and 2840 ± 100 cal yr B.P. but younger than 3370 ± 130 cal yr B.P., 3820 ± 320 cal yr B.P., and 3020 ± 160 cal yr B.P. This surface is assigned an age range of 2800–3000 cal yr B.P.

The shallowest surface, flooding surface 7, extends beneath the chenier plain at 3 mbsl and is manifested as a *Rangia* sp. shell layer (lag) separating muddy, shelly sand below from dark gray mud with plant fragments above. This surface was closely sampled and yielded an age of 2500 to 1800 cal year B.P.

In general, the radiocarbon ages used to constrain the timing of flooding events fall within chronologic order (Table 1). However, there are several anomalously old ages that are out of stratigraphic order. All of these ages come from sand-sized wood/plant fragments extracted from dark gray interdistributary or central-basin mud. We interpret these as reworked material, and, hence, these ages are discounted for the age of flooding-surface isochrons.

DISCUSSION

Allogenic and autogenic flooding surfaces are identified by a change in seismic, litho-, and biofacies within seismic and core data, but they can also be recognized in seismic data as either distinct or subtle truncation surfaces. Depending on location, the flooding surfaces superimpose different depositional environments. A subtle system response can occur for a variety of reasons. Either the back stepping associated with the flooding event was not significant enough to force the system to radically respond; the system responds quickly and progrades back to

TABLE 1. RADIOCARBON AGES

Sample	Lab no.	$\delta^{13}C$ (‰)	^{13}Cct yr B.P.	Error	Cal yr B.P.	Error	Depth (mbsl)	Calib.	Reservoir	Sample type
04cl01_4t8_30t34	OS_50251	–22	4570	45	5250	200	–0.3	Intcal04	0	Wood/plant
04cl01_12_16_108_112	OS-50197	–24	2500	45	2555	185	–3.5	Intcal04	0	Wood/plant
04cl01_16t20_15t20	OS-50204	–2.3	2610	40	2270	120	–3.8	Marine04	0	Foram
04cl01_ 20t 24_ 100t 110	OS-48098	–2.3	3060	40	2835	95	–5.9	Marine04	0	Foram
04cl01_36_40_80t90	OS-50349	–17.4	6820	55	7680	100	–11	Intcal04	0	Wood/plant
04cl01_40t44_0t5	OS-50686	–22.7	6350	130	7225	275	–11	Intcal04	0	Wood/plant
04cl01_40t44_100	OS_50252	–19	6990	55	7820	120	–12	Intcal04	0	Wood/plant
04cl01_44t48_106t108	OS-50701	–25	6890	190	7785	365	–13	Intcal04	0	Wood/plant
04cl01_64_68_1	OS-47096	–27	8060	45	8945	175	–17	Intcal04	0	Peat
04cl01_64_68_96	OS-46748	–26.8	8620	35	9595	65	–18	Intcal98	0	Peat
04cl04_8t12_26t30	25543	–19.6	2735	15	2820	40	–1.5	Intcal04	0	Marsh
04cl04_32t36_70	25538	–4.4	4895	15	5610	40	–9.2	Marine04	–350	*Rangia* sp.
04cl04_40t44_24	25540	–0.8	5455	15	6230	50	–11	Marine04	–350	*Nuculana* sp.
04cl04_56t60_5t7	25544	–26.9	8030	20	8895	115	–16	Intcal04	0	Peat
04cl05_20t25_5t7	OS-53240	–21.5	6680	40	7545	65	–4.9	Intcal04	0	Plant frag.
04cl05_20_25_10_12	OS-53067	–5.01	2860	35	3025	165	–4.9	Marine04	–350	*Rangia* sp.
04cl05_30_35_60	OS-53068	–5.79	6280	35	7135	135	–8.5	Marine04	–350	*Rangia* sp.
04cl05_40_45_8	OS-46750	–25.3	7190	40	8045	105	–11	Intcal98	0	Peat
04cl05_45_50_55_60	OS-53094	–26.2	7130	40	7940	80	–13	Intcal04	0	Peat
04cl05_45_50_83	OS-46981	–27.4	7980	30	8860	140	–13	Intcal04	0	Peat
04cl05_50t55_2t5	OS-53239	–27.7	7720	50	8505	80	–14	Intcal04	0	Plant frag.
04cl06_20t25_12	25534	–7.3	3405	15	3695	85	–5	Marine04	–350	*Rangia*
04cl06_20t25_72	25535	–0.9	4965	15	5660	60	–5.6	Marine04	–350	*Rangia*
04cl06_40_45_106	OS-46749	–27.3	7540	35	8315	95	–12	Intcal98	0	Peat
04cl07_25_30_26	OS-46751	–24.5	6240	30	7135	115	–6.7	Intcal98	0	Peat
04cl07_25_30_30_35	OS-53095	–27.6	6380	35	7335	85	–6.7	Intcal04	0	Plant frag.
04cl07_30t35_55t57	OS-51934	–27.7	6360	40	7300	120	–8.4	Intcal04	0	Plant frag.
04cl07_35_40_70_75	OS-53091	–28.3	12,050	60	13,905	135	–10	Intcal04	0	Plant frag.
04cl09_70	Beta-196917	–6	3530	60	1880	180	–3.1	Marine04	–350	*Rangia* sp.
04cl09_135	Beta-196918	–27	1920	110	3815	325	–3.8	Intcal04	0	Peat
04cl09_3_3_84	OS-53097	–13.7	4490	40	5140	160	–4	Intcal04	0	Plant frag.
04cl10_102_18	25533	–1.7	410	15	390	70	–2.9	Marine04	–350	*Rangia* sp.
04cl10 2 3 70	OS-47876	–6	3140	30	3370	130	–4.8	Marine04	–350	*Rangia* sp.
04cl10 2 3 75 80	OS-48050	–18.8	3830	30	4255	155	–4.9	Intcal04	0	Peat
04cl11_1_3_82	28059	na	2575	15	2715	35	–3.8	Marine04	–350	*Rangia* sp.
04cl11_3_3_40_43	28060	na	5020	15	5750	90	–6.3	Marine04	–350	*Rangia* sp.

Notes: Core locations are shown on Figure 1; core name is the first part of sample name; lab no.—laboratory number, $\delta^{13}C$—carbon-13 isotope fractionation, ^{13}Cct yr B.P.—conventional radiocarbon age corrected for carbon 13 isotope fractionation, Cal yr B.P.—calibrated radiocarbon age in calendar years, Calib—calibration curve used (Stuiver et al., 2004), Reservoir—reservoir used to calibrate radiocarbon age, Sample type—type of material used to obtain radiocarbon date.

original depositional conditions; or the apparent back stepping is an autocyclic event that is manifested as only a very slight environmental change.

The magnitude of depth change exhibited by flooding surfaces associated with back-stepping deltaic-estuarine environments is an indicator of water depth. Delta-plain facies, which tend to preserve organic plant and wood material, occur at sea level, whereas central-basin mud deposition occurs at a range of water depths. An isochron that extends from the deltaic environment into the estuarine environment and shows basinward deepening likely indicates the relative elevation change of depositional environments. Flooding surfaces within a fluvial-deltaic–estuarine–marine depositional system, by definition, will shallow and amalgamate landward or in the direction of diminishing accommodation space. Additionally, flooding surfaces also amalgamate with diminished sedimentation rates, and flooding-surface resolution decreases accordingly. Flooding surfaces are best delineated in high-sedimentation environments such as bay-head deltas as opposed to the central basin.

Seven estuary-wide flooding surfaces coincident with seismic and lithofacies changes are identified within Calcasieu Lake. All of these surfaces shallow in a landward direction. The ages of the flooding surfaces are constrained to ca. 8900–8500 cal yr B.P., 8300–8000 cal yr B.P., 8000–7900 cal yr B.P., 7200 cal yr B.P., 5800–5600 cal yr B.P., 3000–2800 cal yr B.P., and 2500–1800 cal yr B.P. The fact that several isochrons within Calcasieu Lake display 2–4 m of depth range indicates that the depth of the estuary remained relatively constant until present day. This also indicates that the depth of the estuary is not necessarily related to ability of the deltaic system to fill accommodation space but may rather be a function of the wind, wave, and/or tidal conditions.

The integration of several data sets provides insight into likely causal mechanisms for the stratigraphic architecture of Calcasieu Lake. A comparison of the area flooded to radiocarbon-dated surfaces and regional sea level shows that the basal or bayline peat ages fall on the regional sea-level curve (Fig. 10A). The basal/bayline peat in the lower estuary at ~17–15 mbsl and is dated at 8900–8500 cal yr B.P. (cores 04cl01 and 04cl04). This records initial flooding of Calcasieu Lake. The bayline then stepped landward ~20 km to the present-day central-basin location at 12–15 mbsl between 8300–8000 cal yr B.P. (cores 04cl06 and 04cl05). By ca. 8000–7900 cal yr B.P., central-basin deposits

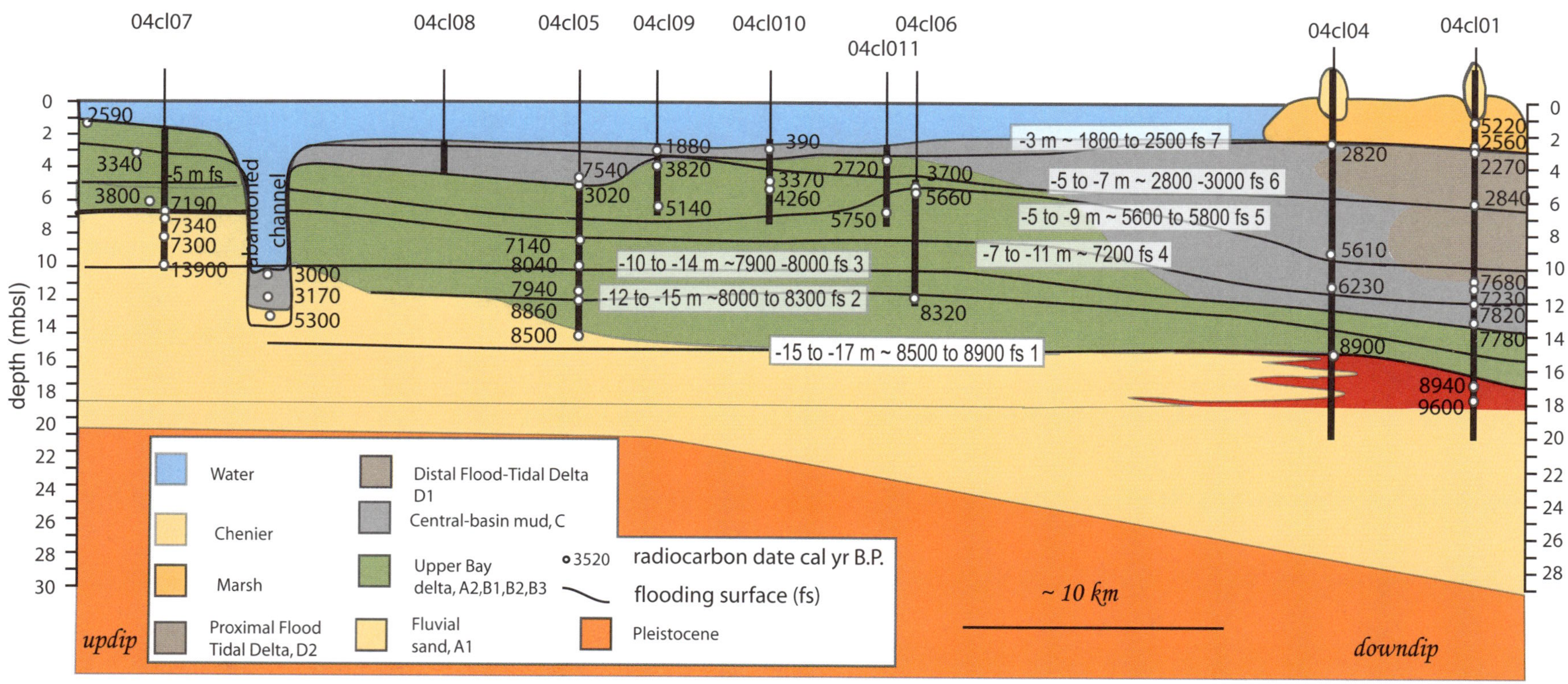

Figure 9. Dip cross section of the Calcasieu incised valley showing interpretation of back-stepping depositional environments and flooding surfaces. Radiocarbon ages are also shown. See Figure 1 for location.

were deposited in the study area at the southern portion of the modern estuary, which coincides with a deltaic back step of more than 10 km. At ca. 7200 cal yr B.P. (6–8 m, core 04cl07), the bayline migrated to the Prien Lake–Lake Charles area, landward of Calcasieu Lake (Fig. 2). The observed back stepping suggests that the fluvial-delta-estuary system responds to eustatic sea-level rise by way of internal thresholds that result in rapid environmental reorganization.

The nonbasal or non–bayline peat or shell ages fall below the regional sea-level curve (Fig. 10A). The peat facies indicate deposition at sea level but show subsequent valley-fill compaction effects, which result in the surfaces lying deeper than expected based on the regional sea-level curve. The shell ages originate from *Rangia cuneata* and assemblages of foraminifera, which were likely deposited in an estuary or shallow-marine environment at depths of a few meters below sea level. The fact that the nonbasal and non–bayline peats reside at depths above the curve for the area inundated by Calcasieu Lake (Fig. 10A) implies that the system was able to outpace sea-level rise at those times by filling available accommodation space.

Sedimentation rates calculated using age-depth relationships within individual cores show a general decrease ca. 5600 cal yr B.P. from 2.0–3.4 mm/yr to 0.3–0.9 mm/yr. These rates roughly mimic the change in regional sea-level rise from 2.9–5.2 mm/yr to 0.5–2.1 mm/yr (Figs. 10B, 10C, and 10D). Figures 10B, 10C, and 10D show sedimentation rates of individual cores from Prien Lake, middle and upper Calcasieu Lake (estuary), and the lower estuary of Calcasieu Lake (Fig. 2). The diminution of sedimentation rate occurs in two intervals (ca. 7200 cal yr B.P. and ca. 5600 cal yr B.P.) in a landward-stepping manner. At 7200 cal yr B.P., the flooding event is manifested as a sedimentation rate decrease in the lower, middle, and upper estuary (Figs. 10C and 10D) as

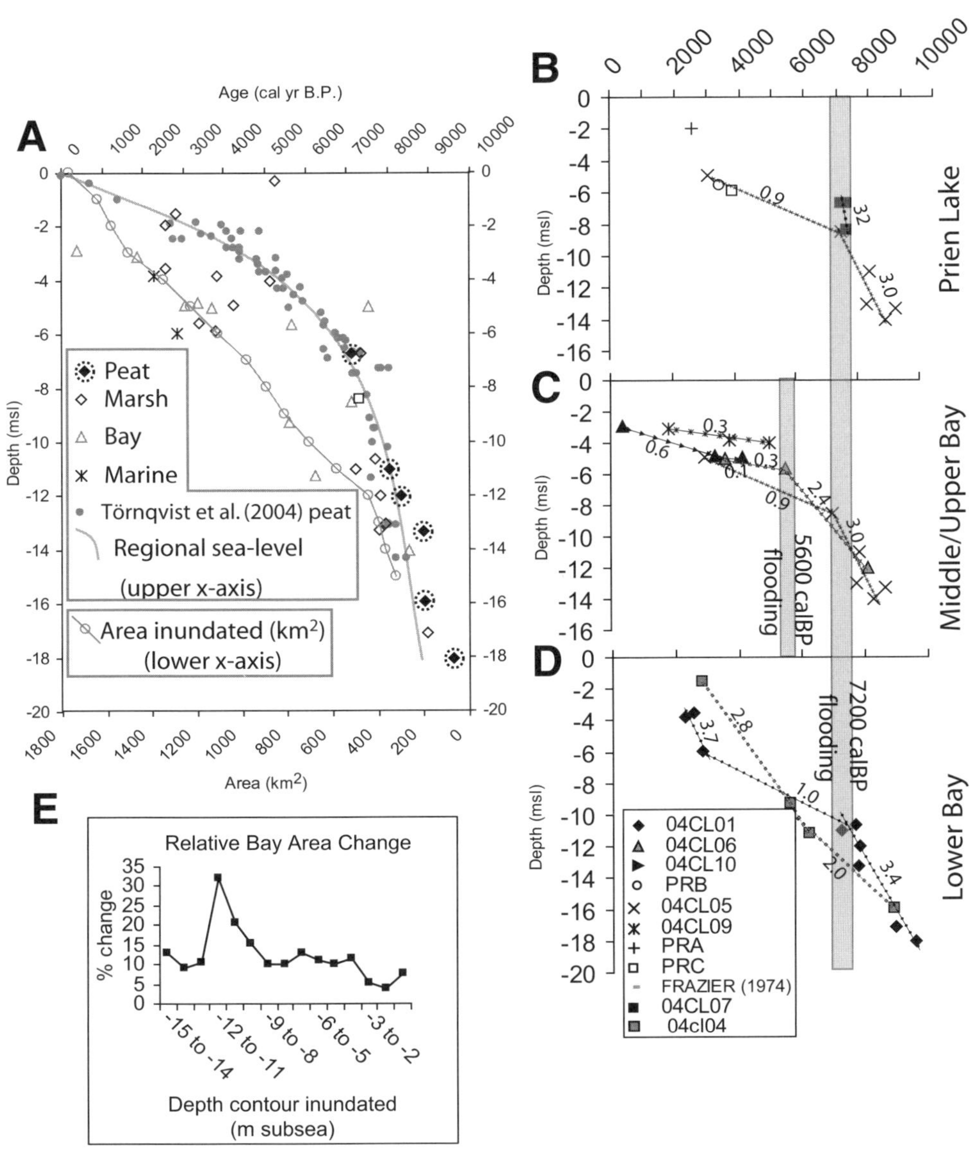

Figure 10. (A) Area of cumulative flooding, radiocarbon samples, and generalized regional sea-level curve (derived from Törnqvist et al., 2004). (B) Age-depth plots of cores located in Prien Lake and linear regression–derived sedimentation rates (mm/yr). (C) Age-depth plots of cores from the middle and upper estuary and linear regression–derived sedimentation rates (mm/yr). (D) Age-depth plots of cores from the lower estuary and linear regression–derived sedimentation rates (mm/yr). The changes in sedimentation rate are highlighted with the gray boxes. The 7200 cal yr B.P. flooding event occurs in the Prien Lake, upper-middle, and lower portions of the fluvial-estuarine system. (E) Relative estuarine area change illustrated by depth contour inundated with % estuarine area increase (% change).

well as peat formation (water-table rise; Fig. 10B) in the area of Prien Lake (Fig. 2). The 5600 cal yr B.P. event is observed in the middle to upper portion of the estuary as a sedimentation rate decrease (Fig. 10C). In the lower estuary, tidal delta and inlet deposition commenced around this time (Fig. 9). The lack of evidence for a 5600 cal yr B.P. event in the Prien Lake area may be due to the paucity of data rather than lack of an event.

In order to quantify the potential effects of antecedent topography on stratigraphic architecture, we compared the area of estuary flooded to sea-level rise. Figure 10E shows the percent change of the estuary area with each successive meter of flooding. If antecedent topography influences the elevation and timing of flooding surfaces, then a particular flooding surface should correlate with a peak in estuarine area addition. When the contour interval of 12–11 m was flooded, the estuary area was increased by 30%. The subtle flooding surface at the level marked by the 12 m bayline in the upper estuary suggests that the system's response was not environment altering but rather a continuation of bay-head delta deposition (Fig. 9). This, in turn, suggests that flooding of antecedent topography is not an important control of the flooding history in Calcasieu Lake.

Sediment supply variation, as controlled by climate, is an important factor in estuarine evolution. If sedimentation is high, the valley fills at the rate of sea-level rise. If sedimentation is low, the valley floods to increase the area of the estuary. All other factors being equal, the balance between sediment supply and sea-level rise controls the flooding history and, therefore, evolution of the estuary. The fact that the Calcasieu system maintained a bay-head delta in the present-day estuary until 3000 cal yr B.P. (5 mbsl, ~1 m subbottom) suggests that the system was fundamentally different than present and had much higher sediment flux into the estuary in the past. This warrants additional study but suggests that sediment supply was much higher prior to 3000 cal yr B.P.

The drainage basin presently resides in a humid setting. One would predict relatively less sediment yield in this type of climate setting versus a more semiarid to arid setting (Langein and Schumm, 1958; Schumm, 2003). Alternatively, a fluvial-estuarine system situated in a climate that oscillates between humid and arid at high frequency would generate and transport more sediment to the basin (Langein and Schumm, 1958; Perlmutter et al., 1998; Schumm, 2003).

Paleogeographic reconstructions (Fig. 11) illustrate the environmental changes that occurred in Calcasieu Lake during the Holocene. The first significant event occurred between 8900 and 8500 cal yr B.P. and resulted in the bay-head delta shifting landward at least 28 km (Figs. 11A and 11B). At 8300 to 8000 cal yr B.P., another rapid event resulted in a 20 km landward shift in the delta (Figs. 11C and 11D). The 8300–8000 cal yr B.P. event is correlated to a rapid sea-level rise at this time (Törnqvist et al., 2005). This rise is believed to have originated from the melting of the Laurentide Ice Sheet. For a few hundred years, the estuary setting remained stable until the first incursion of a central-basin environment, which occurred between 8000–7900 cal yr B.P. (Fig. 11E). At ca. 7200 cal yr B.P., the locus of deposition shifted landward to the northern limits of the study area (Prien Lake–Lake Charles area; Fig. 11F). Between 5800–5600 cal yr B.P., sedimentation rates dramatically diminished, and central-basin muds were deposited over much of the present-day lake (Figs. 10D, 11G, and 11H). At this time, the first tidal inlet was formed in the present-day lower estuary environment. Between 5600 and 3000 cal yr B.P., the bay-head delta prograded and filled in much of the estuary. At the same time, the first chenier ridges were formed (Gould and McFarlan, 1959). At ca. 3000–2800 cal yr B.P., the chenier plain began to prograde and further restricted estuarine circulation (Fig. 11I). During this time interval, the bay-head delta stepped landward 20 km (Fig. 11J). Between 2500–1800 cal yr B.P., the bay-head delta shifted landward of the study area, and a central-basin setting existed throughout much of present day Prien Lake, Lake Charles, and Calcasieu Lake area (Figs. 11K and 11L).

Without additional updip cores, we are not able to reconstruct the estuary setting subsequent to ca. 1600 cal yr B.P. We speculate that the deltaic-fluvial system must have prograded to the northern limit of present-day Calcasieu Lake and retrograded to the northern limit of Lake Charles in order to deposit and create the modern river-lake-estuary morphology (Fig. 11M). Based on work in other estuaries, we further speculate that the progradation may have occurred between 1000–400 cal yr B.P., with rapid retrogradation at ca. 400 cal yr B.P. (900 cal yr B.P. flooding surface in Sabine Lake and 390 cal yr B.P. *Rangia* layer in Calcasieu Lake).

FUTURE CHANGES

Relative sea-level rise across the west Louisiana coastal region has increased 5–15-fold in the past century (Penland et al., 1989; Shinkle and Dokka, 2004). This translates to extreme wetland degradation and marine inundation of vast coastal areas. We utilize the reconstructions of the Holocene evolution to foresee the potential response of the Calcasieu river-lake system to natural and/or anthropogenic relative sea-level rise. Assuming a magnitude of sea-level rise of 0.5–1 m over the next century (5 mm/yr to 10 mm/yr average rates as measured in tidal gauge and benchmark releveling studies; Penland et al., 1989; Shinkle and Dokka, 2004), the scenario of environment retreat is similar to the middle to early Holocene. A 20 km back-stepping event is not unrealistic and would transform the city of Lake Charles into a bayside town with increased tidal energy and saltwater intrusion similar to the setting that existed 1800–1600 yr ago (Fig. 12).

CONCLUSIONS

1. The Calcasieu River–Calcasieu Lake depositional system experienced multiple episodes of dramatic back stepping and landward translation of depositional environments and estuarine shorelines during the Holocene. These back-stepping events occurred within a maximum of a

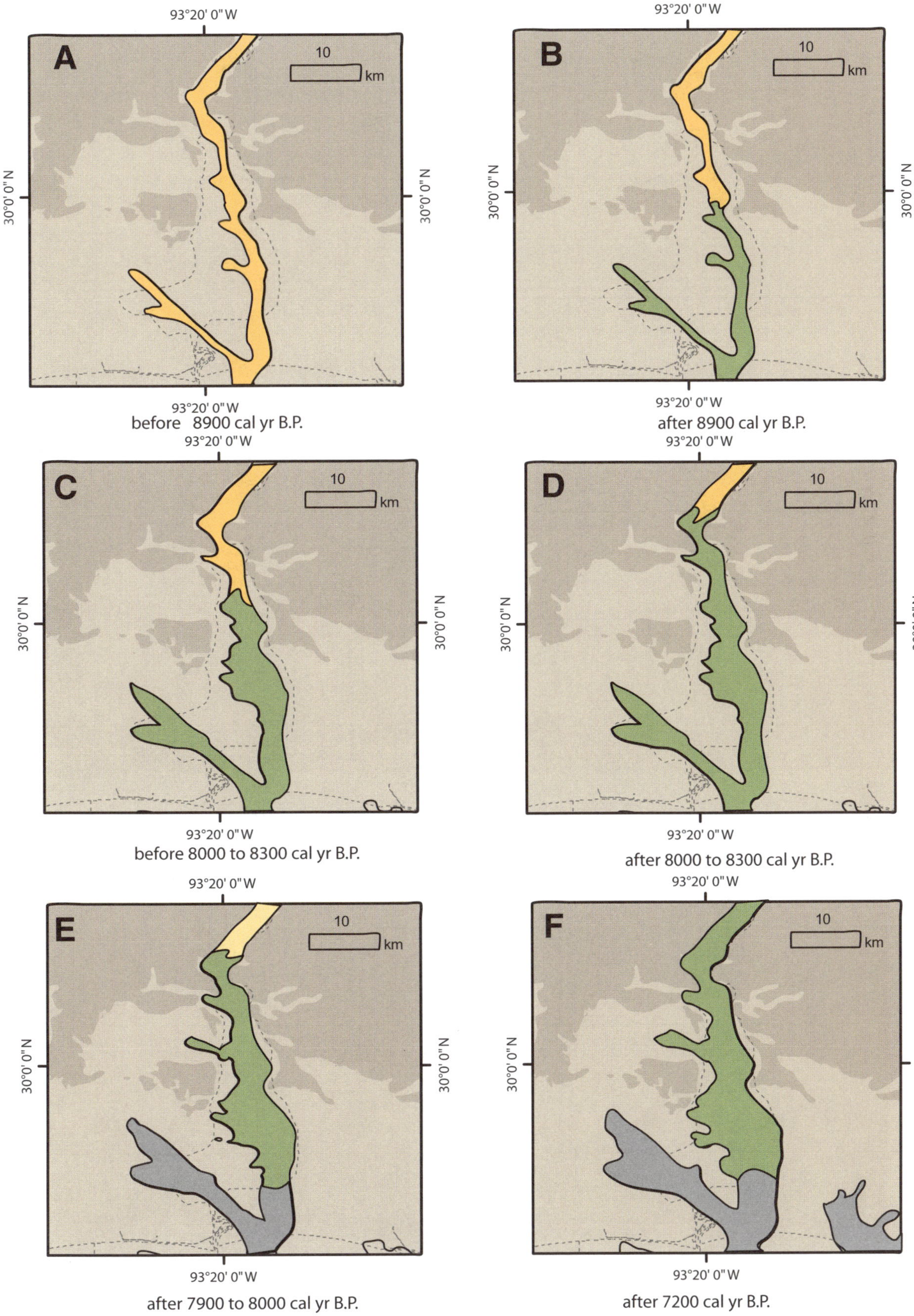

Figure 11 (*on this and following two pages*). Paleogeographic maps showing environmental changes before and after flooding events.

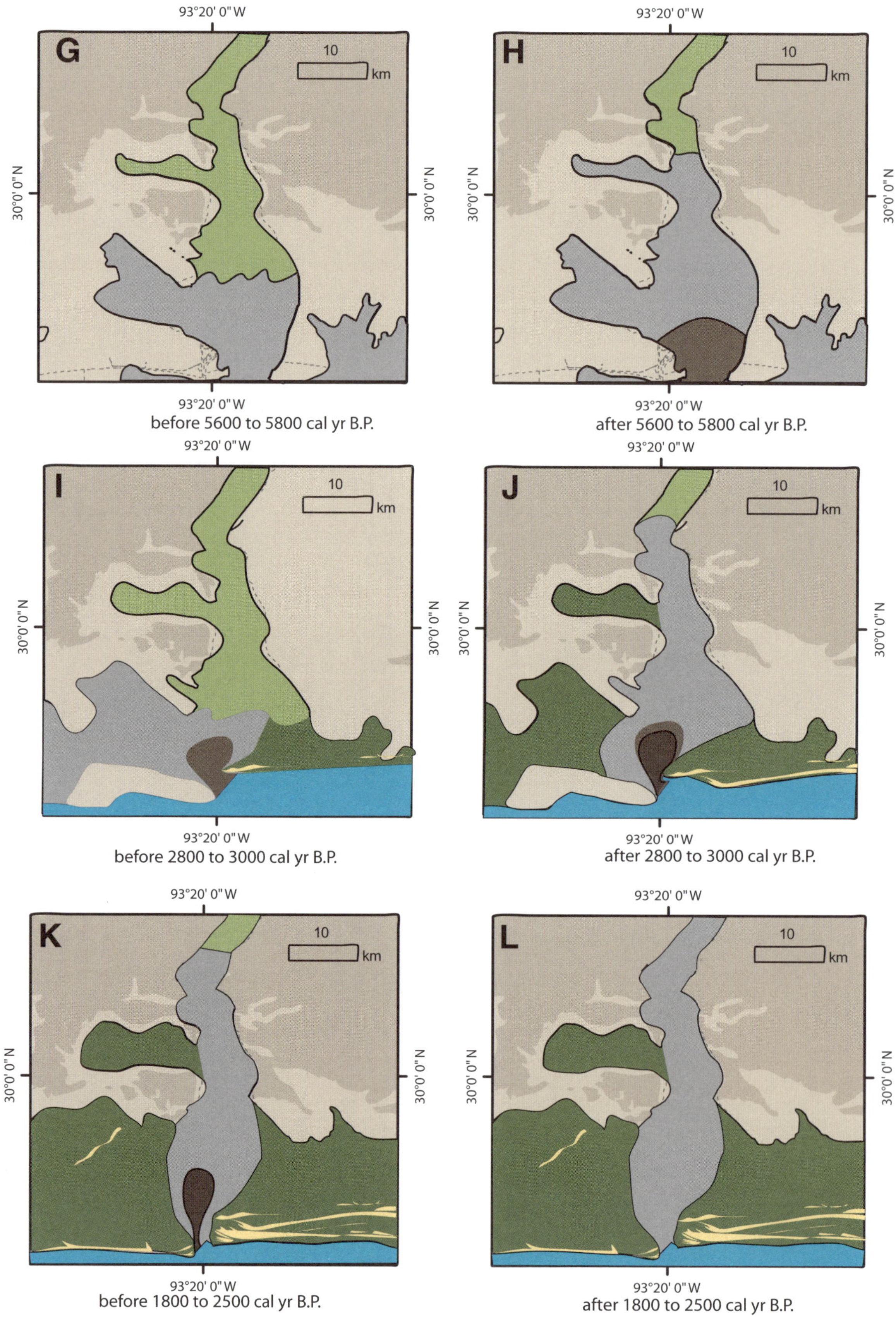

Figure 11 (*continued*).

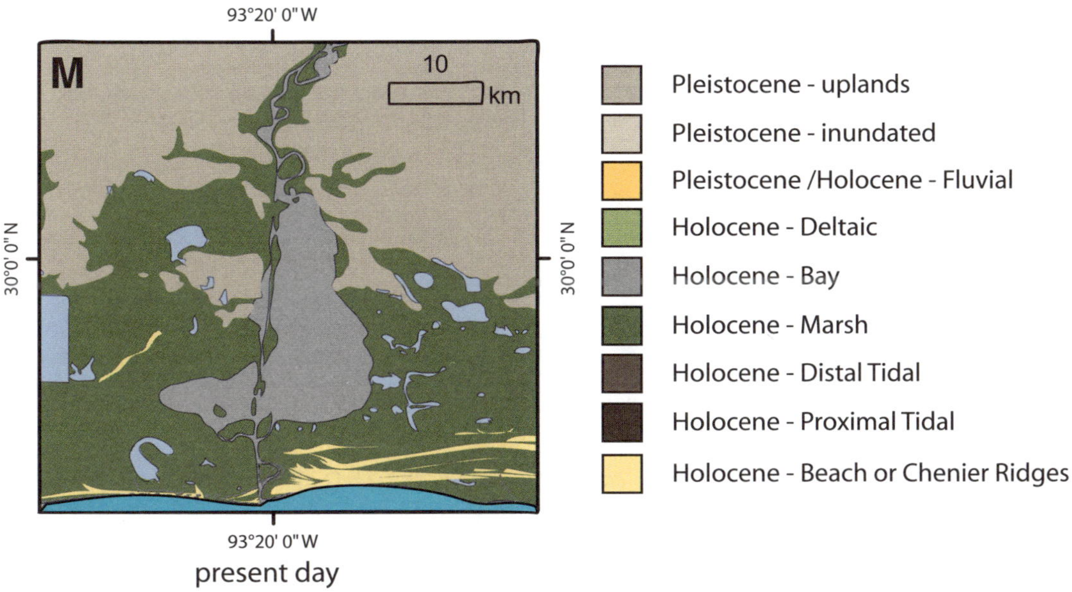

Figure 11 (*continued*).

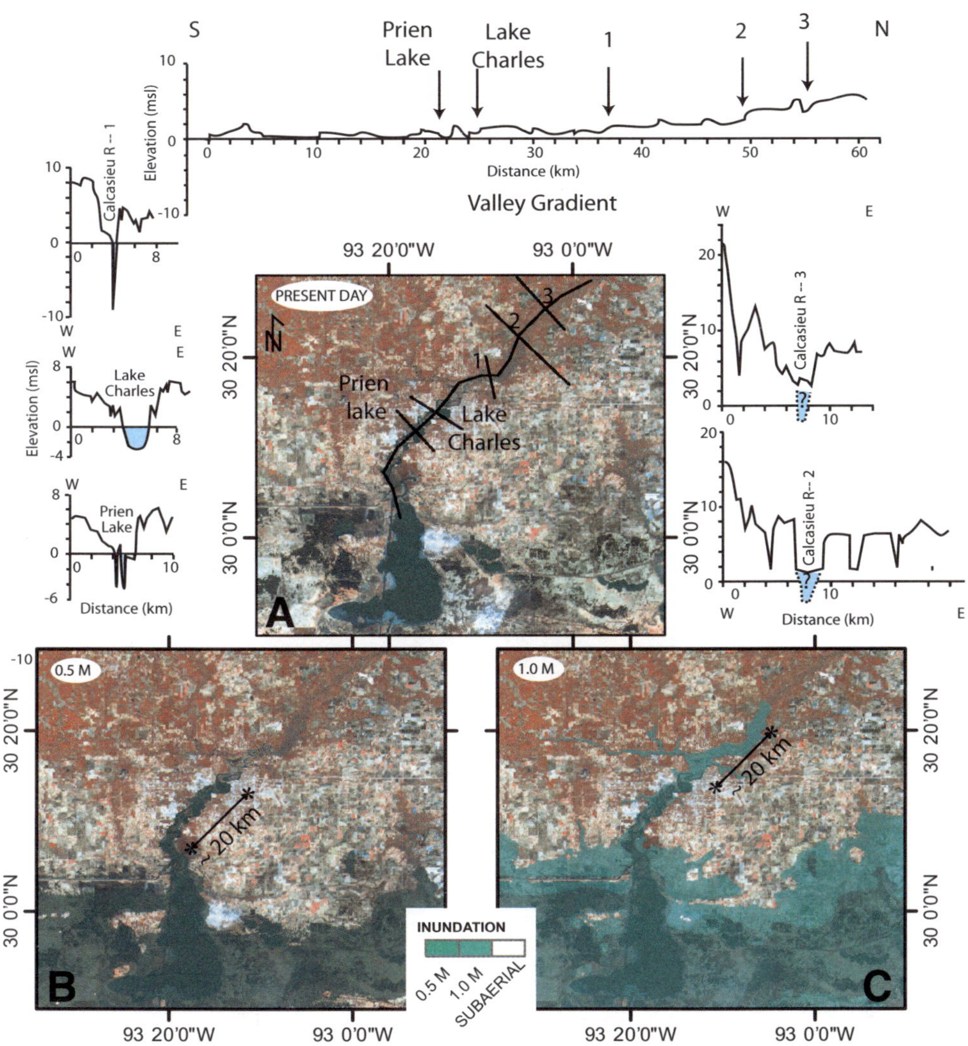

Figure 12. Landsat images of Calcasieu system showing areas that would be inundated by 0.5 and 1.0 m relative sea-level rise.

few hundred years and resulted in tens of kilometers of landward translation of the estuarine shoreline. These events occurred at 8900–8500 cal yr B.P., 8300–8000 cal yr B.P., 8000–7900 cal yr B.P., 7200 cal yr B.P., 5800–5600 cal yr B.P., 3000–2800 cal yr B.P., and 2500–1800 cal yr B.P.

2. The correlation of several flooding events to other estuaries along the Texas coast provides evidence for regional rather than local forcing mechanisms. Regional forcing mechanisms potentially include eustatic rise and/or climate change and associated changes in sediment supply.
3. The meter-scale vertical resolution of regional sea-level curves is likely above the resolution of sea-level changes that caused flooding events. Rather, centimeter- to decimeter-scale sea-level fluctuations can cause significant environmental reorganization in low-gradient fluvial-estuarine systems. Thus, subtle, but perhaps rapid, changes in rate of sea-level rise cannot be ruled out as a cause of observed changes in the estuarine setting.
4. Regional climate change affecting sediment supply could cause periods of diminished sedimentation and environmental reorganization. Additional work is needed to determine if observed flooding events coincide with climatic events.
5. Present accelerated rates of relative sea-level rise translate to future dramatic environmental reorganization within the Calcasieu fluvial-deltaic-estuarine system.

ACKNOWLEDGMENTS

The authors would like to acknowledge the funding for this study from National Science Foundation (NSF) grant EAR-0107-650, Rice Gulf Research Industry Consortium, the David Worthington Fellowship Fund, Ed Picou Student Research Grant, and a Geological Society of America Student Research Grant. We would like to thank reviewers S. Nichol and M. Fearn for thoughtful comments and constructive criticism. Thanks are due to Chip Anderson, Jessica Maddox, Patrick Taha, Alex Simms, and Craig Meyer for data collection assistance.

REFERENCES CITED

Anderson, J.B., Rodriguez, A.R., Abdulah, K., Fillon, R., Banfield, L., McKeown, H., and Wellner, J.S., 2004, Late Quaternary stratigraphic evolution of the northern Gulf of Mexico margin, *in* Anderson, J.B., and Fillon, R., eds., A Synthesis: Late Quaternary Stratigraphic Evolution: Tulsa, Oklahoma, Society for Sedimentary Geology (SEPM), p. 1–24.

Anderson, J.B., Rodriguez, A.B., Milliken, K., and Taviani, M., 2008, this volume, The Holocene evolution of the Galveston estuary complex, Texas: Evidence for rapid change in estuarine environments, *in* Anderson, J.B., and Rodriguez, A.B., eds., Response of Upper Gulf Coast Estuaries to Holocene Climate Change and Sea-Level Rise: Geological Society of America Special Paper 443, doi: 10.1130/2008.2443(06).

Andrews, J., 1981, A Field Guide to Shells of the Texas Coast: Houston, Texas, Gulf Publishing Company, Monthly Field Guide Series 176 p.

Bailey, R.G., 1980, Description of the Ecoregions of the United States: Washington, D.C., U.S. Department of Agriculture Miscellaneous Publication 1391, 77 p.

Bernard, H.A., 1950, Quaternary Geology of Southeast Texas [Ph.D. dissertation]: New Orleans, Louisiana State University, 164 p.

Boesch, D., Mehta, A., Morris, J., Nuttle, W., Simenstad, C., and Swift, D., 1994, Scientific assessment of coastal wetland loss, restoration, and management in Louisiana: Journal of Coastal Research, v. 20, special issue, p. 1–103.

Byrne, J.V., LeRoy, D.O., and Riley, C.M., 1959, The chenier plain and its stratigraphy: Southwestern Louisiana Gulf Coast Association of Geologic Societies Transactions, v. 9 p. 237–260.

Coast 2050 (Louisiana Coastal Wetland Conservation and Restoration Task Force and the Wetlands Conservation and Restoration Authority), 1998, Coast 2050: Toward a Sustainable Coastal Louisiana: An Executive Summary: Baton Rouge, Louisiana, Louisiana Department of Natural Resources, 12 p.

Coleman, J.M., 1966, Recent Coastal Sedimentation: Baton Rouge, Louisiana, Louisiana State University, Central Louisiana Coast, Coastal Studies Institute Technical Report 29, 84 p.

Coleman, J.M., and Smith, W.G., 1964, Lake recent rise of sea level: Geological Society of America Bulletin, v. 75, p. 833–840, doi: 10.1130/0016-7606(1964)75[833:LRROSL]2.0.CO;2.

Day, J.W., Jr., Shaffer, D.P., Britsch, L.D., Reed, D.J., Hawes, S.R., and Cahoon, D.R., 2000, Pattern and process of land loss in the Mississippi Delta, a spatial and temporal analysis of wetland habitat change: Estuaries, v. 23, p. 425–438, doi: 10.2307/1353136.

Doering, J.A., 1956, Review of Quaternary surface formations of Gulf Coast region: Bulletin of the American Association of Petroleum Geologists, v. 40, no. 8, p. 1816–1862.

Fearn, M.L., 1995, Louisiana's Cajun Prairie: Holocene History of a Southern Grassland [Ph.D. thesis]: Baton Rouge, Louisiana State University, 280 p.

Fisk, H.N., 1948, Geological Investigation of the Lower Mermentau River Basin and Adjacent Areas in Coastal Louisiana, Appendix II: Mermentau River, Louisiana, U.S. Army Corps Engineers, Mississippi River Commission, 145 p.

Forbes, M.J., 1988, Hydrologic Investigations of the Lower Calcasieu River, Louisiana: U.S. Geological Service Water-Resources Investigations Report 87-4173, 61 p.

Frazier, D.E., 1974, Depositional Episodes: Their Relationship to the Quaternary Stratigraphic Framework in the Northwest Portion of the Gulf Basin: Austin, Texas, Bureau of Economic Geology, Geologic Circular 74-1, 28 p.

Goldenberg, S.B., Landsea, C.W., Mestas-Nunez, A.M., and Gray, W.M., 2001, The recent increase in Atlantic hurricane activity: Causes and implications: Science, v. 293, p. 474–479, doi: 10.1126/science.1060040.

Gould, H.R., and McFarlan, E., 1959, Geologic history of the chenier plain, southwestern Louisiana: Gulf Coast Association of Geological Societies Transactions, v. 9, p. 261–272.

Howe, H.V., Russell, R.J., McGuirt, J.H., Craft, B.C., and Stephenson, M.B., 1935, Reports on the Geology of Cameron and Vermillion Parishes: New Orleans, Louisiana, Louisiana Department of Conservation, 242 p.

Huh, O.K., Roberts, H.H., Rouse, L.J., and Rickman, D.A., 1991, Fine-grained sediment transport and deposition in the Atchafalaya and chenier plain, *in* Kraus, N.C., Gingerich, K.J., and Kriebel, D., eds., Coastal Sediment '91: Seattle, Washington, Proceedings of a Specialty Conference on Quantitative Approaches to Coastal Sediment Processes, 25 June 1991, p. 817–830.

Kaczorowski, R.T., and Gernant, R.E., 1980, Stratigraphy and coastal processes of the Louisiana chenier plain and Louisiana chenier plain microfossil associations, *in* Kaczorowski, R.T., and Gernant, R.E., eds., The Sedimentary Environments of the Louisiana Coastal Plain: Lafayette, Gulf Coast Association of Geological Societies Field Guide, p. 31–67.

Kane, H., 1959, Late Quaternary Geology of Sabine Lake and Vicinity, Texas and Louisiana: Gulf Coast Association of Geological Societies (GCAGS) Transactions, v. 9, p. 225–235.

Kemp, G.P., 1987, Mud Deposition at the Shoreface: Wave and Sediment Dynamics on the Chenier Plain of Louisiana [Ph.D. dissertation]: New Orleans, Louisiana State University, 147 p.

Langein, W.B., and Schumm, S.A., 1958, Yield of sediment in relation to mean annual precipitation: Eos (Transactions, American Geophysical Union), v. 39, p. 1076–1084.

Le Blanc, R.J., 1949, Recent and Pleistocene Geology of the Calcasieu Entrenched Valley System of Southwest Louisiana: Unpublished Shell

Oil Company Report 128, http://www.searchanddiscovery.net/documents/leblanc/index.htm (last accessed 30 April 2008).

Lee, J.M., Wiseman, W.J., Jr., and Kelly, F.J., 1990, Barotropic, subtidal exchange between Calcasieu Lake and the Gulf of Mexico: Estuaries, v. 13, p. 258–264, doi: 10.2307/1351916.

McCave, I.N., Bryant, R.J., Cook, H.F., and Coughanowr, C.A., 1986, Evaluation of a laser-diffraction-size analyzer for use with natural sediments: Journal of Sedimentary Petrology, v. 56, p. 561–564.

McFarlan, E., 1961, Radiocarbon dating of lake Quaternary deposits, south Louisiana: Geological Society of America Bulletin, v. 72, p. 129–158, doi: 10.1130/0016-7606(1961)72[129:RDOLQD]2.0.CO;2.

Michener, W.K., Blood, E.R., Bildstein, K.L., Brinson, M.M., and Gardner, L.R., 1997, Climate change, hurricanes and tropical storms, and rising sea level in coastal wetland: Ecological Applications, v. 7, p. 770–801, doi: 10.1890/1051-0761(1997)007[0770:CCHATS]2.0.CO;2.

Mitsch, W.J., and Gosselink, J.G., 2000, Wetlands (3rd edition): New York, John Wiley and Sons, 934 p.

Morton, R.A., Bernier, J.C., and Barras, J.A., 2006, Evidence of regional subsidence and associated interior wetland loss induced by hydrocarbon production, Gulf Coast region, USA: Environmental Geology, v. 50, p. 261–274, doi: 10.1007/s00254-006-0207-3.

NCDC, 1994, Time Bias–Corrected Divisional Temperature-Precipitation-Drought Index: Documentation for dataset TD-9640, 12 p. (Available from DBMB, NCDC, NOAA, Federal Building, 37 Battery Park Ave., Asheville, North Carolina 28801-2733, http://www.cdc.noaa.gov.)

Nelson, H.F., and Bray, E.E., 1970, Stratigraphy and history of the Holocene sediments in the Sabine-High Island area, Gulf of Mexico, *in* Morgan, J.P., ed., Deltaic Sedimentation Modern and Ancient: Society for Economic Paleontologists and Mineralogists (SEPM) Special Publication 15, p. 48–77.

Nichol, S.L., Boyd, R., and Penland, S., 1992, Hydrology of a Wave-Dominated Estuary: Lake Calcasieu, Southwest Louisiana: Gulf Coast Association of Geological Societies, v. 42, p. 835–844.

Nichol, S.L., Boyd, R., and Penland, S., 1996, Sequence stratigraphy of a coastal-plain incised valley estuary: Lake Calcasieu, Louisiana: Journal of Sedimentary Research, v. 66, p. 847–857.

Parker, R.H., 1960, Ecology and distributional patterns of marine macro-invertebrates, northern Gulf of Mexico, *in* Shepard, F.P., Phleger, F.B., and van Andel, T.H., eds., Recent Sediments, Northwest Gulf of Mexico: Tulsa, Oklahoma, American Association of Petroleum Geologists, p. 302–337.

Penland, S., Ramsey, K.E., McBride, R.A., Moslow, T.F., and Westphal, K.A., 1989, Relative Sea Level Rise and Subsidence in Louisiana, and the Gulf of Mexico: Louisiana Geological Survey Coastal Geology Technical Report, 65 p.

Perlmutter, M.A., Radovich, B.J., Matthews, M.D., and Kendall, C.G.St.C., 1998, The Impact of High-Frequency Sedimentation Cycles on Stratigraphic Interpretation, Predictive High Resolution Sequence Stratigraphy: Norwegian Petroleum Society No. 8: Elsevier Science, p. 141–170.

Phleger, F.B., 1960, Sedimentary patterns of microfaunas in northern Gulf of Mexico, *in* Shepard, F.P., Phleger, F.B., and van Andel, T.H., eds., Recent Sediments, Northwest Gulf of Mexico: Tulsa, Oklahoma, American Association of Petroleum Geologists, p. 267–301.

Pielke, R.A., Jr., Landsea, C., Mayfield, M., Laver, J., and Pasch, R., 2005, Hurricanes and global warming: Bulletin of the American Meteorological Society, v. 86, no. 11, p. 1571–1575, doi: 10.1175/BAMS-86-11-1571.

Schumm, S.A., 2003, The Fluvial System: New Jersey, Blackburn Press, 338 p.

Simms, A.R., Anderson, J.B., Rodriguez, A.B., and Taviani, M., 2008, this volume, Mechanisms controlling environmental change within an estuary: Corpus Christi Bay, Texas, USA, *in* Anderson, J.B., and Rodriguez, A.B., eds., Response of Upper Gulf Coast Estuaries to Holocene Climate Change and Sea-Level Rise: Geological Society of America Special Paper 443, doi: 10.1130/2008.2443(08).

Shinkle, K.D., and Dokka, R.K., 2004, Rates of Vertical Displacement at Benchmarks in the Lower Mississippi Valley and the Northern Gulf Coast: Silver Spring, Maryland, National Oceanic and Atmospheric Administration (NOAA) Technical Report NOS/NGS 50, 135 p.

Sperazza, M., Morre, J.N., and Hendrix, M.S., 2004, High-resolution particle size analysis of naturally occurring very fine-grained sediment through laser diffractometry: Journal of Sedimentary Research, v. 74, no. 5, p. 736–743, doi: 10.1306/031104740736.

Stuiver, M., Reimer, P.J., and Reimer, R.W., 2005, CALIB 5.0: WWW program and documentation.

Thomas, M.A., and Anderson, J.B., 1994, Sea-level controls on the facies architecture of the Trinity/Sabine incised-valley system, Texas continental shelf, *in* Dalrymple, R.W., Boyd, R., and Zaitlin, B.A., eds., Incised-Valley Systems: Origin and Sedimentary Sequences: Society for Economic Paleontologists and Mineralogists (SEPM) Special Publication 51, p. 63–82.

Tornqvist, T.E., Gonzalez, J.L., Newsom, L.A., van der Borg, K., de Jong, A.F.M., and Kurnik, C.W., 2004, Deciphering Holocene sea-level history on the U.S. Gulf Coast: A high-resolution record from the Mississippi Delta: Geological Society of America Bulletin, v. 116, no. 7–8, p. 1026–1039.

Törnqvist, T.E., Bick, S.J., Gonzalez, J.L., van der Borg, K., and de Jong, A.F.M., 2005, Tracking the sea-level signature of the 8.2 ka cooling event: New constraints from the Mississippi Delta: Geophysical Research Letters, v. 31, L23309, p. 1–4.

Turner, R.E., 1997, Wetland loss in the northern Gulf of Mexico: Multiple working hypothesis: Estuaries, v. 20, p. 1–13, doi: 10.2307/1352716.

Winker, C.D., 1979, Late Pleistocene Fluvial-Deltaic Deposition, Texas Coastal Plain and Shelf [M.A. thesis]: Austin, University of Texas at Austin, 187 p.

Wright, S.S., 1980, Seismic Stratigraphy and Depositional History of Holocene Sediments on the Central Texas Gulf Coast [M.A. thesis]: Austin, University of Texas at Austin, 157 p.

Zaitlin, B.A., Dalrymple, R.W., and Boyd, R., 1994, The stratigraphic organization of incised-valley systems associated with relative sea-level change, *in* Dalrymple, R.W., Boyd, R., and Zaitlin, B.A., eds., Incised-Valley Systems: Origin and Sedimentary Sequences: Society for Economic Paleontologists and Mineralogists (SEPM) Special Publication 51, p. 63–82.

Manuscript Accepted by the Society 10 March 2008

Printed in the USA

The Geological Society of America
Special Paper 443
2008

Tracking the Holocene evolution of Sabine Lake through the interplay of eustasy, antecedent topography, and sediment supply variations, Texas and Louisiana, USA

K.T. Milliken
John B. Anderson
Department of Earth Science, Rice University, 6100 Main Street, MS 126, Houston, Texas 77005, USA

Antonio B. Rodriguez
Institute of Marine Sciences, University of North Carolina at Chapel Hill, 3431 Arendell Street, Morehead City, North Carolina 28557, USA

ABSTRACT

The Sabine-Neches fluvial-estuary system is composed of deposits that represent fluvial, deltaic, central-basin, tidal inlet/delta, and chenier depositional systems. The Holocene deposits and associated environmental changes preserved in the drowned Sabine-Neches alluvial valley provide a valuable analog for present and future environmental changes. These deposits are bounded by flooding surfaces that record episodes of dramatic environmental reorganization during the Holocene. The most dramatic environmental changes are manifested as stratigraphic back stepping in which central-basin sediments overlie deltaic sediments. The magnitude of flooding varies from a few tens of kilometers to less pronounced back stepping followed by rapid progradation.

Initial flooding of the onshore Sabine-Neches incised valley occurred around 9800 cal yr B.P., and by ca. 8900 cal yr B.P., a vast bayhead delta occupied the southern half of the valley. This delta backstepped up the valley during the relatively rapid sea-level rise of the early to middle Holocene, and by ca. 7100 cal yr B.P., it occupied the entire valley. After ca. 7100 cal yr B.P., the bayhead delta shifted up the valley again, and a central-basin setting existed in the lower half of the valley. The middle basin expanded episodically between ca. 5500 cal yr B.P. and 1700 cal yr B.P., and a brief period of delta growth occurred ~300 yr ago.

Controlling mechanisms for flooding surface formation include sea-level rise, changes in the antecedent topography of the incised valley, and sediment supply variations. Antecedent topography was influential in controlling estuarine evolution between ca. 7800 and 7500 cal yr B.P., when an extensive fluvial terrace was inundated. The fact that some flooding surfaces appear to be synchronous, within a few centuries, in several estuaries across the northern Gulf of Mexico suggests a eustatic rather than local control. Flooding events at ca. 8900 cal yr B.P. and ca. 8400–8000

Milliken, K.T., Anderson, J.B., and Rodriguez, A.B., 2008, Tracking the Holocene evolution of Sabine Lake through the interplay of eustasy, antecedent topography, and sediment supply variations, Texas and Louisiana, USA, *in* Anderson, J.B., and Rodriguez, A.B., eds., Response of Upper Gulf Coast Estuaries to Holocene Climate Change and Sea-Level Rise: Geological Society of America Special Paper 443, p. 65–88, doi: 10.1130/2008.2443(05). For permission to copy, contact editing@geosociety.org. ©2008 The Geological Society of America. All rights reserved.

cal yr B.P. were likely caused by rapid, sub–meter-scale sea-level rise events. Sediment supply variations controlled by climatic forcing appear to have been the main cause of other flooding events. Unfortunately, the Holocene climate record for the east Texas–west Louisiana coastal region is poorly documented, and a direct relationship to central and western Texas climate records may be complex. So the exact nature of climate control on sediment flux to the estuary system remains elusive.

Keywords: Quaternary, climate change, incised valley, eustasy, Gulf of Mexico.

INTRODUCTION

Recent land loss and subsidence studies along the chenier plain of southwest Louisiana have documented dramatic environmental changes to the area during the last century (Barras et al., 1994; Turner, 1997; Coast 2050, 1998). Conversion of marsh to open water has constituted the most obvious change. A comparison of aerial photos from 1956 and 1998 of the low-lying areas of the southern portion of the Sabine alluvial valley shows an ~23% (70 km^2) conversion from predominantly subaerial vegetation to open-water ponds (Fig. 1). Proposed causes of the changes include canal construction, ship-channel and intracoastal waterway construction and enlargement, fishing and industry canal maintenance, petroleum and/or groundwater extraction, and hurricane impacts (Turner, 1997; Day et al., 2000; Morton et al., 2006). Regardless of the immediate cause(s), future relative sea-level rise projections (3–15 mm/yr; Penland et al., 1989; Shinkle and Dokka, 2004) foretell more dramatic change. The potential impacts include salinization of groundwater supplies, prairie to wetland conversion, rising water-table levels, and shoreline retreat.

The rate at which low-lying coastal areas are flooded depends on four variables: eustatic sea-level rise, subsidence, sediment supply, and relief of the preexisting landscape (antecedent topography). Holocene eustatic sea-level rise in the Gulf of Mexico has been constrained by multiple studies that suggest smooth but diminishing rates of sea-level rise through time (Törnqvist et al., 2004; Milliken et al., this volume). Long-term (millennial-scale) subsidence has also been constrained by dating sea-level indicators, such as basal salt marsh peat and beach ridges (Milliken et al., this volume). During the Holocene, relative sea-level rise occurred at rates as high as those predicted for the next century (IPCC, 2001). Thus, the sediment preserved beneath Sabine Lake (within the Sabine-Neches alluvial valley) records the evolution of the Sabine-Neches river-estuary system during the Holocene. This paper documents and describes the Holocene flooding history and stratigraphic architecture for the Sabine-Neches fluvial and estuary system following the nomenclature of Zaitlin et al. (1994). A comparison of the results with those of companion studies of sediments in other estuaries along the west Louisiana to central Texas coast (Rodriguez et al., 2005; Milliken et al., this volume; Anderson et al., this volume; Maddox et al., this volume; Simms et al., this volume) provides a means to isolate the relative effects of eustasy, subsidence, antecedent topography, and sediment supply (climate) on coastal inundation.

As the rate of relative sea-level rise increases, inundation of low-lying coastal areas can be offset by increased rates of aggradation. Aggradation is dependent on terrigenous and organic sediment supply to the coast (Baumann et al., 1984; Jervey, 1988; Schumm, 2003). If sediment supply is less than the rate of relative sea-level rise, flooding of low-lying coastal areas occurs. The amount of sediment supplied to estuaries via fluvial processes is dependent upon drainage-basin characteristics, including drainage-basin area, topography and gradient, substrate, vegetation patterns, and discharge. Of these, vegetation patterns and discharge are sensitive to climate change and are, therefore, most likely to influence coastal flooding over century to millennial time scales. Comparison to the Holocene depositional system process-response provides a context for predicted future environmental changes.

GEOLOGICAL SETTING

Two Texas rivers (Sabine and Neches) with roughly equal drainage areas (25,300 km^2 and 25,900 km^2, respectively; Handbook of Texas Online, 2008) terminate in an estuary called Sabine Lake. Sabine River discharge from 1971 to 2002 averaged 250 ± 130 m^3/s near Ruliffe, Texas (U.S. Geological Survey [USGS] 08030500 station). Neches River discharge over the same period averaged 190 ± 90 m^3/s near Evadale, Texas (USGS 08041000 station).

Topographic relief of the Sabine-Neches drainage basin ranges from sea level at the head of Sabine Lake to +250 m in northeast Texas. The drainage basin has a subtropical, subhumid climate, weak seasonal fluctuations in precipitation, and strong seasonal temperature variation (Bull, 1991; Fig. 2). Vegetation in the majority of the basin is dominated by mixed pine-hardwood forests with wetland vegetation near the coast (Frost, 1949; Bezanson, 2000).

Sabine Lake is a freshwater to brackish-water estuary that covers an area of 243 m^2, and water depth is on average 2 m. The estuary receives an average of 1.3 km^3/mo freshwater inflow (Powell and Green, 1994), with a residence time of 0.6 mo (U.S. EPA, 1999). Prevailing winds blow from the southeast during the summer and from the northwest during the winter (Lohse, 1955). Sabine Lake has a microtidal range of 0.33 m (National

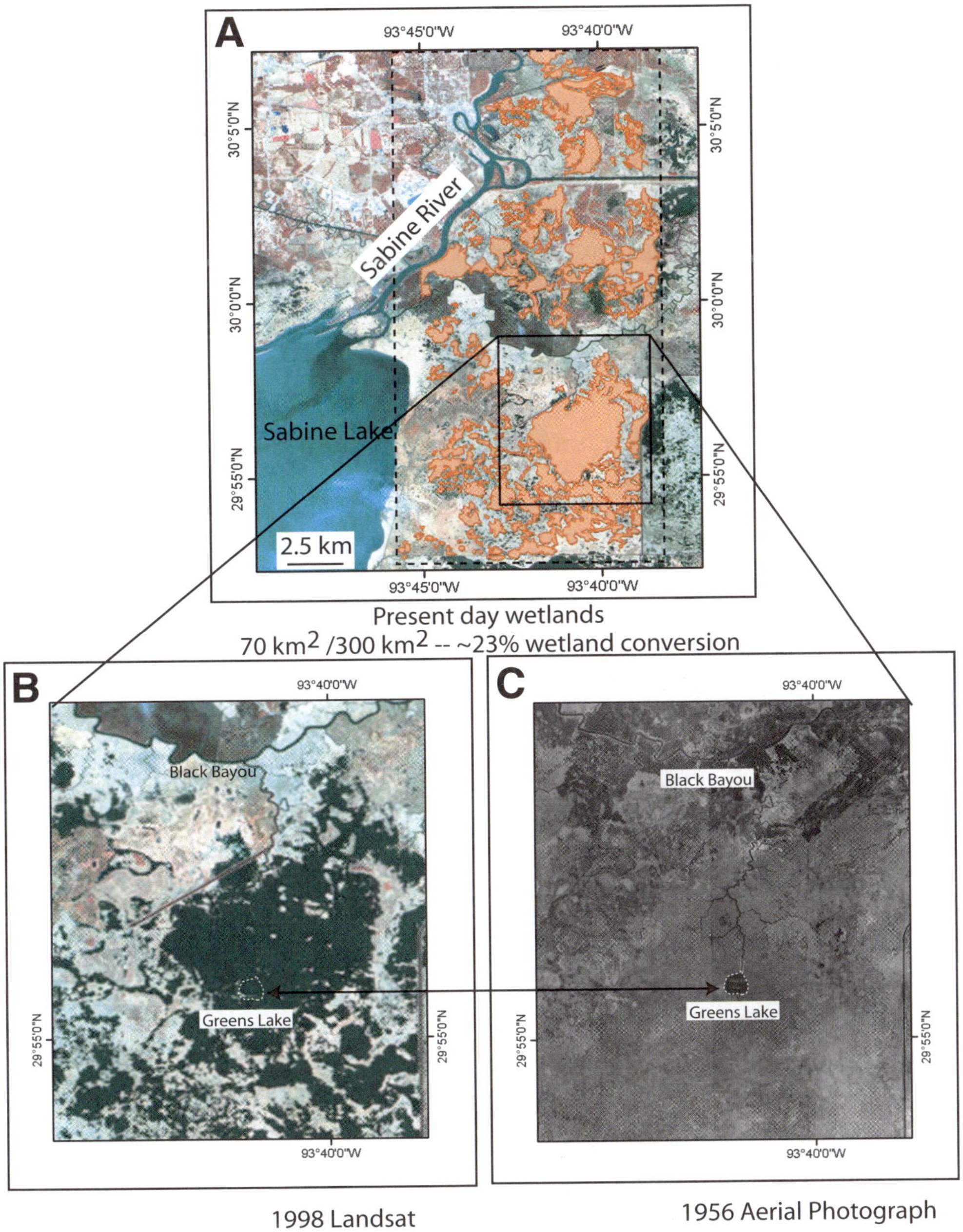

Figure 1. (A) Modern Landsat image focusing on the wetlands northeast of Sabine Lake. (B) A 1998 DOQQ (digital orthoquarter quad) image showing inset area. (C) A 1956 historic aerial photo of the area. The area around Greens Lake was subaerial in 1956 but converted to a large pond by 1998. Approximately 23% of the wetland area has been converted to open water during the past 60 yr.

Oceanic and Atmospheric Administration [NOAA], Sabine Pass tide station).

METHODS

High-resolution seismic data (54 km) were acquired in Sabine Lake aboard the R/V *Jenny* in 2003 using an Edgetech Chirp 2–15 kHz subbottom profiler system. These data augmented 45 km of seismic data acquired in 1991 using a 4 kHz Boomer system aboard the R/V *Matagorda* (Fig. 3; Anderson et al., 1991). Seismic reflections are resolvable at the meter scale, and penetration depths are up to 100 ms or 75 m (1500 m/s assumed velocity).

Core data from Sabine Lake include two cores, 6 and 20 m long, obtained using a push/rotary-drill rig; two vibracores up to 6 m long, and four push cores up to 1 m long. A truck-mounted Geoprobe system was used to collect one core (20 m long) in the southern part of the merged Sabine-Neches valley. Core locations were strategically chosen to verify seismic-data interpretations and sample the complete incised-valley fill sequence. Vibracores (12) from Anderson et al. (1991) are integrated in this study and provide material for additional radiocarbon ages. Thirteen cores

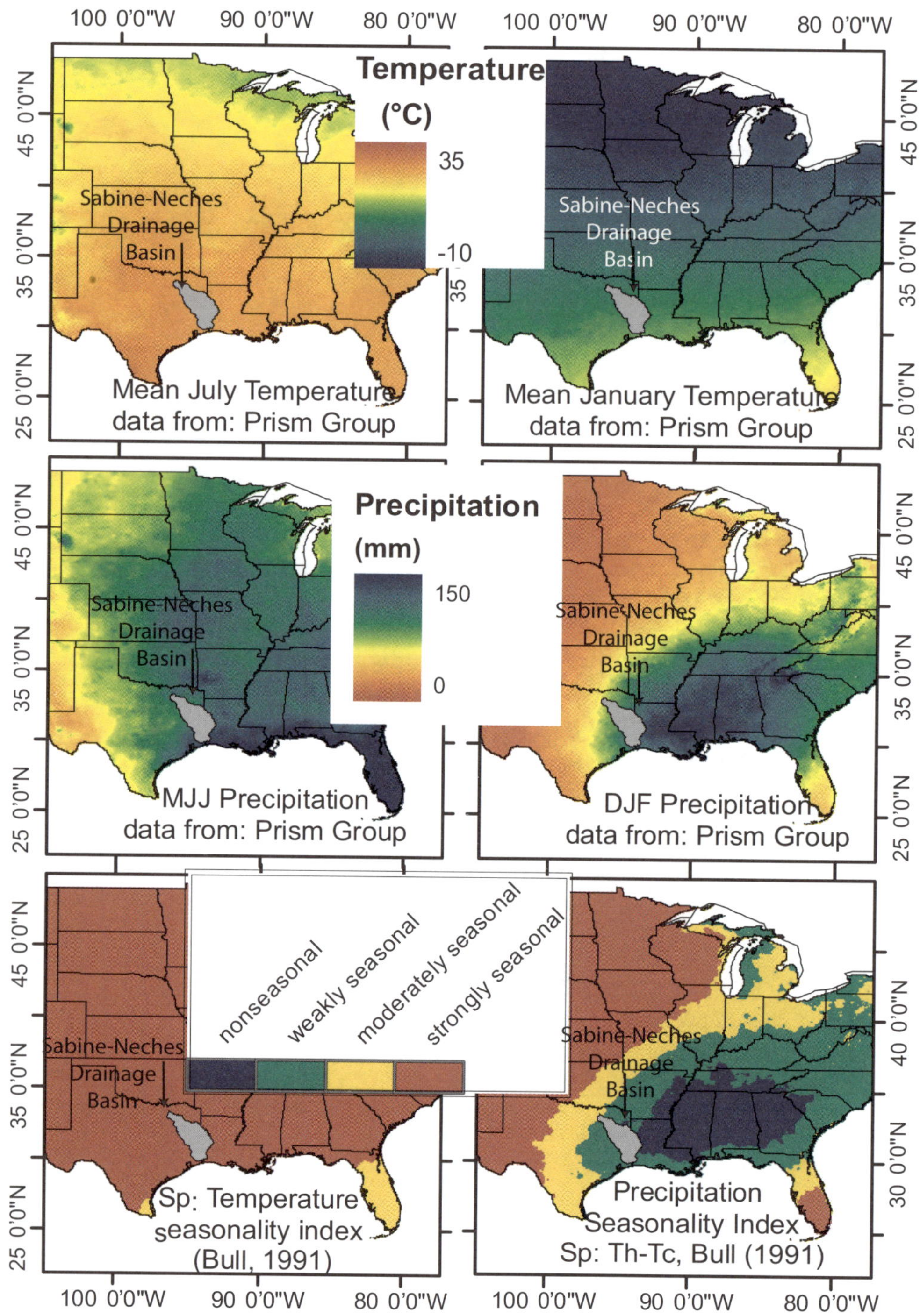

Figure 2. Mean summer and winter temperature as well as May, June, July (MJJ) average precipitation and December, January, February (DJF) average precipitation from 1971 to 2000. Temperature seasonality index and precipitation seasonality index were computed from parameters by Bull (1991). The Sabine River basin is situated in an area with a strongly seasonal temperature index and a weakly to nonseasonal precipitation index.

collected by Morton et al. (2000), and located on the west side of the estuary, were examined at the Texas Bureau of Economic Geology core facility in Austin, Texas.

The cores acquired for this study were cut lengthwise, and each half was designated as either a working or archive half. The cores were stored in a refrigerated facility at 4.4 °C. Cores were cleaned and described immediately after opening; visual inspection included determination of color, grain size, sedimentary structures, nature of contacts, organic type and content, and identification of mollusk shells. Quantitative grain-size analysis was conducted with a Malvern laser particle-size analyzer (McCave et al., 1986; Sperazza et al., 2004) on samples taken every 10–50 cm.

The ages of key bounding surfaces identified from core and seismic data were determined by radiocarbon dating of organic and carbonate material. Humic material was washed from organic-rich mud and muddy peat deposits, whereas carbonate

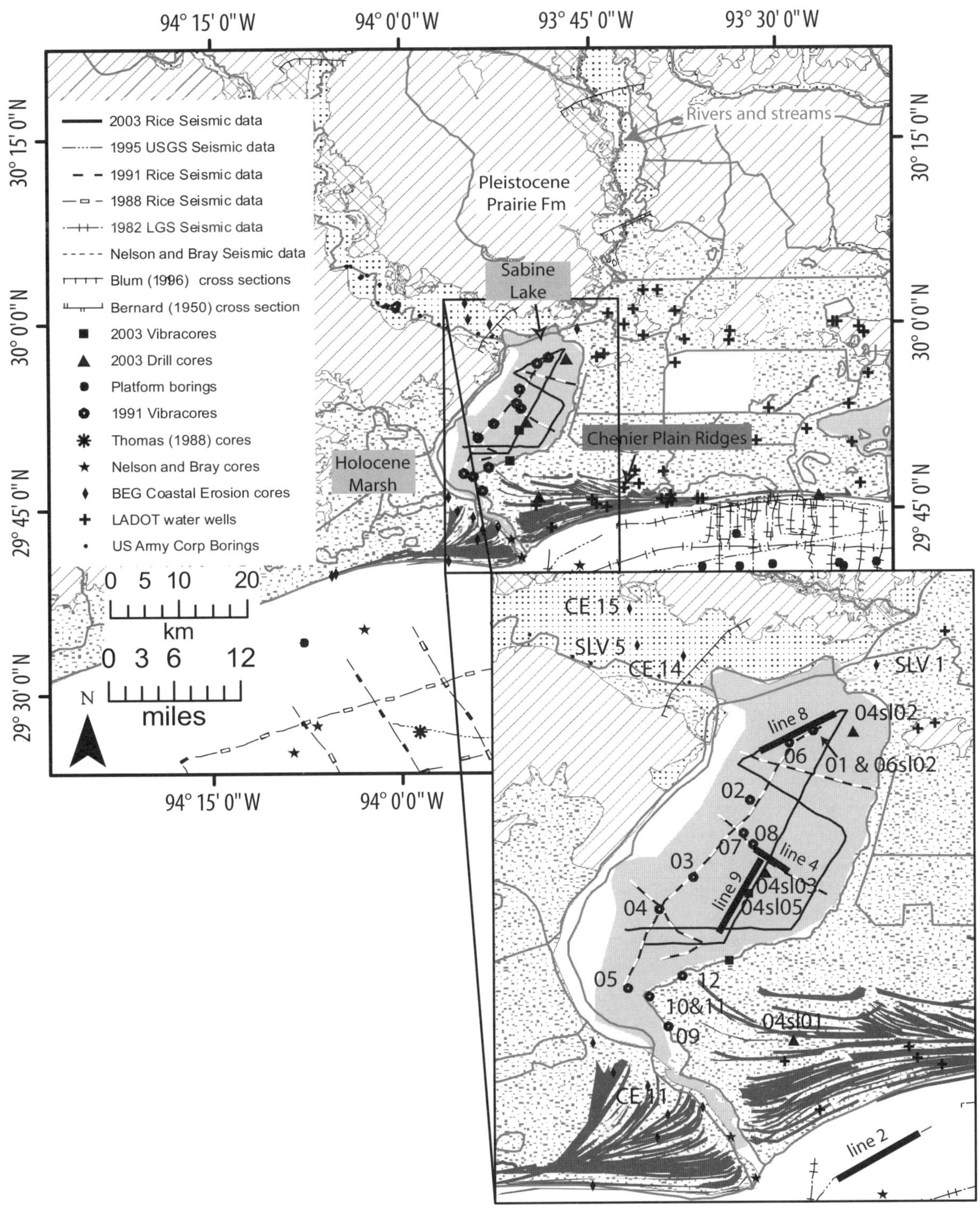

Figure 3. Study area base map with seismic data and core data locations used in this study. 2003 Rice seismic data were collected aboard the R/V *Jenny* for this study; 1995 U.S. Geological Survey (USGS) seismic data were Boomer data obtained from the USGS; 1991 Rice seismic data and 1991 vibracores were published in Anderson et al. (1991); 1988 Rice seismic data and Thomas 88 cores were published in Thomas (1988) and Thomas and Anderson (1994); 1982 LGS seismic data were obtained from the Louisiana Geological Survey; 2003 vibracores and drill cores were collected aboard the R/V *Trinity* for this study; Nelson and Bray cores were published in Nelson and Bray (1970); platform borings were donated by Fugro-McClellan Inc.; Bureau of Economic Geology (BEG) coastal erosion cores were previously published in Morton et al. (2000); LADOT water-well data are well lithological descriptions obtained from the Louisiana Department of Transportation; and the U.S. Army Corps borings are boring descriptions obtained from the U.S. Army Corps of Engineers.

material included mollusk shells and shallow-marine, mixed assemblages of foraminifera. The samples were cleaned and sent to Woods Hole National Ocean and Sciences Accelerator Mass Spectrometry (NOSAMS) and the University of California Irvine Keck (CCAMS) laboratories for radiocarbon dating by accelerator mass spectrometry. The reporting of radiocarbon ages followed the conventions outlined in Stuiver and Polach (1977) and is discussed in detail in Milliken et al. (this volume). The reported ages were calibrated using Calib 2004 (Stuiver et al., 2004). The Northern Hemisphere calibration curve with no assumed reservoir correction was used for ages obtained from wood and plant remains. Attempts were made to measure any radiocarbon reservoir present in Sabine Lake. An articulated *Rangia cuneata* shell embedded in peat was obtained from a core taken from the Sabine Lake area. The *Rangia cuneata* age was 380 yr younger than the peat. This wood-shell pair was therefore discredited as a reservoir indicator for the Sabine-Neches system. Several live *Rangia cuneata* samples were obtained from Sabine Lake, and a live *Donax* sp. sample was obtained from the swash zone of Bolivar Peninsula during the summer 2006. Surf-zone mollusks, such as *Donax* spp., are likely to incorporate CO_2 from water that is well-mixed with the atmosphere, whereas mollusks that reside in deeper water offshore or within estuaries may uptake CO_2 that is not rapidly mixed with the atmosphere (Hogg, 1998). The fraction modern (FM) values of the mollusks were 1.0263 and 1.0549, respectively, which translate to $\Delta^{14}C$ values of 19.3‰ and 47.8‰, respectively. The 1950 (prebomb) ^{14}C atmospheric value has been defined as a fraction modern of 1.0 and age of 0 yr B.P. The fraction modern values of both samples are greater than 1, which indicate that the organisms have incorporated water enriched in ^{14}C due to post-1950 bomb spike. Although the *Donax* sp. organism FM/$\Delta^{14}C$ value is very close to the 2006 atmospheric $\Delta^{14}C$ value (Turnbull et al., 2007), the possibility that the organism incorporated water containing initial bomb ^{14}C ca. 1950 cannot be ruled out without additional sampling. Furthermore, the Sabine Lake *Rangia cuneata* shell contained FM/$\Delta^{14}C$ values of 1.0263/19.3‰. Two explanations for these results include incorporation of ca. 1950 ^{14}C bomb spike or a mixture of modern ^{14}C and partially decayed ^{14}C (reservoir). Again, a time series of samples would help to elucidate the appropriate explanation.

The marine calibration curve was chosen with a reservoir of −175 yr for *Rangia* sp., *Crassostrea* sp., and *Nuculana* sp., whereas the marine calibration curve with no additional reservoir was used for ages extracted from foraminifera. A 350 yr difference in ages of the foraminifera samples would not change our interpretation. Therefore, the calibration curves used for foraminifera were not crucial for this study. Variable reservoir values were utilized for open-marine fauna versus organisms that live in brackish water. The Sabine River drains an area composed of minimal ancient carbonate rocks, whereas the Neches River drainage basin includes no carbonate rocks. Moreover, high precipitation rates regenerate groundwater and surface water supplies frequently. Our assumption is, therefore, that brackish-water organisms should incorporate carbon from a mixture of river water with no reservoir and marine water with a reservoir of ~350 yr (Stuiver et al., 2004). Studies in other estuaries show a diminishing reservoir trend from central Texas to western Louisiana (Milliken et al., this volume).

Published, public, and donated data sets utilized for stratigraphic information include Louisiana and Texas Department of Transportation water well logs and highway and bridge boring information, USGS and Louisiana Geological Survey data, and Fugro-McClellan offshore platform borings (Fig. 3). Previously published data sets were utilized in this study to help map the Pleistocene exposure surface (antecedent topography; Fisk, 1948; Kane, 1959; Anderson et al., 1991). The chenier plain stratigraphy and formation chronology was developed from Gould and McFarlan (1959), McFarlan (1961), Coleman (1966), and Frazier (1974).

RESULTS AND INTERPRETATIONS

Antecedent Topography

The Pleistocene exposure surface, which defines the Sabine-Neches incised valley, up to 37 m below sea level (mbsl) in the study area, is easily recognized in sediment cores (Fig. 4). Holocene sediments are generally soft gray mud and unconsolidated sand, whereas sediment underlying the Pleistocene surface (the top of the Prairie and Deweyville Formations) is stiff mud and sand with orange and tan mottling due to soil-weathering processes (Doering, 1935; Bernard, 1950; Kane, 1959). The Pleistocene surface was indurated while exposed subaerially during the most recent lowstand of sea level.

During the previous sea-level lowstand (marine oxygen isotope stage [OIS] 2), eastern Texas and western Louisiana river valleys extended onto the exposed continental shelf and incised 20–40 m into the inner shelf (Thomas and Anderson, 1994; Anderson et al., 2004). During the subsequent sea-level rise, the rivers aggraded, depositing coarse-grained sand and gravel at the base of the valley. These fluvial deposits grade upward to deltaic sand and mud, central-basin mud, tidal delta sand and mud, and offshore mud, respectively (Rehkemper, 1969; Nelson and Bray, 1970; Thomas and Anderson, 1994). Although, the fluvial part of the section is likely Pleistocene in age (>10,000 yr B.P.), it is included with the Holocene section here, and the Pleistocene surface and incised valley are mapped at the base of the gravel.

Previous work in Galveston Bay (Fig. 5) has shown that the flooding history of the estuary was punctuated by rapid flooding events that occurred when sea level reached the elevation of relatively flat fluvial terraces (Rodriguez et al., 2005). In an attempt to isolate the effect of antecedent topography on the flooding history of Sabine Lake, we measured the cumulative area below sea level (km^2) at 1 m increments of sea-level rise from −30 to 0 m (Fig. 4C). The curve of cumulative area versus sea level reveals any potential terraces as abrupt decreases in slope. Only one possible terrace exists, at a depth of −10 to −12 m. This feature may have genetic links to updip Deweyville fluvial terraces (Fig. 5).

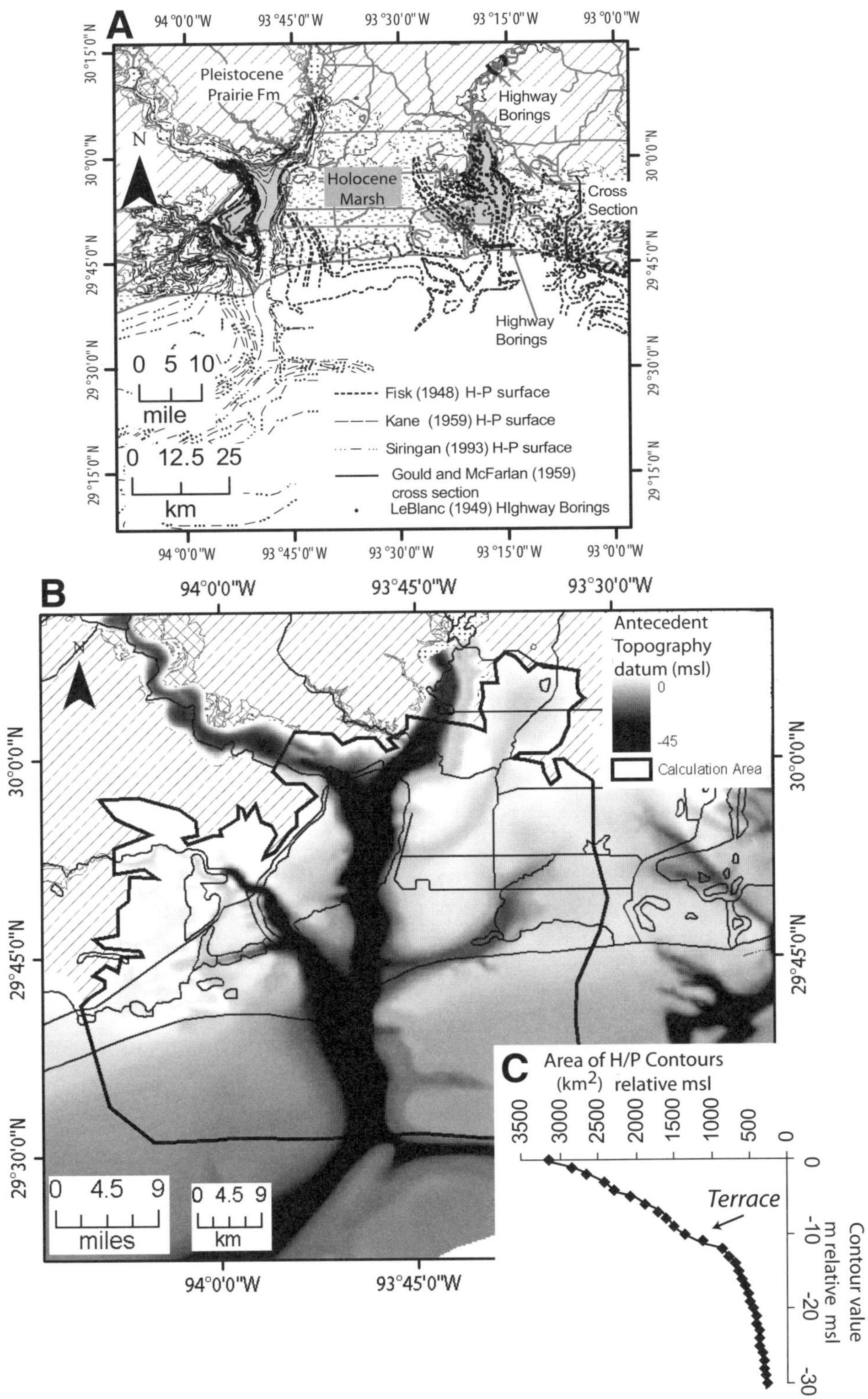

Figure 4. Contour map of the Holocene-Pleistocene (H-P) exposure surface that defines the incised Sabine–Neches river paleovalley. (A) Preexisting data for the Sabine valley, onshore (Kane, 1959) and offshore (Thomas, 1990; Siringan, 1993) areas. (B) Interpolated surface of the antecedent topography; study area outlined in bold. (C) Area flooded with each successive meter rise in sea level. msl—mean sea level.

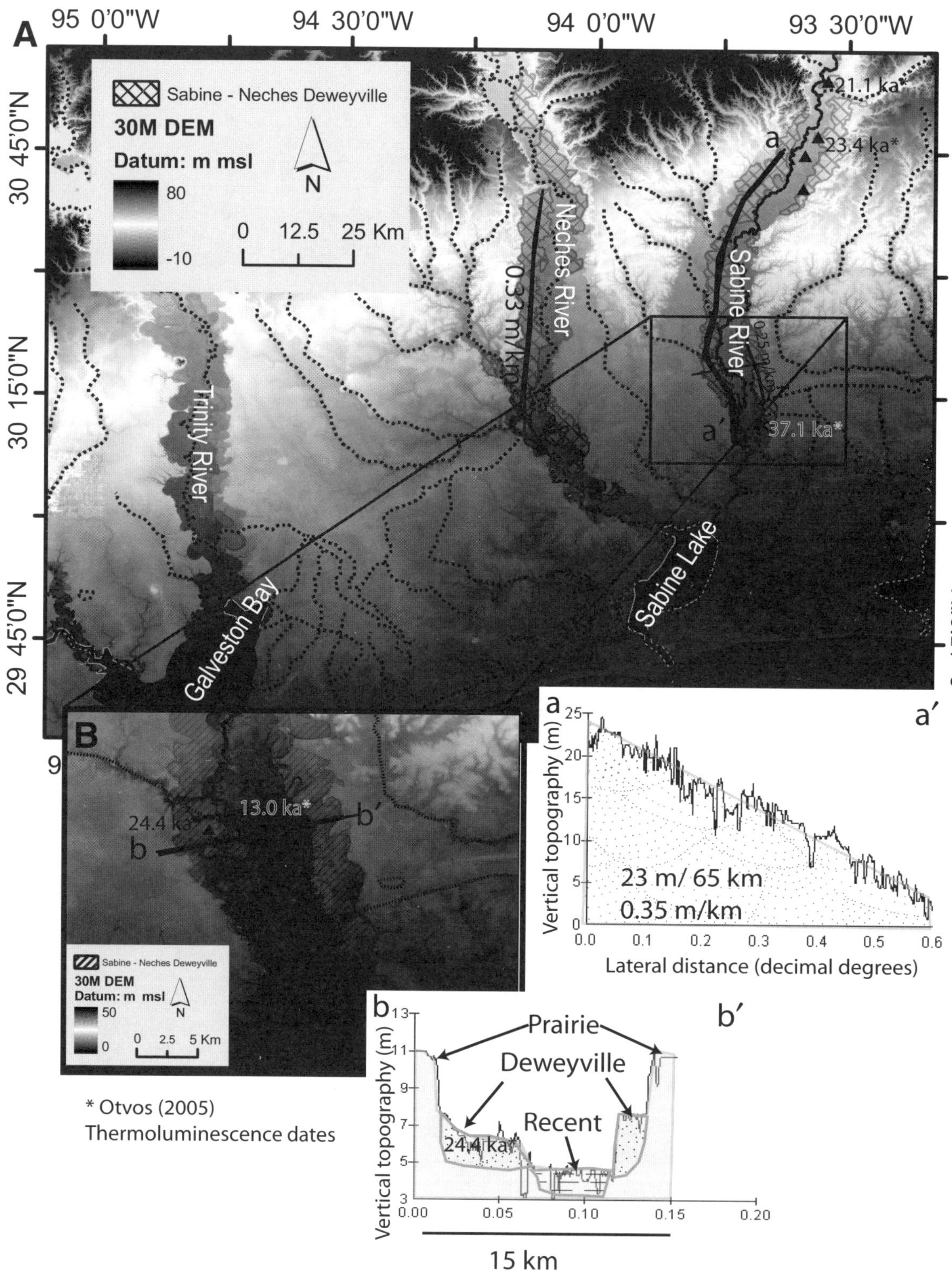

Figure 5. (A) U.S. Geological Survey 30 m digital elevation model (DEM) showing Sabine Lake and the lower part of the Sabine and Neches River basins along with part of the Trinity River basin and Galveston Bay. The Deweyville terraces are outlined with a cross-hatch pattern in the Neches and Sabine River valleys. Cross section a–a′ shows the gradient of the Deweyville Terrace in the Sabine valley. (B) Inset showing Deweyville terrace with large meander scars. Cross section b–b′ shows a strike section across the valley with interpreted Deweyville terraces and Holocene valley fill.

In a classic study, Bernard (1950, p. 59 and p. 62) first named the Deweyville terraces for the town of Deweyville on the west bank of the Sabine River, described them as terraces that occur "…between the level of the present floodplain and the next higher Prairie Terrace," and said they are "…characterized by very large and conspicuous meander scar[s] on the surface." Deweyville terraces were also recognized along the modern Trinity, San Jacinto, Nueces, and Pearl River valleys (Bernard, 1950).

Otvos (2005) published thermoluminescence dates from fluvial sand bodies in the Sabine valley. The ages range from 2.3 to 37.1 ka, suggesting continual or intermittent deposition within the valley. The valley-flanking Deweyville terrace ages range from 21.1 to 37.1 ka, whereas the lowermost floodplain remnant channels range from 2.3 to 13.0 ka (Fig. 5). Using the USGS 30 m digital elevation model (DEM), we identified the Deweyville terraces, calculated a gradient, and extrapolated the gradient beneath the lake to predict the depth of a terrace (Fig. 5). The extrapolated depth of ~5 mbsl is shallower than the observed 10–12 mbsl terrace (Fig. 4C). Either the river gradient that formed the terrace surface was not constant, or transgressive wave ravinement may have formed the terrace beneath Sabine Lake.

Seismic Facies

Seismic facies are distinguished by reflector patterns (laminated, chaotic, or transparent), reflector continuity (continuous or discontinuous), and amplitude characteristics (high, low). A prominent high-amplitude, continuous, subhorizontal reflector (labeled bay line) marks the boundary between fluvial and bayhead delta facies (as described later) and is interpreted to mark a significant acoustic impedance contrast (Figs. 6A and 6B). The bay line ranges in depth from 20 mbsl to 8 mbsl, shallowing landward. Beneath the bay line, ~17–29 m of incised valley fill is not imaged clearly. Based on core data from offshore (Nelson and Bray, 1970; Thomas and Anderson, 1994), the lower valley fill unit (labeled facies A) is likely composed of fluvial sand and gravel. Above the bay line, facies B is characterized by discontinuous, subhorizontal, laminated reflections that are truncated and onlapped by discontinuous, laterally accreting, high-amplitude reflections (Figs. 6A and 6B). Large-scale cut-and-fill (up to 10 m deep and 0.5–1 km wide), lateral accretion, and clinoforms indicate deposition in either deltaic distributary or tidal channels. Clinoforms are also a component of facies B (Fig. 6C), indicating progradation of sedimentary beds.

The top of facies B is sharply truncated and overlain by a low-amplitude, continuous, laminated seismic facies (facies C; Figs. 6A and 6D). In the downdip direction, very shallow channels with laterally accreting subhorizontal reflections (labeled facies D) are similar to seismic facies B and can only be differentiated by lithofacies and surrounding seismic facies (Fig. 6C). Channels on the scale of several meters deep and up to a kilometer wide that grade landward into fluvial facies are likely deltaic distributary channels, whereas cut-and-fill features that grade landward into subparallel horizontal continuous reflections (central basin) likely represent tidal inlet deposits.

Integration of Seismic Facies with Lithofacies

Several lithofacies and biofacies are evident in sediment cores (Fig. 7) and can be linked to the seismic facies. Core data through the section show peat at the base grading upward into variegated tan, gray, and brown mud with soft-sediment deformation features, sandy mud, and gray mud. The variegated mud, which probably represents flood events, is interpreted as a prodelta deposit (lithofacies B3). The peat is a delta-plain deposit (lithofacies A2). The stable carbon isotopic fraction values of all but two of the sampled peats were less than −25 ppt, indicating freshwater marsh deposition (Table 1). The gray mud signifies interdistributary bays (lithofacies B2). The bay line, commonly marked by peat, provides excellent chronological control for the first marine incursion that transformed the fluvial system into an estuary. Deltaic deposits (lithofacies B) are sharply overlain by a *Rangia cuneata* shell layer with soft gray mud above (lithofacies C). Facies C is interpreted to be a middle-basin, and with the shell layer is interpreted to represent a transgressive surface (Figs. 6B, 6C, and 6D).

Figure 6D shows core 04sl01, obtained from the chenier plain, projected onto a seismic line located just offshore, illustrating integration of core facies and seismic stratigraphy. Seismic facies A is correlated to sand and muddy sand within the cores and is interpreted to represent sandy fluvial-channel and point-bar deposits (facies A1). Seismic facies B is interpreted to represent three or more subenvironments within a bayhead delta. The cut-and-fill and lateral accretion observed in seismic profiles correlates to deltaic distributary deposits (facies B1), whereas the subhorizontal laminated reflections correspond to prodelta (lithofacies B3), interdistributary bay (lithofacies B2) or delta-plain (lithofacies A2) deposits. The prograding clinoforms are diagnostic of a delta-front environment and provide evidence for estuary water depths of ~2–3 m (Fig. 6C), similar to present-day Sabine Lake water depths. Vibracores reached up to 5 mbsl and sampled predominantly muddy, very-fine sand with abundant *Rangia cuneata* shells in the top 4 m of the cores. *Rangia cuneata* lives in brackish water (9–18 ppt; LaSalle and de la Cruz, 1985), which exists today in the middle to upper estuary. High-resolution chirp seismic data show subtle seismic facies changes at 2 and 5 mbsl (Fig. 6B). The underlying seismic facies (2–7 mbsl) is transparent with no apparent reflections, whereas the overlying seismic facies (2–5 mbsl) is a slightly higher-amplitude package with hints of subhorizontal to inclined reflections. Integration of the lithofacies and seismic facies suggests a change from central-basin to upper bayhead delta at ~5 mbsl and a return to a central-basin setting at ~2 mbsl.

Quantitative grain-size analysis of core 91sl01 shows muddy very-fine sand with an average of 14% fine sand at or near the central-basin floor (Fig. 7). Core 06sl02, from the same location,

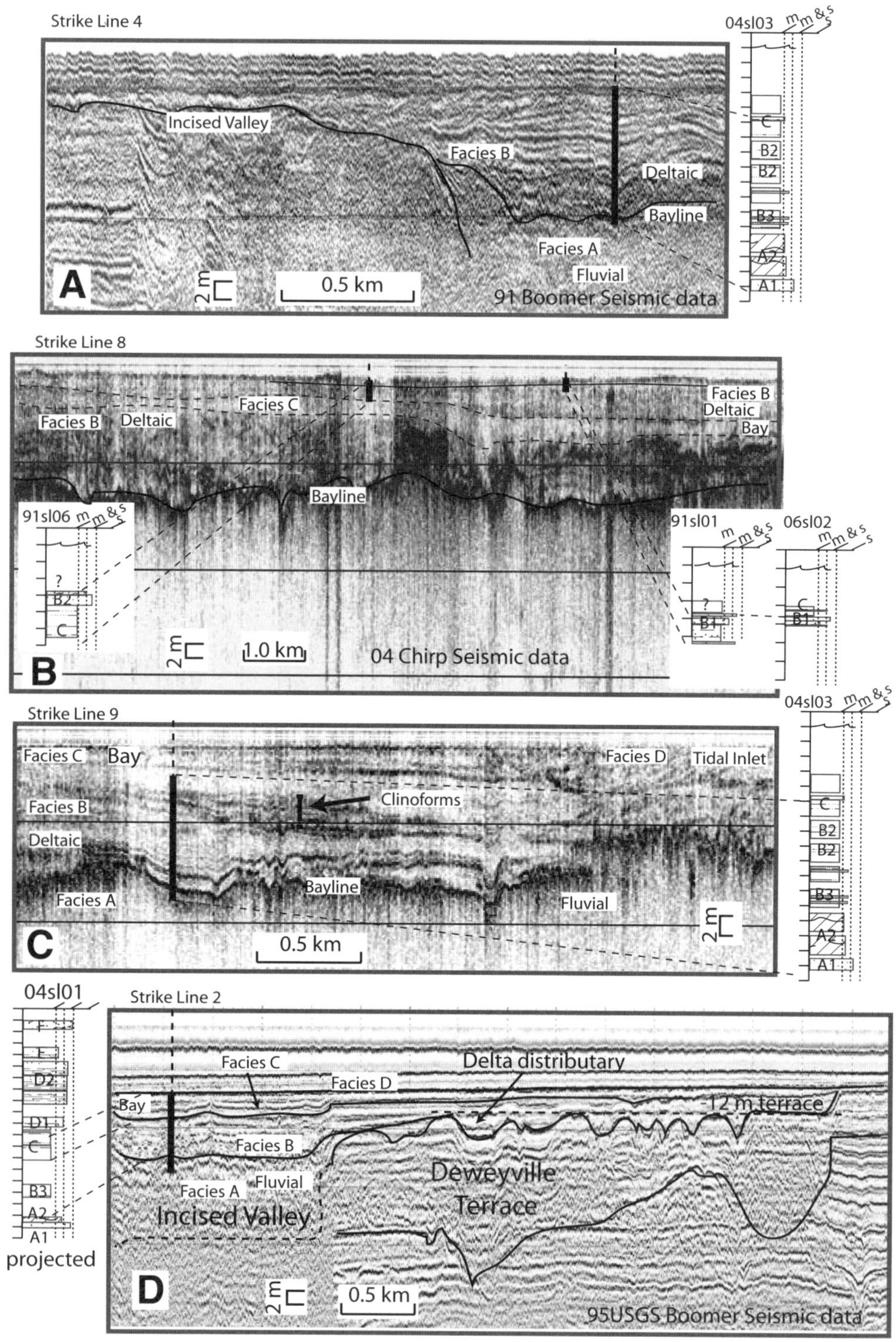

Figure 6. Examples of seismic facies. (A) The bay line surface, which separates facies A (fluvial) below from facies B (deltaic) above. (B) The deltaic seismic facies, characterized by discontinuous, variable-amplitude seismic reflections, is interpreted to reflect variable lithofacies, from sandy mud to muddy sand to predominantly mud, which coincide with deltaic subenvironments. A low-amplitude transparent seismic package overlies facies B (deltaic interval) and is labeled facies C (central basin). (C) The bay line surface separates fluvial (facies A) below from deltaic (facies B) above. Facies B (deltaic), in this example, includes basinward migrating and downlapping seismic reflections that can be described as clinoforms. The clinoforms mark the prograding delta front, and the thickness indicates an average water depth of 2–3 m. The deltaic (facies B) unit is truncated and subsequently overlain by transparent laminated facies C (central basin). Facies C (central basin) laterally changes into facies D (tidal inlet). Facies D is composed of variable-amplitude landward-dipping reflections and is correlated to centimeter-scale interbeds of sand and mud. (D) Core 04sl01 projected onto U.S. Geological Survey seismic line 2 located offshore in 6 m water. The changes in lithofacies correspond to truncational/flooding surfaces on the seismic data. The surfaces separate depositional environments with facies A (fluvial) at the base overlain by facies B (deltaic), which is overlain by central basin (facies C). In the core, tidal delta and/or inlet overlies the central-basin facies, but in the seismic data, shallow-marine facies overlies the central-basin facies. Note the 12 mbsl Deweyville terrace. Locations of seismic lines are shown on Figure 3. msl—mean sea level.

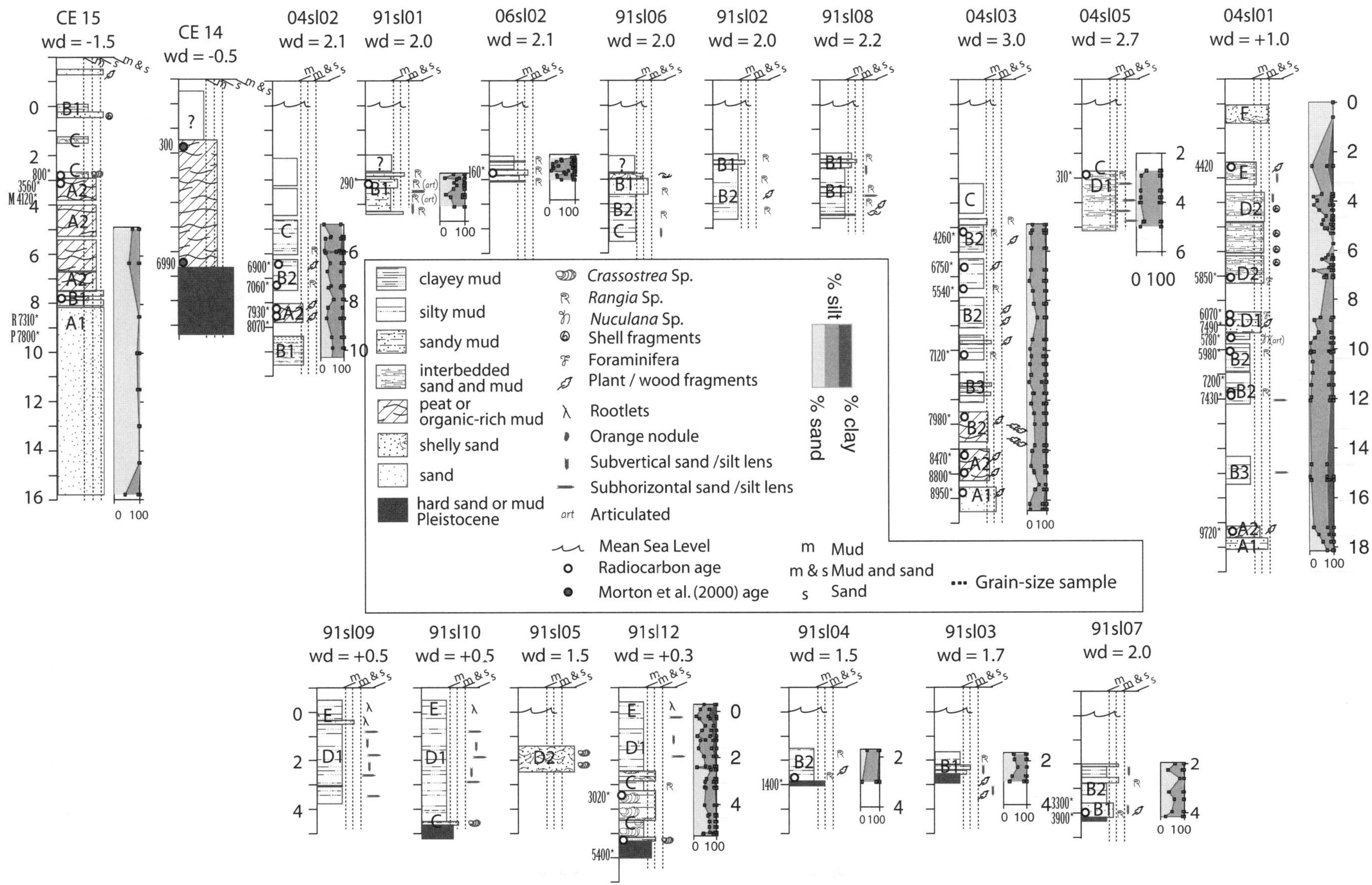

Figure 7. Core logs with grain-size measurements, facies interpretations, structural features, fossils, and radiocarbon ages. Cores are vertically aligned to mean sea level (msl) according to water depth (wd).

TABLE 1. RADIOCARBON DATA AND SAMPLE, AGE, ISOTOPIC, AND DEPTH INFORMATION

Sample	Lab no.	$\delta^{13}C$ (‰)	^{13}C source	^{13}Cct yr B.P.	Error	Cal yr B.P.	Error	Depth (msl)	Calib	Reservoir	Sample type
04sl01_12t16_38t42	OS-50695	–22	m	3970	140	4420	410	–2.8	Intcal04	0	Wood/plant frag
04sl01_24t28_100t105	OS-50203	–3	m	5480	40	5845	105	–7.1	Marine04.14c	0	Foram
04sl01_32t36_10t15	OS-50198	–3	m	5670	50	6070	130	–8.6	Marine04.14c	0	Foram
04sl01_32t36_25t35	OS-50254	–13	m	6590	40	7495	65	–8.8	Intcal04	0	Wood/plant frag
04sl01_32t36_101	25537	–2.06	m	5225	20	5785	85	–9.5	Marine04.14c	–175	*Nuculana* sp.†
04sl01_36t40_26	25536	0.54	m	5415	20	5980	80	–10	Marine04.14c	–175	*Rangia* sp.
04sl01_40t44_96	25539	–0.96	m	6495	20	7205	55	–12	Marine04.14c	–175	*Rangia* sp.
04sl01_40t44_96t96	22972	–3.3	m	6380	20	7430	70	–12	Marine04.14c	–175	*Rangia* sp.
04sl01_60_64_10_15	OS-46681	–25.4	m	8720	40	9720	170	–17	Intcal04	0	Peat
04sl02_25t30_5	OS-50255	–29	m	6060	35	6900	100	–6.5	Intcal04	0	Wood/plant frag
04sl02_25t30_97t99	22972	–3.3	m	6380	20	7065	80	–7.4	Marine04.14c	–175	*Rangia* sp.
04sl02_30_35_18_21	OS-46747	–28.2	m	7110	45	7930	80	–8.1	Intcal04	0	Peat
04sl02_30t35_76	OS-46764	–25.1	*m*	7240	30	8070	90	–8.7	Intcal04	0	Peat
04sl03 20t25 28t30	22974	–3.0	m	3995	25	4260	50	–5.2	Marine04.14c	–175	*Rangia* sp.
04sl03_25t30_35t40	OS-50201	–23	m	5920	55	6760	130	–6.8	Intcal04	0	Wood/plant frag
04s103_25_30_125	28054			5015	15	5540	60	–7.7	Marine04.14c	–175	*Rangia* sp.
04sl03_35t40_77t79	22975	5.1	m	6420	15	7115	60	–10	Marine04.14c	–175	*Rangia* sp.
04sl03_45_50_5	OS-46762	–17.9	*m*	7180	30	7985	45	–13	Intcal04	0	Peat
04sl03_45t50_129	22976	–21.5	m	7670	30	8470	70	–14	Intcal04	0	Wood/plant frag
04sl03_50_55_86t	OS-46746	–28.9	m	7910	45	8795	195	–15	Intcal04	0	Peat
04sl03_50_55_131	OS-46763	–26	m	8070	35	8950	170	–15	Intcal04	0	Peat
04sl05_1o2_5	22977	–4.2	m	470	15	310	10	–2.1	Marine04.14c	–175	*Rangia* sp.
91sl01	28053			470	30	290	40	–2.5	Marine04.14c	–175	*Rangia* sp.
91sab04	Core 4	–6.0	a	1870	100	1620	240	–2.5	Marine04.14c	–175	*Rangia* sp.
91sab07_1	Core 7	–6.0	a	3440	140	3535	355	–4	Marine04.14c	–175	*Rangia* sp.
91sab07_2	Core 7	–6	a	3890	120	4100	330	–4	Marine04.14c	–175	*Rangia* sp.
91s112_3_4	Core 12	–6	a	3040	30	3020	90	–3.4	Marine04.14c	–175	*Crassostrea* sp.
91sab12	Core 12	–6	a	5050	80	5615	215	–5	Marine04.14c	–175	*Rangia* sp.
06sl02	reoccupy91sl01			360	40	160	90	–2.7	Marine04.14c	–175	*Rangia* sp.
CE_15_4_10	OS-48166	–1	m	1080	35	800	100	–3.1	Marine04.14c	–175	*Crassostrea* sp.
CE_15_4_20	OS-48552	–22	m	3330	50	3560	110	–3.2	Intcal04	0	Peat
CE_15_7_120	OS-53096	–25	a	7070	40	7885	85	–8.7	Intcal04	0	Peat
CE_15_7_120	OS-48156	–4	m	6610	40	7315	85	–8.7	Marine04.14c	–175	*Rangia* sp.

Notes: Sample—Sample designation, Lab no.—laboratory number, $\delta^{13}C$—stable carbon isotope value, ^{13}C source—whether the sample ^{13}C isotopic fraction was measured (m) or assumed (a), ^{13}Cct yr B.P.—conventional radiocarbon values, Cal yr B.P.—calibrated radiocarbon age, Depth—depth in meters relative to mean sea level, Calib—Calib05 calibration curve used, Reservoir—reservoir used to correct the ages, and Sample type—sample material.

†Articulated mollusks.

sampled ~70 cm of dark gray mud overlying gray sand and muddy sand. Seismic resolution is not adequate to image a surface at ~2–3 mbsl, but there does appear to be a lithofacies change in this core that represents a change from deltaic deposition to central-basin mud deposition. The lack of historical distributary channels and prodelta mud suggests that the delta is not active today, but the shallow sandy lithofacies sampled in vibracores implies very recent delta deposition in the central basin. Furthermore, Bernard (1950) identified the Sabine River valley alluvial cone as the segment of the river valley where the gradient changes from steep to shallow and where alluvial sediment is deposited. In the Sabine River valley, Bernard (1950) placed the modern-day advancing alluvial cone 20 km landward of Sabine Lake.

Core 04sl01 was obtained from a chenier ridge (Figs. 6D and 7) and shows a facies change at 9 mbsl, with gray sandy mud containing foraminifera and articulated *Nuculana concentrica* (facies D1) overlying dark gray silty mud with bivalves, dominated by *Rangia cuneata* (facies C). This change is interpreted as tidal inlet or distal tidal delta sediments overlying central-basin sediments, and thus it represents a flooding surface. Another facies change occurs at ~7 mbsl, where gray muddy sand with bivalve shells and foraminifera (facies D2) overlies gray sandy mud (Fig. 6D). This surface is interpreted as a flooding surface that places proximal flood tidal delta or tidal inlet sediments over distal tidal delta deposits. A sharp facies contact at ~3 mbsl separates the underlying tidal deposits from dark gray sandy mud with a layer of organic material (facies E), interpreted as a marsh deposit. Above the marsh deposit, there are gray-brown, decimeter-scale interbedded shelly sand and shell beds (facies F), interpreted as a beach or chenier ridge deposit. Other examples of cores that preserve central-basin deposits include core 91sl12, which sampled mud with *Crassostrea* sp. at 3 to 6 mbsl, and core 04sl05, which sampled dark gray silty mud with very fine to fine sand laminations at ~3 to 5 mbsl (Fig. 7).

Chronostratigraphy

Seven estuary-wide flooding surfaces are evident in both seismic and core data, whereas one local surface occurs in the lower bay. The estuary-wide surfaces shallow landward and display a few meters of relief. Radiocarbon ages from above and below the surfaces provide constraints on the timing of surface formation and environmental reorganization (Table 1). We linked these ages to the seismic and core data to determine the chronostratigraphy (Fig. 8). The fact that most of ages, particularly in the lower/older section, are (1) in stratigraphic order and (2) from

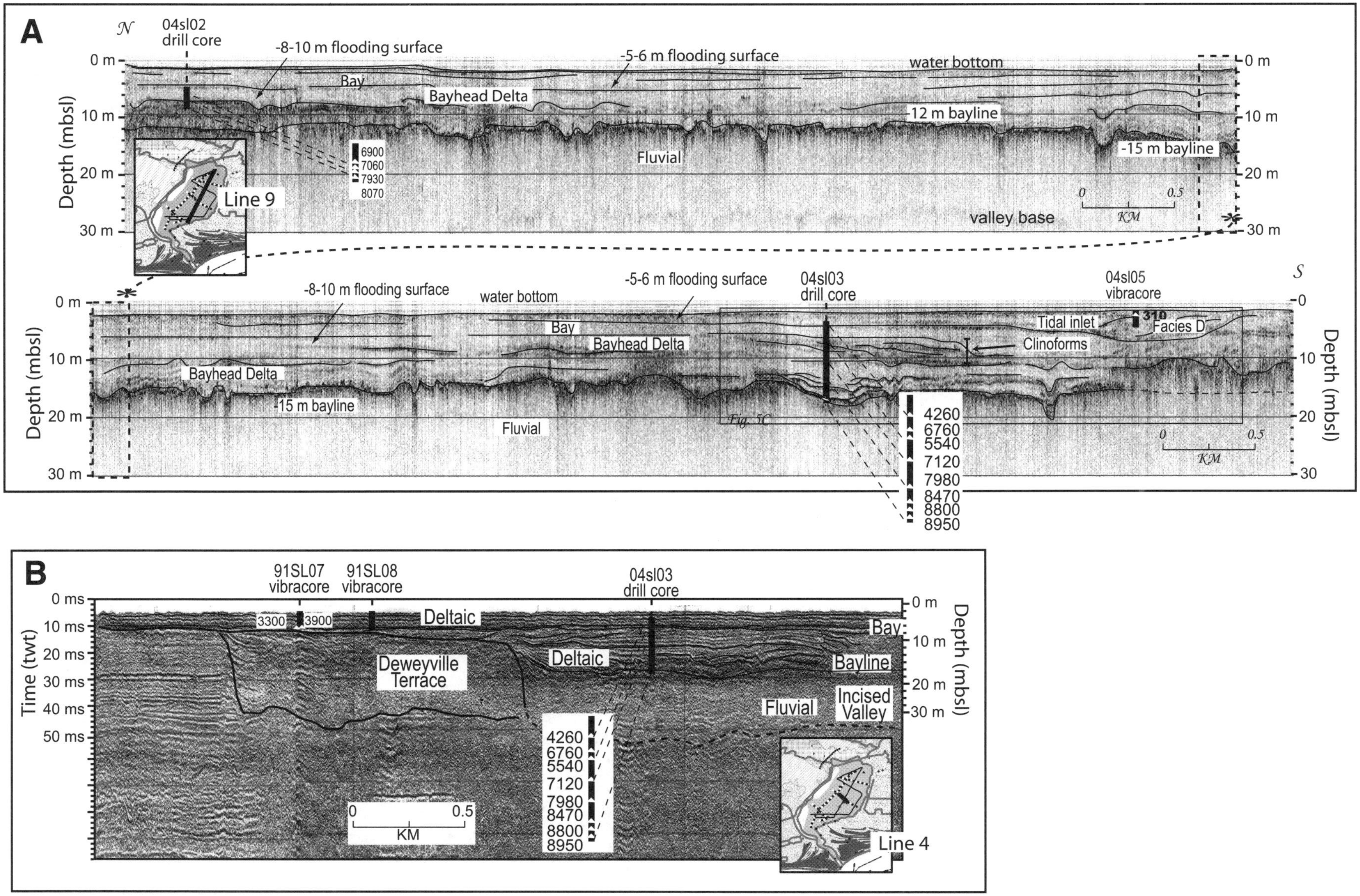

Figure 8. Seismic lines illustrating the integration of the core data and chronostratigraphy obtained from the radiocarbon ages. Core locations, depth sampled, and radiocarbon ages are projected onto the seismic line. Radiocarbon ages in cal yr B.P. constrain the age and depth of the flooding surfaces, which shallow landward. The depositional locus steps landward episodically. (A) Dip-oriented seismic line 9 showing the step in the bay line from 15 m to 12 mbsl and overlying bayhead delta deposits and central-basin deposits. The step in the bay line is constrained to younger than 8470 ± 70 cal yr B.P. but older than 7980 ± 40 cal yr B.P. (8400–8000 cal yr B.P. flooding surface). Delta clinoforms downlap a flooding surface at 10 mbsl (7120 ± 60 cal yr B.P.) and are truncated by 7 mbsl flooding surface (5540 ± 60 cal yr B.P.). The 2–3 m clinoform thickness indicates that the delta was prograding into a 2–3-m-deep central basin. A tidal inlet/delta at the southern portion of the estuary is manifested as shallow-channel landward-dipping bedding. (B) Strike-oriented line 4 from 1991 Rice survey. Deweyville terrace interpretation is from Anderson et al. (1991). Incised valley contains fluvial sediments at the base overlain by deltaic and estuarine deposits. The strike orientation of the line shows four of the six prominent flooding surfaces; 15 mbsl bay line (8900 cal yr B.P.), 10 mbsl (7100 cal yr B.P.), 7 mbsl (5800–5500 cal yr B.P.), and 6 mbsl (4200 cal yr B.P.). twt—two-way travel time.

deltaic facies with relatively high rates of sedimentation suggests that reworking and/or burrowing was not a prominent factor for the time period prior to ca. 5500–5800 cal yr B.P. Following this time period, with commencement of central-basin depositional conditions and lower sedimentation rates, the probability of reworking and burrowing across flooding surfaces increases (Fig. 9).

The bay line separates fluvial deposits below from bayhead delta deposition and is a diachronous surface that steps landward through time in response to relative sea-level rise. This surface defines three steps in Sabine Lake at 15, 12–14, and 8–12 mbsl. The 15 mbsl step, flooding surface 1, formed ca. 8900 cal yr B.P., on the basis of two radiocarbon analyses that yielded ages of 8950 ± 170 cal yr B.P. and 8800 ± 200 cal yr B.P. from below and above the surface, respectively (Fig. 9). Although the 12–14 mbsl step in the bay line, flooding surface 2, was not age constrained directly, it was traced basinward into core 04sl03, where it is recognized as a flooding surface separating interdistributary bay deposits below from prodelta deposits above. From core 04sl03, the 12–14 mbsl flooding surface and associated step in the bay line formed ca. 8000–8400 cal yr B.P., on the basis of two ages from peat and organic material, 8470 ± 70 cal yr B.P. and 7980 ± 80 cal yr B.P. from below and above the surface, respectively (Fig. 9). The stable carbon isotope values from the peat below the surface were less than −25 ppt and greater than −25 ppt above the surface, indicating changing salinity conditions from freshwater to saline marsh. The bayhead delta deposit directly above the 8–12 mbsl step in the bay line, flooding surface 3, was constrained as younger than 7880 ± 80 cal yr B.P. This step extends basinward into core 04sl02, where it is recognized as a flooding surface separating delta plain below from interdistributary bay above. Ages below and above the flooding surface in this core are 7930 ± 80 cal yr B.P. and 7060 ± 80 cal yr B.P., respectively. Additionally, an age obtained from core 04sl01 constrains the age to older than 7430 ± 70 cal yr B.P. An age range of 7500–7800 cal yr B.P. is therefore assigned to this surface.

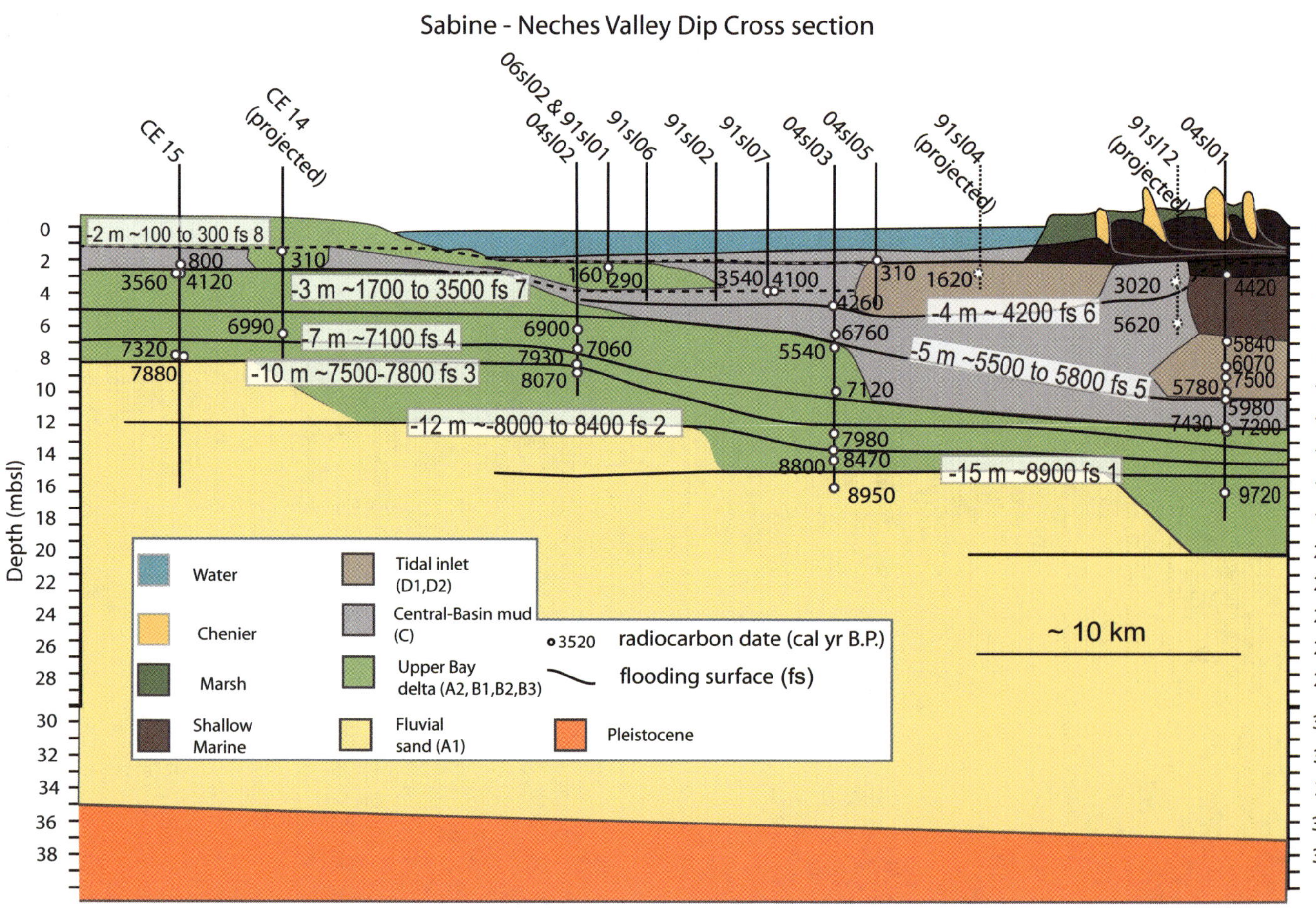

Figure 9. Generalized dip cross section showing flooding surfaces and interpreted ages of these surfaces. The cross section was constructed by integration of core and seismic data. Flooding surfaces are black lines that shallow a few meters in the landward direction and are labeled with interpreted flooding surface ages. Several cores are projected onto the section and represent lateral facies with associated ages. The contact between the fluvial section and the underlying Pleistocene was determined from previous studies as well as seismic data.

The seismic package designated facies B can be subdivided into two packages with a bounding surface at ~10–12 mbsl (flooding surface 4) in the portion of the study area defined by the present-day Sabine lake (Figs. 8A and 8B). Updip, this surface climbs to 7 mbsl, and downdip, this surface occurs at 13 mbsl (Fig. 9). This surface truncates reflections below and acts as a downlap surface for the prograding clinoforms above (Fig. 8A). Five ages constrain the age of this surface to older than 7060 ± 80 cal yr B.P. but younger than 7320 ± 80 cal yr B.P., 7930 ± 80 cal yr B.P., 7120 ± 60 cal yr B.P., and 7200 ± 60 cal yr B.P. The age of this surface is interpreted to be 7100 cal yr B.P.

A fifth flooding surface, at 10 mbsl in the lower estuary to 5 mbsl in the upper estuary, exists as an upward facies change from dark gray central-basin mud with *Rangia cuneata* to muddy sand with foraminifera and *Nuculana concentrica* of a tidal delta or inlet. An age of 5780 ± 80 cal yr B.P. was obtained on an articulated *Nuculana concentrica* immediately below the lithofacies change, and it provides a maximum age for the flooding surface. A landward-shallowing clinoform toplap surface (5–8 mbsl) coincides with a *Rangia* sp. shell in core 04sl03 at 7.7 mbsl. This shell gave an age of 5540 ± 60 cal yr B.P. At approximately the same time, saline estuary conditions of 10–30 ppt, based on the occurrence of *Crassostrea* sp. (Sellers and Stanley, 1988), were established at the location of core 91sl12 (5 mbsl) at 5620 ± 220 cal yr B.P. Additionally, Morton et al. (2000) reported initial flooding of the Holocene-Pleistocene boundary in core CE 11 at 6.6 mbsl and provided an age of 5570 ± 80 cal yr B.P. (Fig. 3). The 5500–5800 cal yr B.P. flooding surface represents incursion of tidal inlet or delta deposition and flooding of adjacent interfluvial areas in conjunction with landward translation of the bayhead delta.

A local flooding surface occurs at 4–6 mbsl, flooding surface 6, in seismic data collected in the lower portion of the estuary (Figs. 6C and 8A). Above this surface, landward-dipping seismic reflections are correlated to interbedded sand and mud in cores. This facies is interpreted as a tidal inlet deposit that overlies central-basin mud. Based on two analyses, the surface is younger than 4260 ± 50 cal yr B.P. but older than 4100 ± 330 cal yr B.P. This surface is assigned an age of 4200 cal yr B.P.

A seventh estuary-wide flooding surface at 2–4 mbsl is recognized as a facies change in core CE 15 from peat below to gray mud with *Crassostrea* sp. above (Fig. 7). It is also marked in core 91sl07 by muddy sand with *Rangia* sp. overlying Pleistocene deposits. The age is constrained as younger than 3560 ± 110 cal yr B.P. and 3540 ± 355 cal yr B.P. but older than 800 ± 100 cal yr B.P. and 1620 ± 240 cal yr B.P. We tentatively assign an age of 1700–3500 cal yr B.P. to this surface (Fig. 9). The surface marks a dramatic environmental change during which the bayhead delta stepped landward of the study area.

The shallowest flooding surface, flooding surface 8, occurs within a meter of the Sabine Lake bottom, or 2–3 mbsl, and emplaces central-basin mud on deltaic sand and muddy sand with *Rangia cuneata* shells. Furthermore, cores in the upper part of the estuary (e.g., core 91sl01) contain articulated *Rangia cuneata* shells in muddy sand at ~2.5–3 mbsl. Intact *Rangia* sp. shells indicate brackish-water conditions. Ages obtained from *Rangia* sp. collected at this surface provide maximum ages of 290 ± 40 cal yr B.P. and 160 ± 90 cal yr B.P. (Fig. 6B). This surface can also be identified in core 04sl05 as a facies change from underlying muddy sand and sand with laminations to dark gray silty mud with a *Rangia cuneata* shell lag. Seismic data show a laterally accreting shallow channel (Fig. 8A) truncated and overlain by transparent seismic facies. This facies change is interpreted as tidal inlet deposits overlain by central-basin mud. Analysis of a *Rangia* sp. sample yielded an age of 300 ± 10 cal yr B.P. for termination of tidal inlet deposition. This surface is therefore assigned an age of 100–300 cal yr B.P. and signifies a relatively recent reorganization of estuarine environments.

Figure 10 shows sedimentation rates for the past few hundred years as determined by radiocarbon (this study) and Pb concentrations (Ravichandran et al., 1995). The upper estuary sedimentation rate for the uppermost 15 cm (post-1963) equals 5.2 mm/yr (Ravichandran et al., 1995). A sedimentation rate of 3.8 mm/yr is derived from ages bracketing the interval 0.7 m subbottom to 1.2 m subbottom using cores 91sl01 and 06sl02.

DISCUSSION

A general stratigraphic section for Sabine Lake consists of fluvial sand and mud overlain by distributary channels and deltaplain deposits, which are in turn overlain by prodelta mud and interdistributary bay mud. Central-basin mud may be interbedded with deltaic sediment but is the capping facies of the succession. Tidal inlet and tidal delta facies occur at ~2–6 mbsl and at the estuary floor in the lower estuary. This generalized stratigraphic

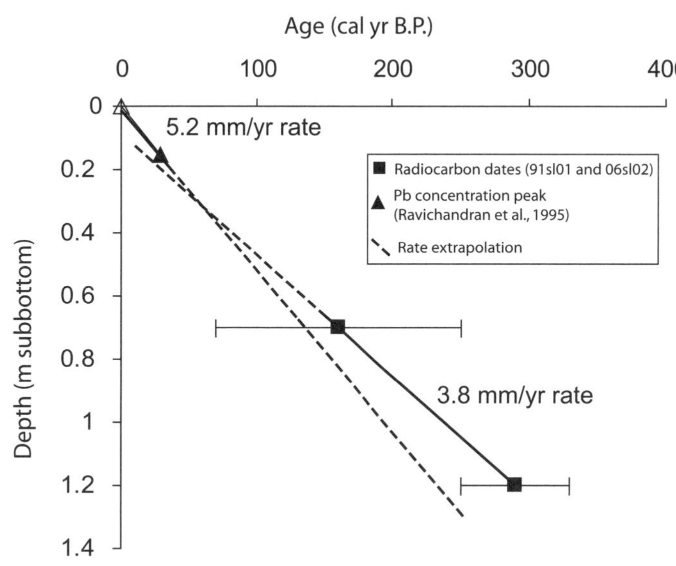

Figure 10. Age-depth graph illustrating historical sedimentation rates for the upper estuary of Sabine Lake.

succession shows a landward shift in facies over the past 9800 cal yr B.P. (Fig. 9).

Seven estuary-wide flooding surfaces and one local lower-estuary flooding surface are recognized in the sediments underlying Sabine Lake. Flooding surfaces occur at 8900 cal yr B.P., 8400–8000 cal yr B.P., 7800–7500 cal yr B.P., 7100 cal yr B.P., 5800–5500 cal yr B.P., 4200 cal yr B.P., and 3500–1700 cal yr B.P. One potential autogenic flooding surface occurs at 4200 cal yr B.P. The estuary-wide surfaces, on average, shallow 2–4 m in the landward direction. This likely indicates paleo–water depths, which are similar to the current depth of Sabine Lake.

Paleogeographic maps illustrate environmental changes that occurred in Sabine Lake before and after significant flooding events (Fig. 11). Prior to ca. 9800 cal yr B.P., fluvial deposition dominated the Sabine-Neches incised valley in the study area. Estuarine depositional environments occurred offshore and were migrating landward under the influence of eustatic sea-level rise (Thomas, 1990; Thomas and Anderson, 1994). Climate reconstructions from central Texas suggest that cool, wetter climates transitioned to warm, drier conditions by 12–15 ka (Musgrove et al., 2001). After ca. 9600 cal yr B.P., the fluvial system stepped landward a minimum of 10 km, and deltaic deposition dominated in the southern portion of the estuary (Figs. 11A and 11B). At ca. 8900 cal yr B.P., the fluvial-deltaic system again stepped landward ~10 km within a century or two (Fig. 11C). Between ca. 8400 to ca. 8000 cal yr B.P., another flooding event caused the fluvial system to step landward several kilometers (Fig. 11D). At this time, tidal deposition commenced in the southwestern portion of the valley. During the time interval between ca. 7800 to ca. 7500 cal yr B.P., a significant flooding event took place in which the bay line stepped landward a minimum of 10–20 km (Figs. 11E and 11F). During this time interval, deltaic deposition dominated most of present-day Sabine Lake.

A flooding surface at 7–12 mbsl was formed ca. 7100 cal yr B.P. in Sabine Lake (Figs. 4C, 8A, and 8B). The delta shifted ~20 km landward, and the area of the estuary increased by 30% (Figs. 9, 11G, and 11H). There are several explanations for this flooding surface and short-term landward retreat of the delta depositional system. The potential causes include: (1) lateral flooding of a Deweyville terrace and significant increase in lateral estuary extent, (2) short-term climatic fluctuation, (3) sea-level perturbation, and/or (4) period of intense storm events. Following the flooding event, the bayhead delta prograded to fill new accommodation space within approximately a century, which suggests that sediment supply was not impacted on the millennial time scale. Recent studies suggest a rapid rise in global sea level sometime between 7600 and 6500 cal yr B.P. (Blanchon and Shaw, 1995; Blanchon et al., 2002; Bird et al., 2007) due to the deglaciation of the Laurentide Ice Sheet (Carlson et al., 2007).

A lack of Holocene climate records from lakes, caves, or archaeological sites in the coastal zone of Texas and Louisiana has hampered Holocene climate reconstructions for the area. Those data that do exist indicate that at about the time the 7100–7300 cal yr B.P. flooding surface formed, multiple localities in central and northeastern Texas and northwestern Louisiana experienced an arid mid-Holocene climate phase (Knox, 1983; Toomey et al., 1993; Bense, 1995; Humphrey and Ferring, 1994; Holliday, 1995), also known as altithermal (Antevs, 1948). This was followed by a shift to more humid conditions ca. 3–4 ka. Evidence for an increase in summer thunderstorms, seasonal precipitation patterns, and summer temperature suggests that the Sabine drainage basin was situated in the prime climatic regime for increased fluvial sediment supply to the estuary (Knox, 1983).

At ca. 5800 to ca. 5500 cal yr B.P., Sabine Lake was largely occupied by a bayhead delta (Fig. 11I). Oyster beds were established along the margins of the estuary (core 91sl12) around 5600 cal yr B.P. and persisted until at least 3000 cal yr B.P. Tidal deposits were also prevalent in the lower portion of the estuary at this time. The 4200 cal yr B.P. flooding resulted in a landward shift in tidal deposits. After 4200 cal yr B.P., bayhead deltas stabilized in the upper reaches of the Neches and Sabine River valleys until at least 3500 cal yr B.P. (Fig. 11K). From 3500 to 1700 cal yr B.P., the Sabine-Neches bayhead delta shifted further landward, beyond the limits of the study area. At ca. 1700 cal yr B.P., brackish-water *Rangia* spp. resided in much of the estuary, but by 800 cal yr B.P., *Crassostrea* spp. were present as far north as the mouth of the modern-day Neches River. At this time, the chenier plain was also forming and gradually restricting estuarine circulation (Figs. 11M and 11N). Sabine Lake oscillated between more brackish and more saline (10–20 ppt) conditions as either precipitation and discharge into the estuary varied or the tidal connection to the Gulf of Mexico varied. The time periods in which oysters lived in the Sabine Lake estuary are similar to time periods (6.1–5.6 ka and 2.4–0.5 ka.) when oysters colonized the Hudson River estuary (Carbotte et al., 2004), suggesting a climate forcing. Modern-day humid conditions allow the brackish-water clam *Rangia cuneata* to thrive in the middle estuary.

The Sabine chenier plain is a series of alternating, prograding mud flats and beach ridges that began to form ca. 2.8 ka (Gould and McFarlan, 1959). The chenier plain adjacent to the Sabine tidal inlet is classified as regressive-ridge and laterally accreted-ridge complexes (Taylor et al., 1996; McBride et al., 2007). The progressive formation of the chenier plain acted as a baffle to marine and estuary water exchange. The estuary became more restricted and was characterized by amplified tidal currents and potentially reduced tidal ranges. The deposition of tidal inlet and tidal delta facies probably was related to chenier plain development.

The occurrence of sandy lithofacies with abundant *Rangia* spp., including articulated *Rangia* specimens constrained to ca. 300 cal yr B.P. at 2–4 mbsl, suggests that just a few centuries ago, a bayhead delta existed in the upper estuary. A facies change at 2 mbsl to dark gray mud deposition marks the back stepping of the delta. Little evidence exists for active delta deposition and progradation presently.

Possible controlling mechanisms responsible for the observed flooding events in Sabine Lake include eustasy, subsidence, flooding of antecedent topography, and variations in sediment

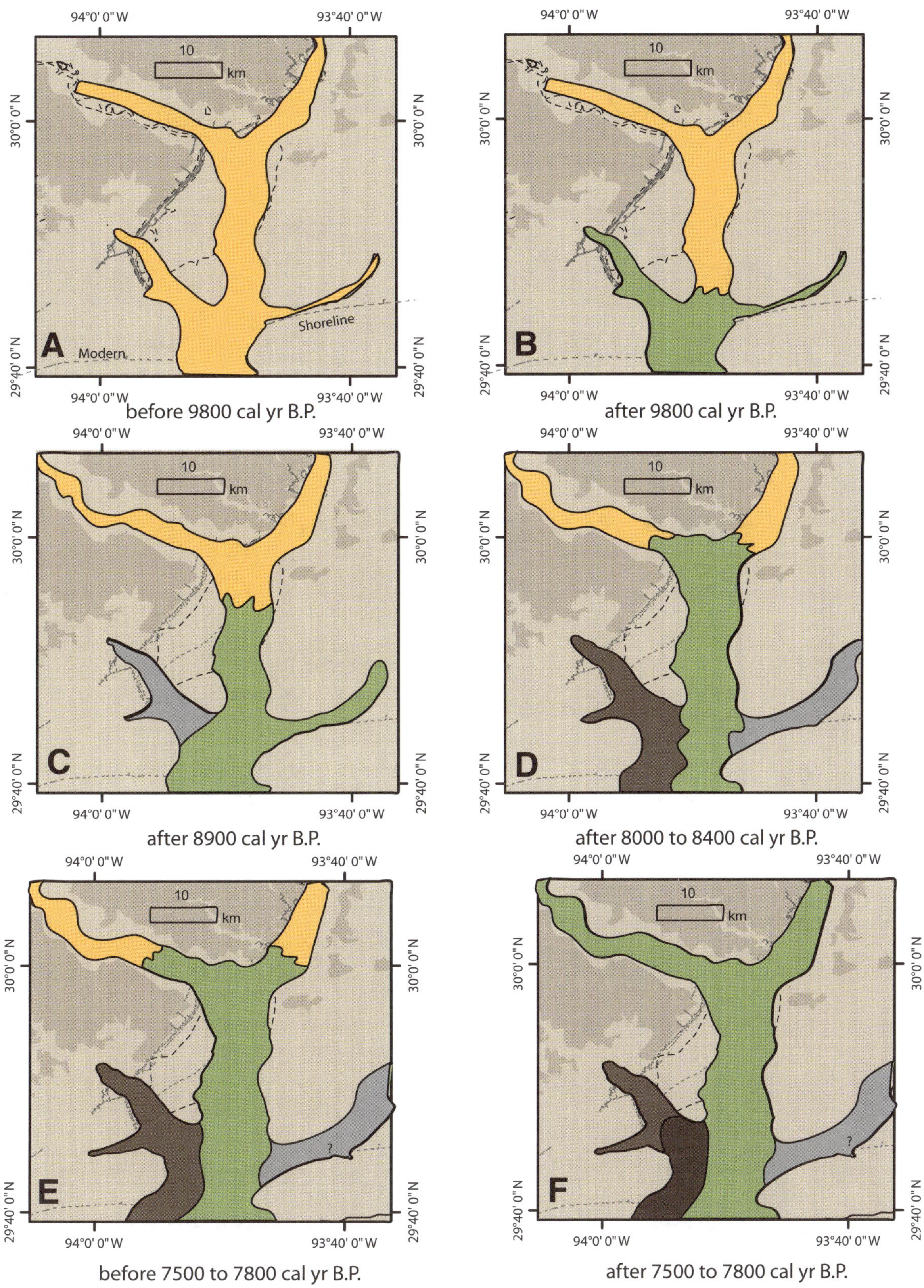

Figure 11 (*on this and following two pages*). Paleogeographic maps illustrating the nature of Holocene environmental change in the Sabine valley. Environments back step and prograde tens of kilometers within a few hundred years.

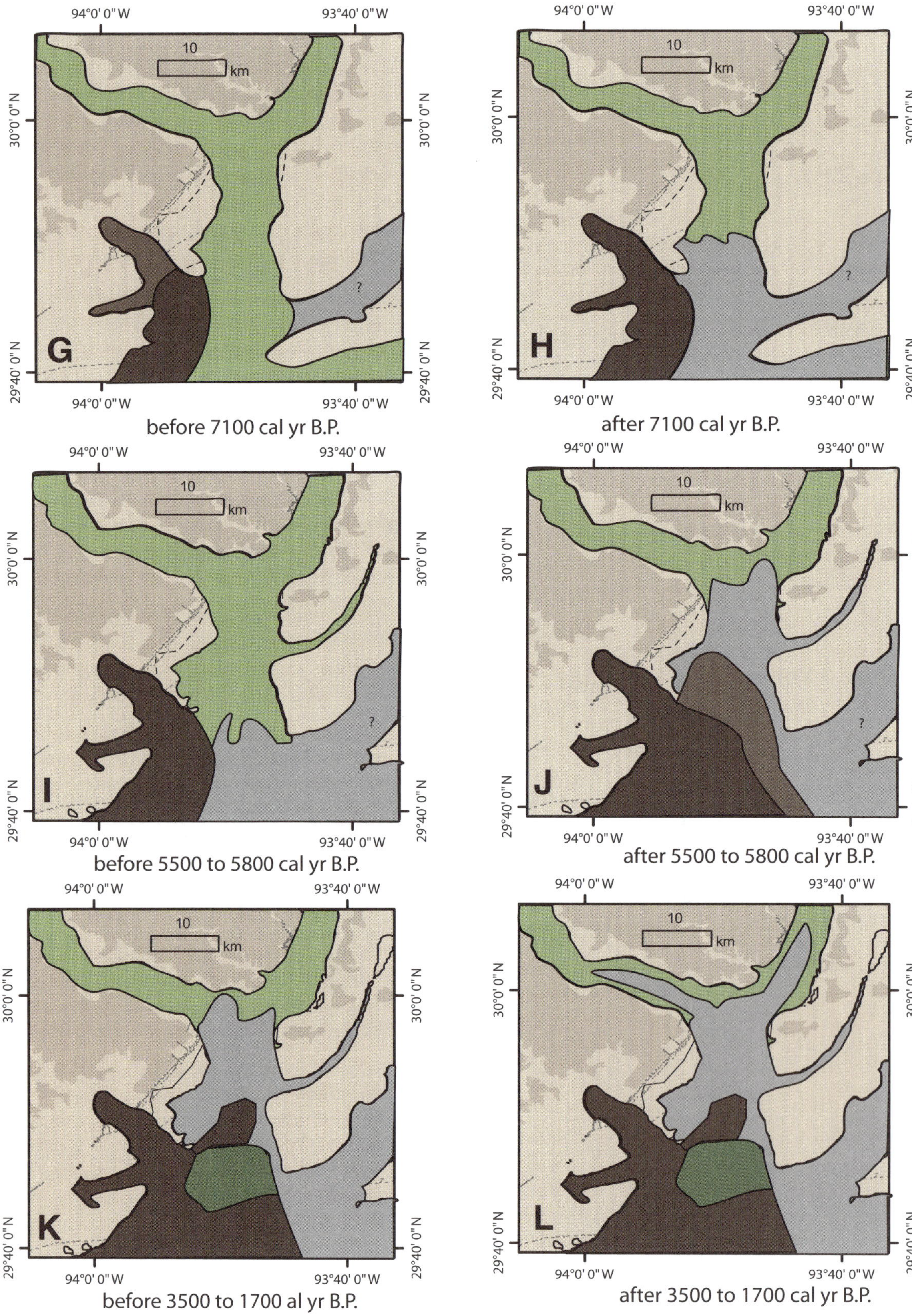

Figure 11 (*continued*).

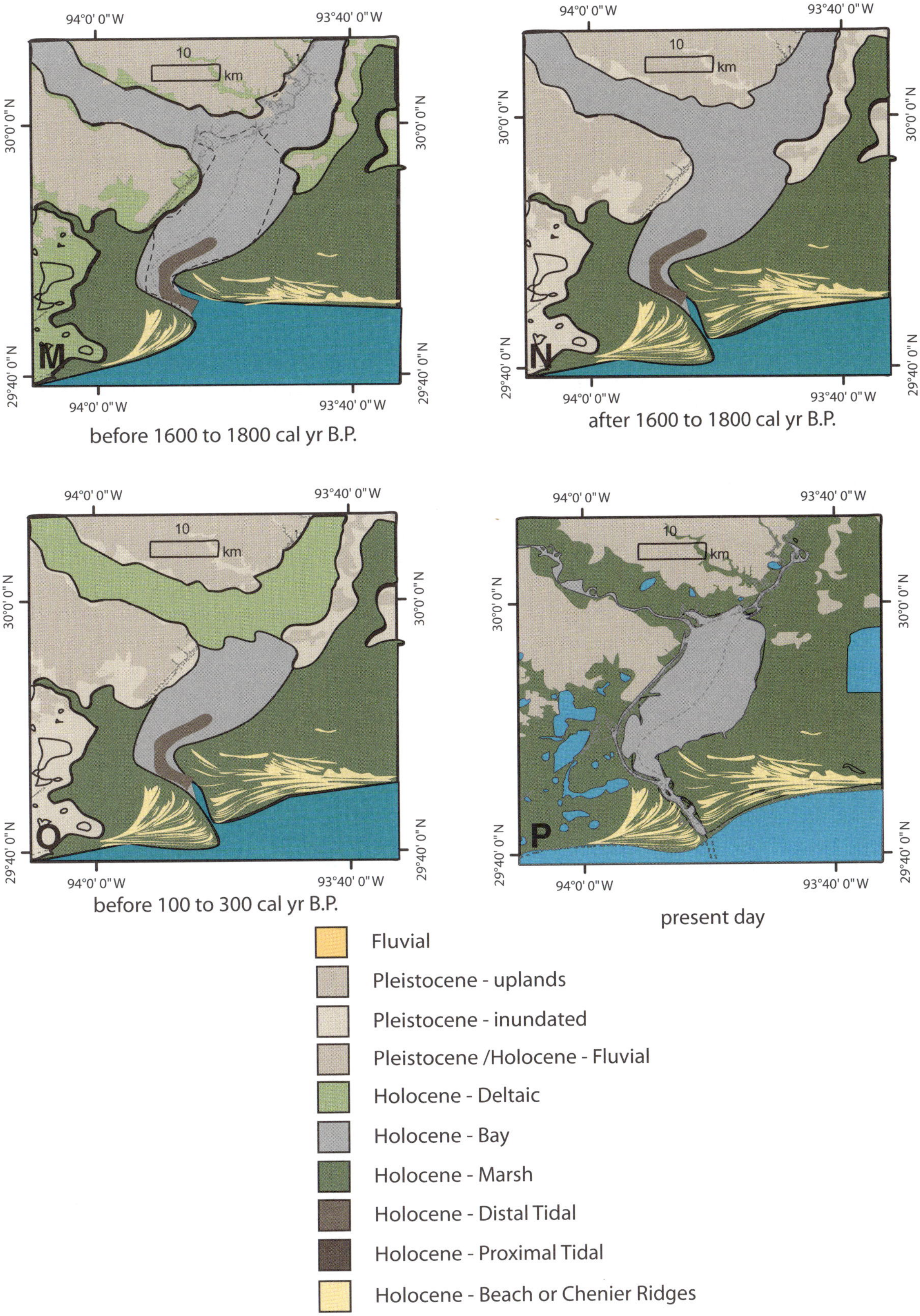

Figure 11 (*continued*).

supply. One model of generating back-stepping uses a threshold response of the fluvial system to the overall relative sea-level rise (Schumm, 2003). Under this scenario, several fluvial systems that have differing internal controls, such as in drainage-basin areal extent and topography, and that have experienced different external climatic regimes should respond asynchronously to similar rates Holocene sea-level rise. Therefore, comparison of the ages of flooding surfaces in several fluvial-estuary systems will underscore or negate the eustatic control on each flooding surface.

While coastal evolution in the Texas-Louisiana region was strongly influenced by rising sea level, the best available Holocene sea-level curves indicate that the rate of rise was continuous, with the exception of rapid rises at ca. 7.6 ka and ca. 8.2 ka (Blanchon and Shaw, 1995; Törnqvist et al., 2004; Milliken et al., this volume). Furthermore, only one flooding surface within the Sabine Lake area corresponds to the elevation of a fluvial terrace. That is the 7100 cal yr B.P. flooding surface. The other flooding surfaces are thus interpreted as having been caused by stepwise decreases in sediment input to the estuary.

Sedimentation rate variations can be used as a proxy for sediment supply changes. Based on this proxy, the Sabine-Neches deltaic system was barely able to keep pace with regional relative sea-level rise throughout most its history. Sedimentation rates in the deltaic facies averaged 1.5–2.8 mm/yr during the period of time in which regional sea-level rise was on the order of 2.1–5.2 mm/yr (8000–12,000 cal yr B.P.) and 0.5–1.5 mm/yr when sea level was rising at 0.8–2.1 mm/yr (4500–8000 cal yr B.P.) (Fig. 12).

Age-depth relationships for basal and bay line peat deposits, age constrained in this study, coincide with the regional relative sea-level curve and show a landward shift in the locus of fluvial-deltaic deposition at ca. 7900 cal yr B.P. (Fig. 12). The delta plain existed in the study area until at least 3500 cal yr B.P., but possibly as recently as 800 cal yr B.P. Sedimentation rates determined from aggrading delta-plain peat deposits average 1.5 mm/yr and potentially kept pace with sea-level rise of 0.8–2.1 mm/yr from 7900 cal yr B.P. to 3500 cal yr B.P. At ca. 3500 cal yr B.P., the bayhead delta stepped landward an unknown distance. The delta began to prograde again by at least 300 cal yr B.P., as indicated by delta-plain peat of this age resting above 800 cal yr B.P. oyster-bearing estuarine deposits. Pleistocene Deweyville terraces crop out updip and constrain the volume of updip Holocene fluvial aggradation. Furthermore, thermoluminescence ages obtained from surface-channel sand bodies of 2.3–9.9 ka (Otvos, 2005) suggest nominal overbank sedimentation during the past two thousand years and therefore minimal floodplain storage of sediment in the lower part of the drainage basin.

Initiation or intensification of agriculture and industry in fluvial drainage basins correlates to sedimentation rate changes (Colman et al., 2002). Historically, a possible sedimentation rate increase coincides with the timing of extensive logging of the East Texas pine forests around the beginning of the twentieth century (Handbook of Texas Online, 2008). Alternatively, the 100–300-yr-old flooding surface could be linked to anthropogenic land-use modifications or ship-channel construction and dredging. Further detailed work is needed to address the potential record of anthropogenic environmental effects to the fluvial-estuary system. Historical measurements of permanent land loss of the coastal wetlands adjacent to the fluvial-estuary system have been quantified as ~11% for the period of 1956 to post-2005 hurricanes (Barras, 2006). The area may now be undergoing an unprecedented, rapid environmental reorganization.

FUTURE IMPLICATIONS

Episodes of rapid environmental reorganization during the Holocene provide a baseline for future urban and landscape planning. During the Holocene, the Sabine-Neches estuary system experienced multiple episodes of dramatic environmental reorganization, which include landward shifts in the bayhead delta of 10 to 20 km within a few centuries. These events occurred while sea level was rising more or less continuously at rates of less than 5.2 mm/yr. This is less than the predicted eustatic rise and considerably less than relative sea-level rise for the end of this century (IPCC, 2001). A recent report indicates that the Louisiana coastal plain may be experiencing subsidence rates on the order to 5–15 mm/yr (Shinkle and Dokka, 2004). This relative rate of sea-level rise is close to the extreme forecasts of the IPCC (2001) predictions. Considering the area of land below 0.5 and 1.0 m elevation (Fig. 13), effects of predicted relative sea-level rise for the area could be catastrophic.

The results of this study also indicate that flooding and associated reorganization of estuarine environments during the Holocene have been influenced by a reduction in sediment supply to the Sabine-Neches estuary system. The type and magnitude of the presumed climatic events that caused these changes remain uncertain however. It is clear that even without a reduction in sediment supply to the estuary system, flooding events similar to those that occurred during the Holocene may occur by the end of this century.

CONCLUSIONS

1. The Sabine River–Neches River–Sabine Lake system experienced several episodes of environmental change during the Holocene. These changes are recorded by landward shifts in environments on the order of 10–20 km within a few centuries.
2. The area of the modern Sabine Lake system was flooded and experienced pronounced reorganization of environments at 8900 cal yr B.P., 8400–8000 cal yr B.P., 7800–7500 cal yr B.P., 7100 cal yr B.P., 5800–5500 cal yr B.P., 4200 cal yr B.P., 3500–1700 cal yr B.P., and 300–100 cal yr B.P.
3. Relative sea-level rise is a likely cause of flooding events at 8900 cal yr B.P. and 8400–8000 cal yr B.P., but climate and/or sediment supply variations cannot be ruled out. Flooding of antecedent topography likely influenced

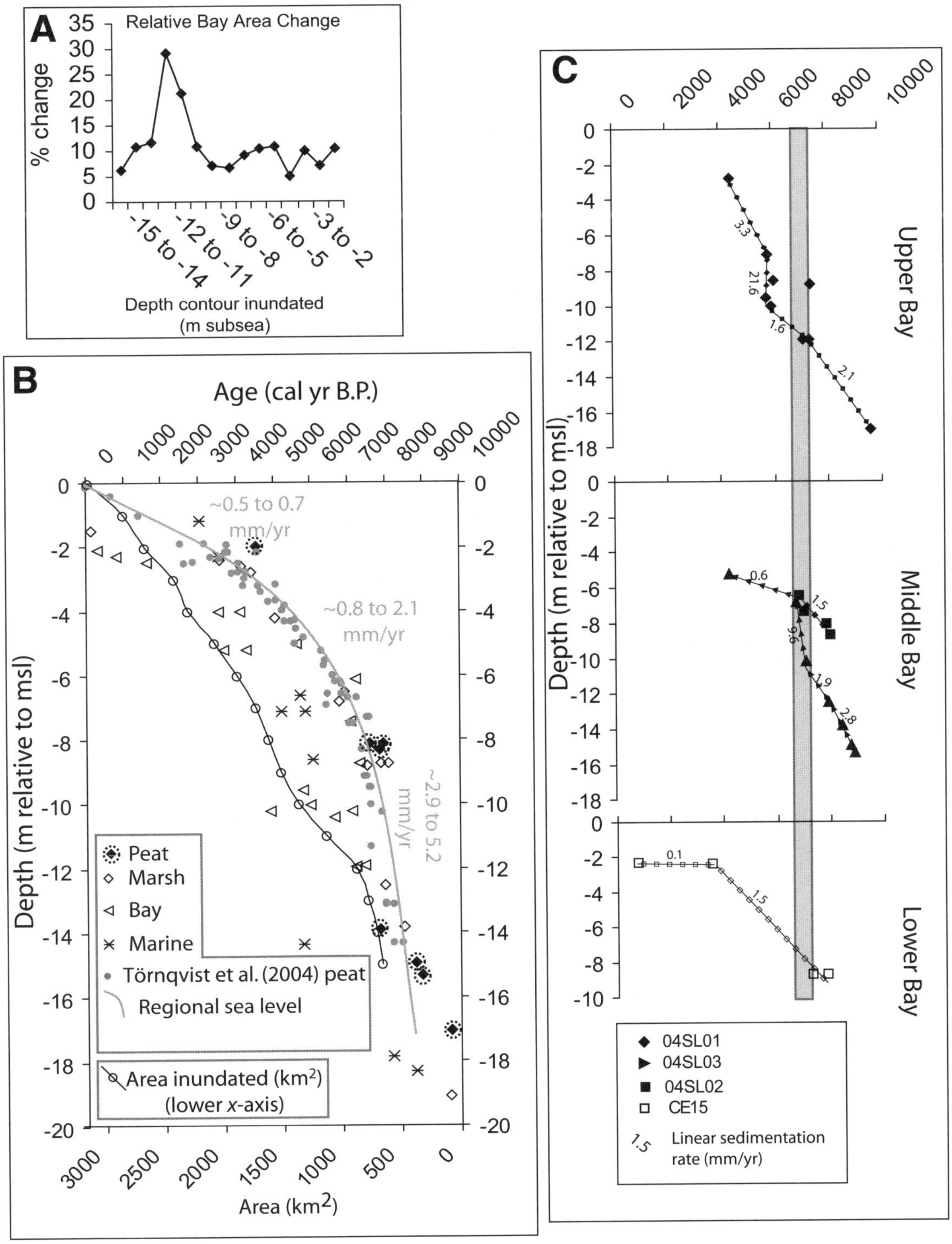

Figure 12. (A) Relative estuarine area change illustrated by depth contour inundated with % estuarine area increase (% change). (B) Area of cumulative flooding, radiocarbon samples, and generalized regional sea-level curve (derived from Törnqvist et al., 2004). Radiocarbon ages (calibrated yr B.P.) are plotted versus depth (mbsl). Generalized regional sea-level curve shows diminishing sea-level rise through time. Radiocarbon ages obtained from basal and bay line peat coincide with the regional sea-level curve (derived from Törnqvist et al., 2004). (C) Separate age-depth plots from cores situated in the upper, middle, and lower estuary and linear regression–derived sedimentation rates (mm/yr). Sedimentation rates change in the upper and middle estuary at ca. 7000 cal yr B.P. The increase in sedimentation rates between 2 and 9 mbsl in the lower-estuary core correspond to tidal delta and tidal inlet deposition. The upper-estuary sedimentation rates diminish to 0.1 mm/yr by at least 3500 cal yr B.P.

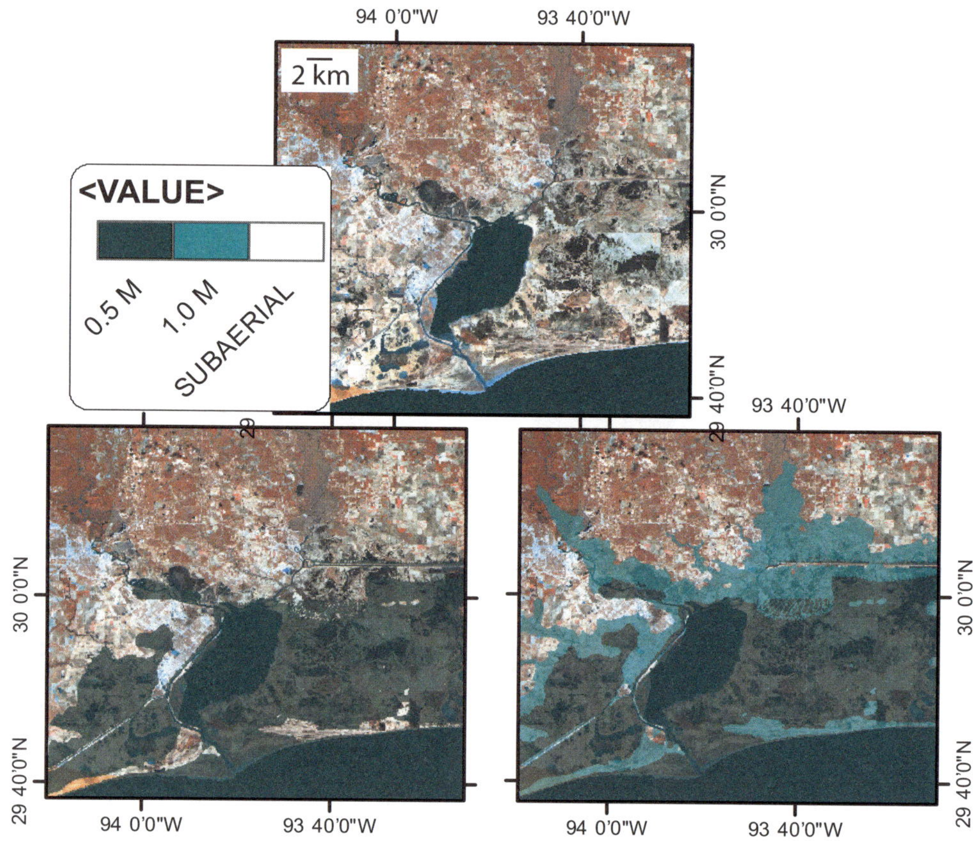

Figure 13. Landsat images showing areas at or below 0.5 and 1.0 m elevation to illustrate the vulnerability of the area to sea-level rise.

the 7800–7500 cal yr B.P. flooding event. Sediment supply variations controlled by climate are assumed to have been the drivers for the 7100 cal yr B.P., 5800–5500 cal yr B.P., 3500–1700 cal yr B.P., and 300–100 cal yr B.P. flooding events. The 4200 cal yr B.P. flooding event may have been a local autocyclic event related to tidal inlet migration, but sediment supply variations cannot be discounted at this time.

4. Past flooding events provide a baseline to measure future environmental changes due to predicted accelerated sea-level rise and subsidence. In the absence of climate change, rapid coastal inundation is realistic and could result in updip saltwater intrusion, increased rates of estuary shoreline ravinement, migration of marsh habitats, and greater impact from hurricanes.

ACKNOWLEDGMENTS

The authors would like to acknowledge Chip Anderson, Z.P. Taha, J.K. Maddox, and C. Meyer for field assistance. Funding for this project was provided through the National Science Foundation (NSF) grant EAR-0107650, the Dave Worthington research grant, and the Anadarko, ExxonMobil, and Conoco Phillips Industry consortium. K.T. Milliken also received student research grants from the Gulf Coast Association of Geological Societies—Ed Picou student research grant 2005 and 2006; and the 2005 Geological Society Student Research Grant. The authors also appreciate reviews by J. Bratton and R. McBride.

REFERENCES CITED

Anderson, J.B., Siringan, F.P., Taviani, M., and Lawrence, J., 1991, Origin and Evolution of Sabine Lake, Texas–Louisiana: Transactions of the Gulf Coast Association of Geological Societies, v. 41, p. 12–16.

Anderson, J.B., Rodriguez, A.R., Abdulah, K., Fillon, R., Banfield, L., McKeown, H., and Wellner, J.S., 2004, Late Quaternary stratigraphic evolution of the northern Gulf of Mexico margin, *in* Anderson, J.B., and Fillon, R., eds., A Synthesis: Late Quaternary Stratigraphic Evolution: Tulsa, Oklahoma, Society for Sedimentary Geology (SEPM), p. 1–24.

Anderson, J.B., Rodriguez, A.B., Milliken, K., and Taviani, M., 2008, this volume, The Holocene evolution of the Galveston estuary complex, Texas: Evidence for rapid change in estuarine environments, *in* Anderson, J.B., and Rodriguez, A.B., eds., Response of Upper Gulf Coast Estuaries to Holocene Climate Change and Sea-Level Rise: Geological Society of America Special Paper 443, doi: 10.1130/2008.2443(06).

Antevs, E., 1948, Climatic changes and pre-white man: In the Great Basin with emphasis on glacial to postglacial times: University of Utah Bulletin, v. 38, p. 169–191.

Barras, J.A., 2006, Land area changes in coastal Louisiana after the 2005 hurricanes—A series of three maps: U.S. Geological Survey Open-File Report 06-1274, 3 sheets.

Barras, J.A., Bourgeois, P.E., and Handley, L.R., 1994, Land loss in coastal Louisiana, 1956–1990: National Wetland Research Center Open-File Report 94-01, 5 p.

Baumann, R.H., Day, J.W., and Miller, C.A., 1984, Mississippi deltaic wetland survival versus coastal submergence: Science, v. 224, no. 4653, p. 1093–1095, doi: 10.1126/science.224.4653.1093.

Bense, J.A., 1995, Archaeology of the Southeastern United States: London, UK, Academic Press, 350 p.

Bernard, H.A., 1950, Quaternary Geology of Southeast Texas [Ph.D. dissertation]: Baton Rouge, Louisiana, Louisiana State University, 165 p.

Bezanson, M.A., 2000, Natural Vegetation Types of Texas and Their Representation in Conservation Areas [M.S. thesis]: Austin, University of Texas, 215 p.

Bird, M.I., Fifiled, L.K., The, T.S., Chang, C.H., Shirlaw, N., and Lambeck, K., 2007, An inflection in the rate of early mid-Holocene eustatic sea-level rise: A new sea-level curve from Singapore: Estuarine, Coastal and Shelf Science, v. 71, p. 523–536, doi: 10.1016/j.ecss.2006.07.004.

Blanchon, P., and Shaw, J., 1995, Reef drowning during the last deglaciation: Evidence for catastrophic sea-level rise and ice-sheet collapse: Geology, v. 23, no. 1, p. 4–8, doi: 10.1130/0091-7613(1995)023<0004:RDDTLD>2.3.CO;2.

Blanchon, P., Jones, B., and Ford, D.C., 2002, Discovery of a submerged relic reef and shoreline off Grand Cayman: Further support for an early Holocene jump in sea level: Sedimentary Geology, v. 147, p. 253–270, doi: 10.1016/S0037-0738(01)00143-9.

Bull, W.B., 1991, Geomorphic Responses to Climate Change: New York, Oxford University Press, 326 p.

Carbotte, S.M., Bell, R.E., Ryan, W.B.F., McHugh, C., Slagle, A., Nitsche, F., and Rubenstone, J., 2004, Environmental change and oyster colonization within the Hudson River estuary linked to Holocene climate: Geo-Marine Letters, v. 24, p. 212–224, doi: 10.1007/s00367-004-0179-9.

Carlson, A.E., Clark, P.U., Raisbeck, G.M., and Brook, E.J., 2007, Rapid Holocene deglaciation of the Labrador sector of the Laurentide Ice Sheet: Journal of Climate, v. 20, p. 5126–5130, doi: 10.1175/JCLI4273.1.

Coast 2050: Louisiana Coastal Wetlands Conservation and Restoration Task Force and the Wetlands Conservation and Restoration Authority, 1998, Coast 2050: Toward a Sustainable Coastal Louisiana: Baton Rouge, Louisiana, Louisiana Department of Natural Resources, 161 p.

Coleman, J.M., 1966, Recent Coastal Sedimentation: Central Louisiana Coast: Baton Rouge, Louisiana University Press, 73 p.

Colman, S.M., Baucom, P.C., Bratton, J.F., Cronin, T.M., McGeehin, J.P., Willard, D., Zimmerman, A.R., and Vogt, P.R., 2002, Radiocarbon dating, chronologic framework, and changes in accumulation rates of Holocene estuarine sediments from Chesapeake Bay: Quaternary Research, v. 57, p. 58–70, doi: 10.1006/qres.2001.2285.s.

Day, J.W., Britsch, L.D., Hawes, S.R., Shaffer, G.P., Reed, D.J., and Cahoon, D., 2000, Pattern and process of land loss in the Mississippi Delta: A spatial and temporal analysis of wetland habitat change: Estuaries, v. 23, no. 4, p. 425–438, doi: 10.2307/1353136.

Doering, J., 1935, Post-Fleming surface formation of coastal southeast Texas and South Louisiana: Bulletin of the American Association of Petroleum Geologists, v. 19, no. 5, p. 651–688.

Fisk, H.N., 1948, Geological Investigation of the Lower Mermentau River Basin and Adjacent Areas in Coastal Louisiana, Appendix II: Mermentau River, Louisiana, U.S. Army Corps Engineers, Mississippi River Commission, 145 p.

Frazier, D.E., 1974, Depositional Episodes: Their Relationship to the Quaternary Stratigraphic Framework in the Northwest Portion of the Gulf Basin: Austin, Texas, Bureau of Economic Geology, Geologic Circular 74-1, 28 p.

Frost, S.L., 1949, The forest regions of Texas: The Texas Journal of Science, v. 1, p. 52–55.

Gould, H.R., and McFarlan, E., 1959, Geologic history of the chenier plain, southwestern Louisiana: Gulf Coast Association of Geological Societies Transactions, v. 9, p. 261–272.

Handbook of Texas Online, 2008, http://www.tshaonline.org/handbook/online/articles.

Holliday, V.T., 1995, Stratigraphy and Paleoenvironments of Late Quaternary Valley Fills on the Southern High Plains: Boulder, Colorado, Geological Society of America Memoir 186, 136 p.

Humphrey, J.D., and Ferring, C.R., 1994, Stable isotopic evidence for latest Pleistocene and Holocene climatic change in north-central Texas: Quaternary Research, v. 41, p. 200–213, doi: 10.1006/qres.1994.1022.

IPCC, 2001, Climate Change 2001: The Scientific Basis: Intergovernmental Panel on Climate Change, http://www.grida.no/climate/ipcc_tar/wg1/index.htm (last accessed 25 April 2008).

Jervey, M.T., 1988. Quantitative geological modeling of siliciclastic rock sequences and their seismic expression, *in* Wilgus, C.K., Hastings, B.S., Kendall, C.G.St.C., Posamentier, H.W., Ross, C.A., and Van Wagoner, J.C., eds., Sea-Level Changes: An Integrated Approach: Society of Economic Paleontology and Mineralogy Special Publication 42, p. 47–69.

Kane, H., 1959, Late Quaternary geology of Sabine Lake and vicinity, Texas and Louisiana: Gulf Coast Association of Geological Societies (GCAGS) Transactions, v. 9, p. 225–235.

Knox, J.C., 1983, Responses of river systems to Holocene climates, *in* Wright, H.E., Jr., ed., Late-Quaternary Environments of the United States: Volume 2. The Holocene: Minneapolis, Minnesota, University of Minnesota Press, p. 26–41.

LaSalle, M.W., and de la Cruz, A.A., 1985, Species Profiles: Life Histories and Environmental Requirements of Coastal Fishes and Invertebrates (Gulf of Mexico)—Common *Rangia*: U.S. Fish Wildlife Service Biological Report 82(11.31), U.S. Army Corps of Engineers Report TR EL-82-4, 16 p.

Lohse, A.E., 1955, Dynamic geology of the modern coastal region, northwest Gulf of Mexico, *in* Hough, J.L., and Menard, H.W., eds., Finding Ancient Shorelines: Society for Economic Paleontologists and Mineralogists (SEPM) Special Publication 3, p. 99–104.

Maddox, J., Anderson, J.B., Milliken, K., Rodriguez, A.B., Dellapenna, T.M., and Giosan, L., 2008, this volume, The Holocene evolution of the Matagorda and Lavaca estuary complex, Texas, USA, *in* Anderson, J.B., and Rodriguez, A.B., eds., Response of Upper Gulf Coast Estuaries to Holocene Climate Change and Sea-Level Rise: Geological Society of America Special Paper 443, doi: 10.1130/2008.2443(07).

McBride, R.A., Taylor, M.J., and Byrnes, M.R., 2007, Coastal morphodynamics and chenier-plain evolution in southwestern Louisiana, USA: A geomorphic model: Geomorphology, v. 88, p. 367–422.

McCave, I.N., Bryant, R.J., Cook, H.F., and Coughanowr, C.A., 1986, Evaluation of a laser diffraction size analyzer for use with natural sediment: Journal of Sedimentary Petrology, v. 56, p. 561–564.

McFarlan, 1961, Radiocarbon dating of late Quaternary deposits, south Louisiana: Geological Society of America Bulletin, v. 72, p. 129–158, doi: 10.1130/0016-7606(1961)72[129:RDOLQD]2.0.CO;2.

Milliken, K.T., Anderson, J.B., and Rodriguez, A.B., 2008, A new composite Holocene sea-level curve for the northern Gulf of Mexico, *in* Anderson, J.B., and Rodriguez, A.B., eds., Response of Upper Gulf Coast Estuaries to Holocene Climate Change and Sea-Level Rise: Geological Society of America Special Paper 443, doi: 10.1130/2008.2443(01).

Morton, R.A., Paine, J.G., and Blum, M.D., 2000, Responses of stable bay-margin and barrier-island systems to Holocene sea-level highstands, western Gulf of Mexico: Journal of Sedimentary Research, v. 70, no. 3, p. 478–490, doi: 10.1306/2DC40921-0E47-11D7-8643000102C1865D.

Morton, R.A., Bernier, J.C., and Barras, J.A., 2006, Evidence of regional subsidence and associated intertidal wetland loss induced by hydrocarbon production, Gulf Coast region, USA: Environmental Geology, v. 50, p. 261–274, doi: 10.1007/s00254-006-0207-3.

Musgrove, M., Banner, J.L., Mack, L.E., Combs, D.M., James, E.W., Cheng, H., and Edwards, R.L., 2001, Geochronology of late Pleistocene to Holocene speleothems from central Texas: Implications for regional paleoclimate: Geological Society of America Bulletin, v. 113, no. 12, p. 1532–1543, doi: 10.1130/0016-7606(2001)113<1532:GOLPTH>2.0.CO;2.

Nelson, H.F., and Bray, E.E., 1970, Stratigraphy and history of the Holocene sediments in the Sabine–High Island area, Gulf of Mexico, *in* Morgan, J.P., ed., Deltaic Sedimentation Modern and Ancient: Society for Economic Paleontologists and Mineralogists (SEPM) Special Publication 15, p. 48–77.

Otvos, E.G., 2005, Numerical chronology of Pleistocene coastal plain and valley development; extensive aggradation during glacial low sea-levels: Quaternary International, v. 135, no. 1, p. 91–113, doi: 10.1016/j.quaint.2004.10.026.

Penland, S., Ramsey, K.E., McBride, R.A., Moslow, T.F., and Westphal, K.A., 1989, Relative Sea Level in Louisiana and the Gulf of Mexico: Baton Rouge, Louisiana Geological Survey Coastal Geology Technical Report No. 3, 65 p.

Powell, G.L., and Green, A.W., 1994, Freshwater inflows to Texas bays and estuaries: Ecological relationships and methods for determination of needs, *in* Longley, W.L., ed.: Austin, Texas, Water Development Board and Texas Parks and Wildlife Department, 386 p.

Ravichandran, M., Baskaran, M., Santschi, P.H., and Bianchi, T.S., 1995, Geochronology of sediments in the Sabine-Neches estuary, Texas, USA: Chemical Geology, v. 125, no. 3–4, p. 291–306, doi: 10.1016/0009-2541(95)00082-W.

Rehkemper, J.L., 1969, Sedimentology of Holocene estuarine deposits, Galveston Bay, *in* Lankford, R.R., and Rogers, J.J.W., eds., Holocene Geology of the Galveston Bay Area: Houston, Houston Geological Society, p. 12–52.

Rodriguez, A.B., Anderson, J.B., and Simms, A.R., 2005, Terrace inundation as an autocyclic mechanism for parasequence formation: Galveston estuary, Texas, USA: Journal of Sedimentary Research, v. 75, p. 608–620, doi: 10.2110/jsr.2005.050.

Schumm, S.A., 2003, The Fluvial System: Caldwell, New Jersey, Blackburn Press, 338 p.

Sellers, M.A., and Stanley, J.G., 1988, Species Profiles: Life Histories and Environmental Requirements of Coastal Fishes and Invertebrates (Gulf of Mexico)—American Oyster: U.S. Fish Wildlife Service Report FWS/OBS-82/11.64, U.S. Army Corps of Engineers Report TR EL-82-4, 15 p.

Shinkle, K., and Dokka, R.K., 2004, Rates of Vertical Displacement at Benchmarks in the Lower Mississippi Valley and the Northern Gulf Coast: National Oceanic and Atmospheric Administration (NOAA) Technical Report 50, 135 p.

Simms, A.R., Anderson, J.B., Rodriguez, A.B., and Taviani, M., 2008, this volume, Mechanisms controlling environmental change within an estuary: Corpus Christi Bay, Texas, USA, *in* Anderson, J.B., and Rodriguez, A.B., eds., Response of Upper Gulf Coast Estuaries to Holocene Climate Change and Sea-Level Rise: Geological Society of America Special Paper 443, doi: 10.1130/2008.2443(08).

Siringan, F.P., 1993, Coastal lithosome evolution and preservation during an overall rising sea level: East Texas Gulf Coast and continental shelf [Ph.D. thesis]: Houston, Rice University, 226 p.

Sperazza, M., Morre, J.N., and Hendrix, M.S., 2004, High-resolution particle size analysis of naturally occurring very fine-grained sediment through laser diffractometry: Journal of Sedimentary Research, v. 74, no. 5, p. 736–743, doi: 10.1306/031104740736.

Stuiver, M., and Polach, H.A., 1977, Discussion: Reporting of ^{14}C data: Radiocarbon, v. 19, p. 355–363.

Stuiver, M., Reimer, P.J., and Reimer, R.W., 2004, CALIB 5.0: www program and documentation, http://calib.qub.ac.uk/calib/ (last accessed 29 April 2008).

Taylor, M.J., Byrnes, M.R., and McBride, R.A., 1996, Form/process relationships and geomorphic evolution of the southwestern Louisiana chenier plain: Transactions of the Gulf Coast Association of Geological Societies, v. 46, p. 413–422.

Thomas, M.A., 1990, The Impact of Long-Term and Short-Term Sea Level Changes on the Evolution of the Wisconsinan–Holocene Trinity/Sabine Incised Valley System, Texas Continental Shelf [Ph.D. thesis]: Houston, Rice University, 247 p.

Thomas, M.A., and Anderson, J.B., 1994, Sea-level controls on the facies architecture of the Trinity/Sabine incised-valley system, Texas continental shelf, *in* Dalrymple, R.W., Boyd, R., and Zaitlin, B.A., eds., Incised-Valley Systems: Origin and Sedimentary Sequences: Society for Economic Paleontologists and Mineralogists (SEPM) Special Publication 51, p. 63–82.

Toomey, R.S., Blum, M.D., and Valastro, S.J., 1993, Late Quaternary climates and environments of the Edwards Plateau, Texas: Global and Planetary Change, v. 7, p. 299–320, doi: 10.1016/0921-8181(93)90003-7.

Törnqvist, T.E., Gonzalez, J.L., Newsom, L.A., van der Borg, K., de Jong, A.F.M., and Kurnik, C.W., 2004, Deciphering Holocene sea-level history on the U.S. Gulf Coast: A high-resolution record from the Mississippi Delta: Geological Society of America Bulletin, v. 116, no. 7–8, p. 1026–1039, doi: 10.1130/B2525478.1.

Turnbull, J.C., Lehman, S.J., Miller, J.B., Sparks, R.J., Southon, J.R., and Tans, P.P., 2007, A new high precision $^{14}CO_2$ time series for North American continental air: Journal of Geophysical Research, v. 112, p. D11310, doi: 10.1029/2006JD008184.

Turner, R.E., 1997, Wetland loss in the northern Gulf of Mexico, multiple working hypotheses: Estuaries, v. 20, no. 1, p. 1–13, doi: 10.2307/1352716.

U.S. EPA, 1999, Ecological Condition of Estuaries in the Gulf of Mexico: Gulf Breeze, Florida, U.S. Environmental Protection Agency, Office of Research and Development, National Health and Environmental Effects Research Laboratory, Gulf Ecology Division, Report EPA 620-R-98-004, 80 p.

Zaitlin, B.A., Dalrymple, R.W., and Boyd, R., 1994, The stratigraphic organization of incised-valley systems associated with relative sea-level change, *in* Dalrymple, R.W., Boyd, R., and Zaitlin, B.A., eds., Incised-Valley Systems: Origin and Sedimentary Sequences: Society for Economic Paleontologists and Mineralogists (SEPM) Special Publication 51, p. 63–82.

Manuscript Accepted by the Society 10 March 2008

Printed in the USA

The Geological Society of America
Special Paper 443
2008

The Holocene evolution of the Galveston estuary complex, Texas: Evidence for rapid change in estuarine environments

John B. Anderson
Department of Earth Science, Rice University, 6100 Main Street, MS 126, Houston, Texas 77005, USA

Antonio B. Rodriguez
Institute of Marine Sciences, University of North Carolina at Chapel Hill, 3431 Arendell Street, Morehead City, North Carolina 28557, USA

K.T. Milliken
Department of Earth Science, Rice University, 6100 Main Street, MS 126, Houston, Texas 77005, USA

Marco Taviani
Istituto di Scienze Marine, Consiglio Nazionale delle Ricerche, via Gobetti 101, Bologna, 40129, Italy

ABSTRACT

Seismic data and sediment cores from the Galveston estuary complex were used to reconstruct the evolution of the estuary during the Holocene. These data show that the estuary complex has had a history of rapid and dramatic change in response to (1) sea-level rise across the irregular topography of the ancestral Trinity River valley and (2) changes in climate, which regulated sediment supply to the estuary. In general, the valley morphology consists of a deep incision near the center and broad, terraced flanks. As sea level rose during the Holocene and flooded the valley, the shape of the estuary changed from narrow and deep to wide and rounded. As sea level rose to the elevation of the relatively flat fluvial terraces, these areas were flooded rapidly, resulting in expansions in bay area and dramatic reorganization of bay environments. The most notable changes were up-valley shifts in the bayhead delta of tens of kilometers in a few centuries. Radiocarbon ages indicate that these events took place ca. 9600, ca. 8500, and between ca. 7700 and 7400 yr B.P. The early flooding events occurred when sea level was rising rapidly (average 4.2 mm/yr; Milliken et al., this volume, Chapter 1), perhaps episodically. The ca. 8200 yr B.P. flooding surface corresponds to a prominent terrace at –14 m. The ca. 7700–7400 yr B.P. flooding episode was the most dramatic in terms of its impact on the estuary setting. Following this event, the area of the estuary increased by ~30%. This flooding event occurred as the rate of sea-level rise was starting to decrease. The level of this flooding surface also corresponds to a terrace at ~–10 m, but the magnitude of flooding is too large to be explained entirely by flooding of this terrace. At the same time, Matagorda Bay to the west and Sabine Lake to the east experienced similar dramatic flooding events. This

Anderson, J.B., Rodriguez, A.B., Milliken, K.T., and Taviani, M., 2008, The Holocene evolution of the Galveston estuary complex, Texas: Evidence for rapid change in estuarine environments, *in* Anderson, J.B., and Rodriguez, A.B., eds., Response of Upper Gulf Coast Estuaries to Holocene Climate Change and Sea-Level Rise: Geological Society of America Special Paper 443, p. 89–104, doi: 10.1130/2008.2443(06). For permission to copy, contact editing@geosociety.org.
©2008 The Geological Society of America. All rights reserved.

event occurred when the climate of east-central Texas was in transition from cool and moist to warm and dry, and the vegetation cover of the region was undergoing a reduction in forest and an increase in grasslands. Hence, the ca. 7700–7400 event was likely amplified by a reduction in sediment supply to the estuary triggered by this regional climatic change coupled with an increase in sediment accommodation space caused by flooding of a terrace. Following the ca. 7700–7400 yr B.P. flooding event, the estuary setting changed relatively little as the rate of sea-level rise decreased to less than 2.0 mm/yr. By ca. 2600 cal yr B.P., the modern Trinity bayhead delta had begun to form. Circa 1600 cal yr B.P., the delta experienced a phase of rapid growth. This more recent episode of delta growth may have resulted from an increase in the rate of sediment supply, perhaps associated with human occupation and agriculture in the drainage basin.

More recent anthropogenic changes include accelerated subsidence due to groundwater and hydrocarbon extraction. These changes are occurring at rates that approach those that occurred during prior flooding events of the Holocene. Thus, the Galveston estuary complex could be on the verge of another flooding event that would mainly impact the Trinity bayhead delta and low-lying areas around the delta.

Keywords: estuary, sea-level rise, climate change, flooding.

INTRODUCTION

Galveston estuary complex borders on one of the world's largest population centers and petrochemical complexes, which has led to significant alteration of the bay environment and its adjacent watersheds. Most of Galveston Bay's western shoreline has been developed, resulting in the loss of virtually all of its fringing wetlands. Two rivers empty into the bay, the Trinity River and the San Jacinto River (Fig. 1). Both rivers have been dammed, but the impact of these dams on sediment supply to the estuary remains problematic (Phillips et al., 2004; White et al., 2002). Within the past century, Galveston Bay and adjacent areas have experienced high rates of subsidence caused by groundwater and hydrocarbon extraction (Paine, 1993; Morton et al., 2006).

The increased rate of subsidence caused by subsurface fluid extraction has placed considerable stress on wetlands (White and Morton, 1997; White et al., 2002; Morton et al., 2006). However, the full impact of these changes may not have occurred yet, as the rate of relative sea-level rise is expected to increase in the next several decades. What will be the impact of accelerated relative sea-level rise and reduced sediment input to the Galveston Bay complex? One way to address this question is to constrain how the bay complex responded to past changes in the rate of relative sea-level rise and sediment supply, changes that were of a similar magnitude to those that are expected to occur during the next few decades to centuries. This paper presents results from an investigation aimed at reconstructing the Holocene evolution of the Galveston estuary complex that will hopefully raise public awareness of the sensitivity of the bay to sea-level rise and climate change and will provide a foundation for predicting future changes in the estuary complex.

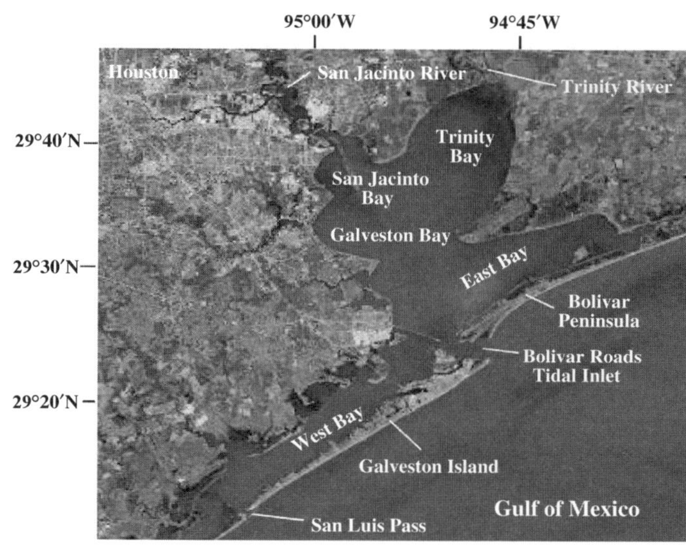

Figure 1. Aerial photo of the Galveston estuary complex showing key geographic locations.

SETTING

The Galveston estuary complex includes Galveston Bay, Trinity Bay, San Jacinto Bay, East Bay, and West Bay (Fig. 1). These estuaries are situated landward of two prominent coastal barriers, Bolivar Peninsula and Galveston Island. Galveston Bay is 48 km long and has a variable width that reflects the highstand fluvial geomorphology that was flooded to create the bay (Morton and McGowen, 1980; Rodriguez et al., 2005). The depth of the bay averages 2–3 m, and astronomical tides in the bay are 0.3 m.

Prevailing winds are from the southeast, with occasional strong northerly winds that are associated with passing cold fronts. Wind tides of up to 1.0 m above and below mean tide occur during strong wind events.

Fluvial sediment supply to Galveston Bay comes from the Trinity and San Jacinto Rivers, and to a lesser degree the many small streams that enter the estuary. The Trinity River drains an area of 46,500 km^2 and has an annual discharge of 7.1×10^9 m^3. The San Jacinto River has an annual discharge of 2.4×10^9 m^3. The geology of both drainage basins consists of Tertiary-age strata that are, for the most part, compacted but poorly lithified.

The Trinity River has formed a prominent bayhead delta that occupies the upper portion of Trinity Bay (Fig. 2). The modern delta plain covers an area of 16 km^2 and is characterized by a distributary channel network and associated mouth bars. In the past two centuries, the delta has prograded toward the east, sealing off the upper part of Trinity Bay to form Lake Anahuac (Fig. 2). Since the early 1900s, the Trinity River mouth has been artificially maintained as a navigation channel, forcing sediment to bypass the delta plain, except during major floods. This has resulted in the extension of the river mouth into the estuary, which has resulted in sediment bypass of the delta front and mouth bars.

The San Jacinto River formed a much smaller delta that has been removed by channel dredging, maintenance, and subsidence. Both the San Jacinto and Trinity Rivers have been dammed, but the impact of sediment flux to the estuary may not have been significantly reduced (Phillips et al., 2004). The entire Galveston estuary complex experiences significant bay line erosion, which contributes to the sediment supply of the estuary (Paine and Morton, 1986). In addition, tidal currents deliver sand from the Gulf of Mexico to the estuary complex. Most of this sand is deposited in two tidal deltas, the Bolivar tidal delta located between Galveston Island and Bolivar Peninsula, and San Luis Pass tidal delta, located at the west end of Galveston Island (Fig. 1). Construction of jetties at the Bolivar Roads tidal inlet has significantly reduced the amount of sediment moving into the estuary through the inlet (Siringan and Anderson, 1993).

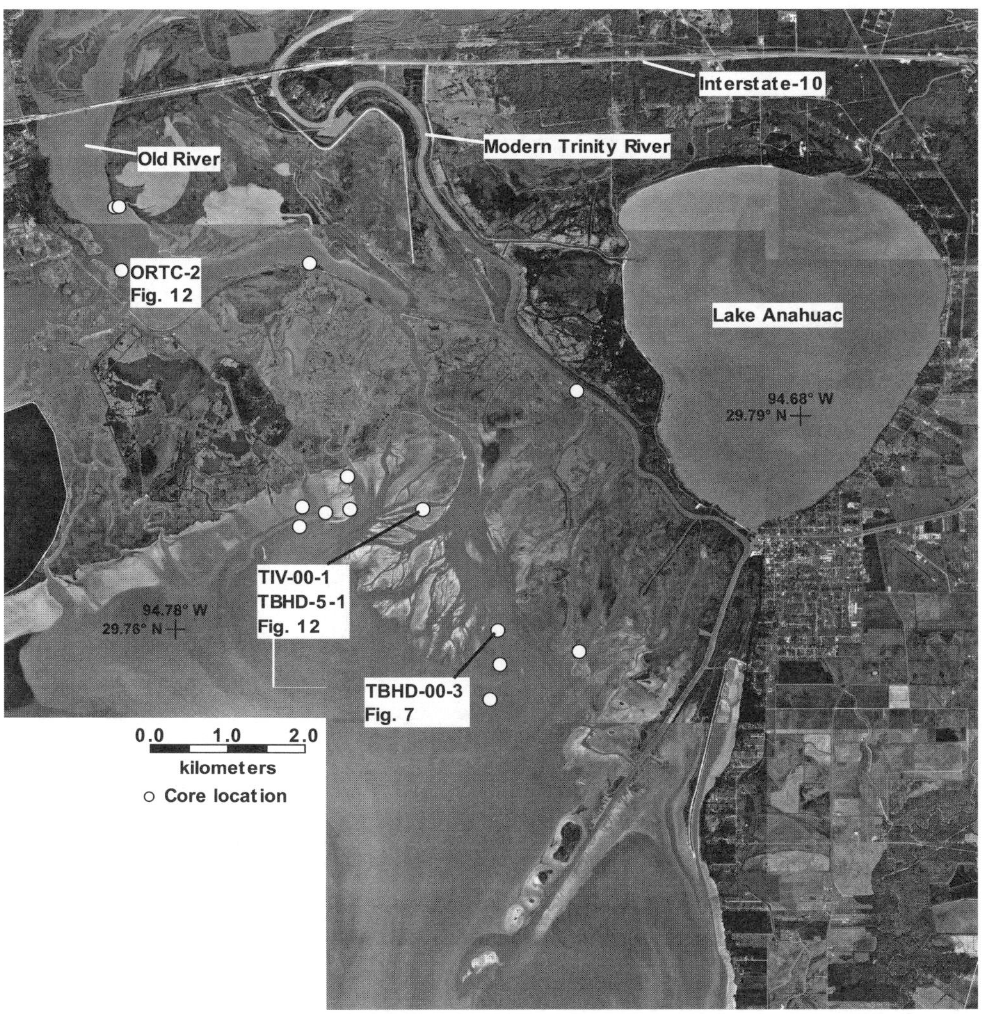

Figure 2. Aerial view of the Trinity River bayhead delta showing locations of cores presented in Figures 7 and 12.

METHODS

Two seismic data sets were used for this investigation. The Pleistocene surface was mapped using minisparker data collected by the U.S. Geological Survey (Fig. 3). These data provided penetration to a subbottom depth of 300 ms. The sedimentary fill of the valley was examined using seismic data acquired aboard the Rice University research vessels R/V *Matagorda* and R/V *Lone Star*. These data were acquired using an EG&G miniboomer and a CAP 6000 Chirper. Onboard processing included band-pass filtering and automatic gain control. These data were augmented by boomer data collected by Williams et al. (1979) in the southern part of the bay.

Sediment cores were collected with a pneumatic-hammer corer and with a rotary-drill rig. Seven drill cores were acquired along the valley axis using push coring to a subbottom depth of up to 20 m from the Rice University drilling barge R/V *Trinity* (Fig. 3). The core sites were selected to sample the central part of the valley, a strategy that generally worked, except in the cases of sites TV-99-1 and TV-93-3, which were off the center of the valley.

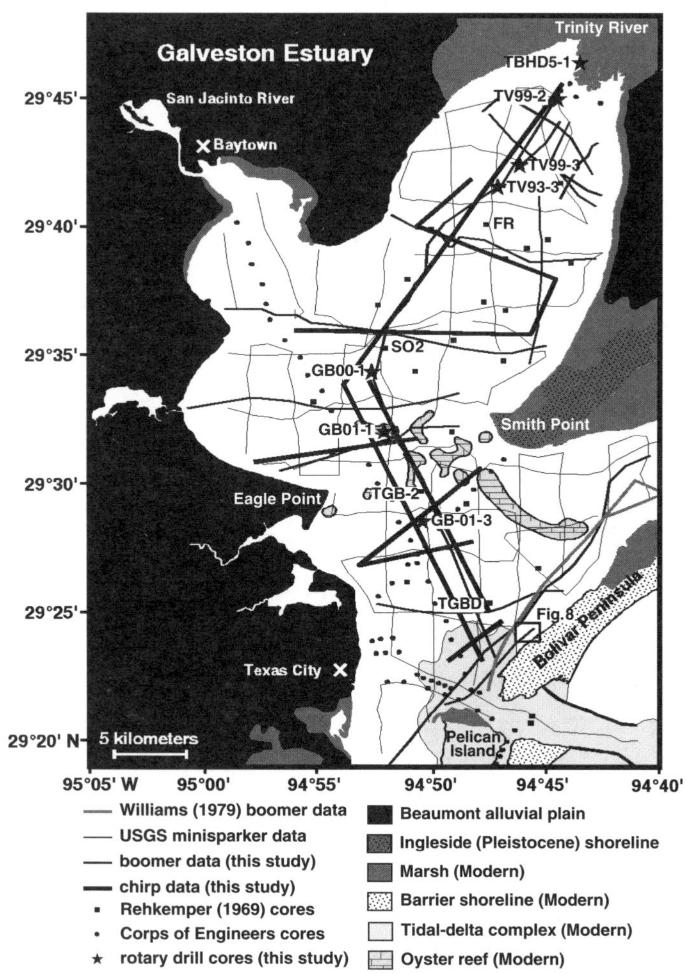

Figure 3. The data set used for this study includes a dense grid of seismic data collected over nearly three decades, borings from the U.S. Army Corps of Engineers, and several long rotary-drill cores collected by Rehkemper (1969) and as part of this study.

Cores were cut and described for color, lithology, sedimentary structures, and macrofossil content. Radiocarbon dating was conducted on 28 samples using the accelerator mass spectrometry (AMS) facilities at the University of Arizona and Woods Hole Oceanographic Institution. An additional five samples were dated by standard radiocarbon analysis at Beta Analytic. When possible, analyses were conducted on articulated bivalves or on individual wood and plant fragments and on basal peat samples. These data are presented here in both conventional radiocarbon yr B.P. and calibrated radiocarbon yr B.P. (Table 1). These new data were augmented with radiocarbon dates from prior studies by Rehkemper (1969) and Siringan and Anderson (1993). All radiocarbon ages were calibrated using Calib 5.0 software, which accounts for atmospheric variations in radiocarbon over time by comparing sample dates against a calibrated data set (Calib 5.2, 14 March 2005). See Milliken et al. (this volume, Chapter 1) for further details concerning the calibration of radiocarbon ages.

RESULTS

Valley Morphology

One key to reconstructing the evolution of Galveston estuary complex was mapping the incised Trinity and San Jacinto paleovalleys that were flooded to create the estuaries (Fig. 4). This antecedent fluvial geomorphology was an important factor in the flooding history of the valleys. The structure map of the valley shown in Figure 4 was modified from an earlier study by Smyth et al. (1988). It is based on a dense grid of seismic data and constrained by long sediment cores and borings that penetrated the Pleistocene surface.

During the falling limb of sea level in the Pleistocene (120–20 ka), when the Trinity and San Jacinto valleys were being incised, the morphology of the valleys changed from that of broad meander belts to narrower and deeper valleys. The fall in sea level was episodic in nature, and this resulted in the formation of terraces in the upper part of the valleys (Blum et al., 1995; Morton et al., 1996; Rodriguez et al., 2005). The two most prominent terraces occur −14 m and −10 m below mean sea level throughout much of the bay (Fig. 4). During the maximum lowstand, the Trinity River incised a 35–40-m-deep valley near the present coast. The valley extended across the continental shelf and merged with the Sabine River valley ~40 km offshore of the current shoreline (Thomas and Anderson, 1994; Simms et al., 2006).

During the initial rise of sea level following the Last Glacial Maximum, the ancestral Trinity River maintained a shelf-margin delta until ca. 15,000 cal yr B.P., when rising sea level overtook the delta (Anderson, 2005). From ca. 15,000 cal yr B.P. until ca. 10,000 cal yr B.P., deposition shifted updip, resulting in backstepping bayhead delta, middle-bay, and lower-bay facies within

TABLE 1. RADIOCARBON DATING RESULTS

Lab code	Sample name	Sample depth (cm)	Material	Conventional ^{14}C age (yr B.P.; 1σ)	Calibrated ^{14}C age (cal. yr B.P.; 1σ)
Beta-163927	TBHD-5-1	174–178	Wood	810 ± 60	720 ± 40
Beta-163928	TBHD-5-1	308–313	*R. cuneata*	1250 ± 70	1170 ± 100
AA39383	TBHD-5-1	419–421	Wood	1728 ± 37	1670 ± 25
Beta-163929	TBHD-5-1	554–556	*C. virginica*	4180 ± 60	4250 ± 100
AA38437	TBHD-5-1	641–644	*R. cuneata*	5276 ± 75	5645 ± 70
AA38439	TBHD-5-1	725–730	Wood	6232 ± 55	7185 ± 30
AA39386	TIV-00-1.1	810–820	Wood	6457 ± 62	7360 ± 40
AA39381	TIV-00-1.3	1165–1166	Wood	7169 ± 81	7985 ± 55
AA39403	TIV-00-1.4	1340–1342	Wood	7738 ± 60	8495 ± 50
OS-34795	TV-99-2-17′	746–751	*R. cuneata*	6690 ± 35	7225 ± 35
OS-34717	TV-99-2-17′	816–818	Wood	6900 ± 35	7714 ± 33
OS-34716	TV-99-2-26′	978–982	Wood	7050 ± 40	7885 ± 50
OS-34797	TV-99-3-0′	439–441	*C. virginica*	4130 ± 35	4185 ± 50
OS-34796	TV-99-3-15′	809–813	*R. cuneata*	6530 ± 35	7035 ± 55
Beta-135431	TV-99-3-15′	864–865	*M. mitchelli*	6860 ± 60	7370 ± 50
Beta-135429	TV-99-3-24′	1015–1024	Peat	8030 ± 150	8840 ± 195
AA39400	GB-00-1.1	941–943	*C. virginica*	6620 ± 64	7150 ± 80
AA39395	GB-00-1.1	957–960	Wood	6454 ± 51	7375 ± 50
OS-34715	GB-00-1.1	1026–1029	Wood	6900 ± 35	7715 ± 35
AA39384	GB-00-1.2	1126–1133	Wood	7413 ± 53	8295 ± 35
39032	GB-00-1.2	1288–1298	Wood	7660 ± 40	8420 ± 30
AA39385	GB-00-1.3	1638–1641	*R. cuneata*	8502 ± 57	8970 ± 65
39034	GB-00-1.3	1679–1682	Wood	8330 ± 45	9311 ± 45
AA44774	GB-01-1	1460–1464	Wood	7803 ± 70	8545 ± 95
AA44773	GB-01-1	1781–1785	Wood	8702 ± 61	9645 ± 95
AA44771	GB-01-3	1252–1254	*M. lateralis**	7275 ± 100	7735 ± 95
AA44772	GB-01-3	1507–1511	Peat	7666 ± 70	8430 ± 40
AA44770	GB-01-3	1654–1658	Peat	8020 ± 120	8865 ± 170

the offshore valley. These facies are bounded by prominent flooding surfaces that record landward shifts in environments (Thomas and Anderson, 1994). The most pronounced back-stepping events involved the ancestral Trinity bayhead delta (Rodriguez et al., 2005). Although the age of the deltas was poorly constrained by Thomas and Anderson (1994), they did suggest that the updip shifts in valley-fill facies and associated environments were caused by episodic increases in the rate of sea-level rise. This interpretation is consistent with published sea-level curves for the Gulf of Mexico (e.g., Nelson and Bray, 1970; Frazier, 1974; Otvos, 2004), and global eustatic curves (e.g., Fairbanks, 1989; Bard et al., 1990, 1996; Toscano and Macintyre, 2003) for the pre-Holocene and early Holocene flooding events. Later work on the inner shelf yielded radiocarbon dates that indicate the two youngest deltas are between 10,000 and 8400 yr old (Rodriguez et al., 2004). In all, these early Holocene bayhead deltas shifted landward ~60 km during this time interval, an average rate of 3.7 km/century. However, the actual flooding surfaces may reflect much faster rates of change.

Recently published sea-level curves indicate that the early Holocene delta back-stepping events within the offshore Trinity valley took place when sea level was rising fairly continuously at an average rate of 4.2 mm/yr (Toscano and Macintyre, 2003; Törnqvist et al., 2004; Milliken et al., this volume, Chapter 1). This raises the question: What caused the step-wise updip shifts in the location of the deltas? One explanation is flooding of variable antecedent topography (Rodriguez et al., 2005). The other explanation is a decrease in sediment supply to the delta. Unfortunately, poor age control and lack of detail about the nature of antecedent topography on the inner shelf have prevented us from resolving this issue. In this study, we focus on the estuarine part of the valley, where the causes of apparently abrupt changes in bay evolution can be examined.

Trinity–San Jacinto Valley Fill

Sediment cores that penetrated the base of the Trinity–San Jacinto valley sampled stiff (high shear strength), red to yellow-brown clay with calcareous nodules (Rehkemper, 1969). Cores from the margins of the valley penetrated Pleistocene fluvial deposits (Deweyville and Beaumont Formations), including point-bar sand and floodplain clay. A sharp contact, both in cores and seismic records, separates Pleistocene deposits from the younger valley-fill succession (Figs. 4 and 5).

The overall valley-fill succession includes fluvial sands and gravels at the base of the valley that were deposited during the lowstand and initial rise in sea level. The fluvial deposits are overlain by organic-rich mud and bayhead delta deposits, which are, in turn, overlain by estuarine deposits. Cores collected in the southern part of the estuary sampled well-sorted sand and sand-and-mud couplets, interpreted as flood-tidal delta deposits, and shell-rich sand and shell hash interpreted as tidal-inlet deposits (Siringan and Anderson, 1993). This overall succession records the flooding history of the estuary and evolution of Bolivar and Galveston coastal barriers and Bolivar tidal inlet. The following is a brief description of these facies.

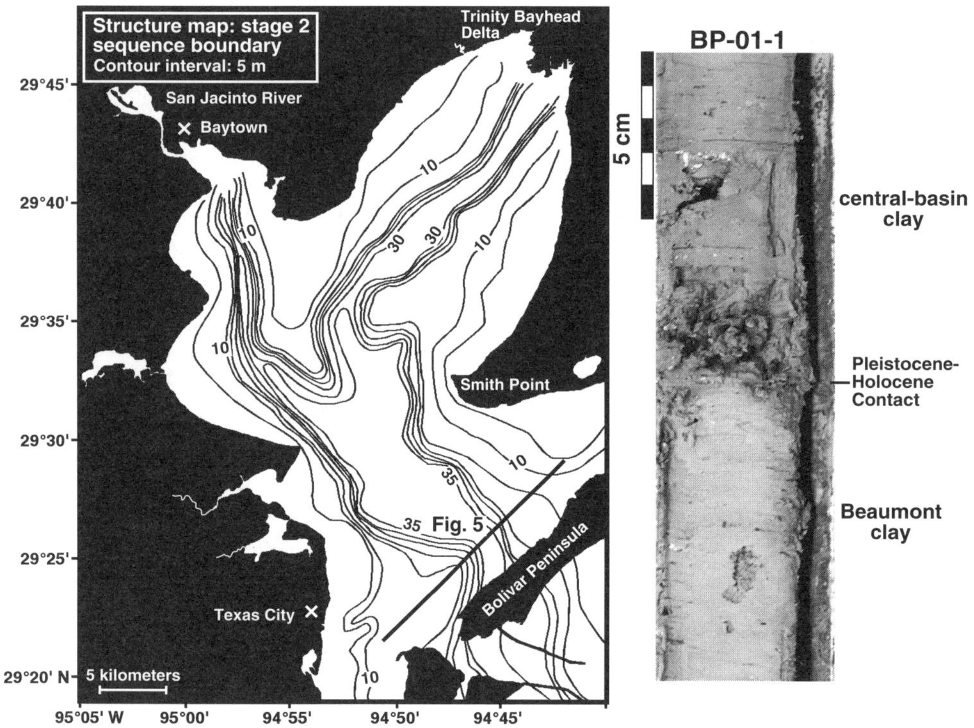

Figure 4. Structure map of the late Pleistocene lowstand surface showing the geomorphology of the San Jacinto and Trinity incised valleys (modified from Smyth et al., 1988). The inset core photograph illustrates the contact between Pleistocene Beaumont clay and Holocene estuary deposits.

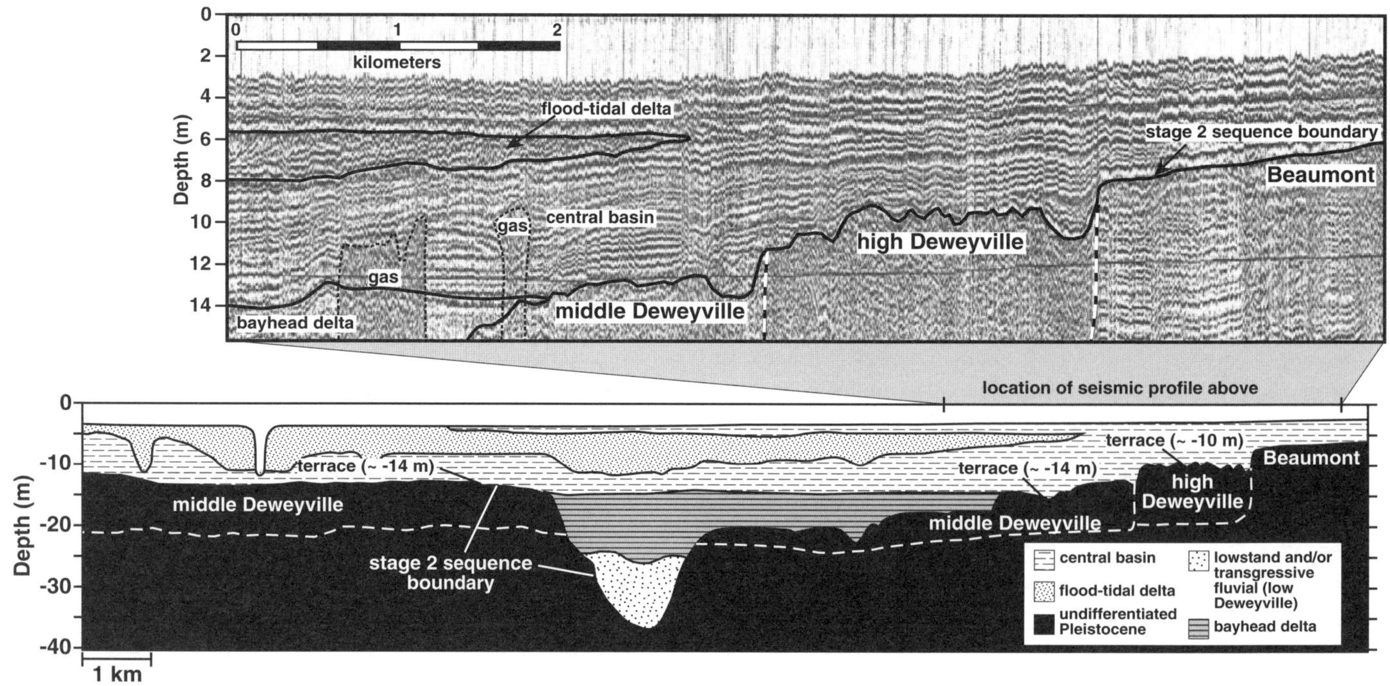

Figure 5. Cross section of Galveston Bay based on seismic data and sediment cores. Also shown is a representative seismic section showing fluvial terraces and onlapping acoustically laminated bay deposits. The deepest part of the valley was cut during the maximum lowstand in sea level. See Figure 4 for location.

Fluvial Facies

Fluvial deposits are confined to the deepest part of the valley and are recognized in seismic records by channels with chaotic seismic facies that cut into subparallel, high-amplitude reflections interpreted as floodplain deposits (Fig. 6). These seismic facies have been used to map the distribution of fluvial deposits within the bay (Smyth et al., 1988). The fluvial section averages 10 m in thickness. Although several drill cores penetrated the fluvial section, only a few undisturbed cores were obtained. Those cores recovered gravelly sand, sand, and mud containing reworked (Cretaceous and Tertiary) foraminifera (Rehkemper, 1969). Channel deposits also contain calcareous nodules reworked from Pleistocene deposits.

Marsh Facies

Marsh deposits consist of medium to dark gray, organic-rich silty clay and peat. Macrofauna are rare and consist of barnacles, and oyster and other bivalve shell fragments that occur in the organic-rich clay. The microfauna within these clays include rare benthic foraminifera, including *Ammonia beccarii*, *Elphidium gunteri*, *Elphidium delicatulum*, and *Trochammina* spp. (Wantland, 1969). The combined fauna indicate brackish rather than freshwater conditions (Rehkemper, 1969).

Trinity Bayhead Delta Facies

Seismic profiles from offshore of the modern Trinity bayhead delta image a seismic facies that consists of subparallel, medium-amplitude, horizontal and oblique reflections (Fig. 6). Channels, with width to depth ratios similar to modern distributary channels, are common and are generally characterized by concave, moderate to strong reflectors. The modern delta has prograded into water depths of 3–4 m, so the angle of clinoforms is quite gentle and difficult to resolve in seismic records.

A detailed sedimentological study of the Trinity delta by McEwen (1969) revealed a fluvial-dominated delta succession composed of prodelta clay grading upward into distal bar and sandy mouth-bar facies capped by organic-rich delta-plain deposits (Fig. 7). Cores that sampled the prodelta recovered greenish-gray mud with thin silt layers and scattered *Rangia cuneata* shells. The distal bar consists of interbedded sand and mud with minor bioturbation. The mouth bar consists of quartz sand with variable amounts of mud matrix, and it is highly bioturbated. Delta-plain deposits include *Rangia*-bearing mud in interdistributary bays and organic-rich mud and peat. The foraminiferal assemblage is dominated by *Ammonia beccarii*, with lesser numbers of *Elphidium* spp., *Miliammina fusca*, and *Ammotium salsum* (Brann, 1969).

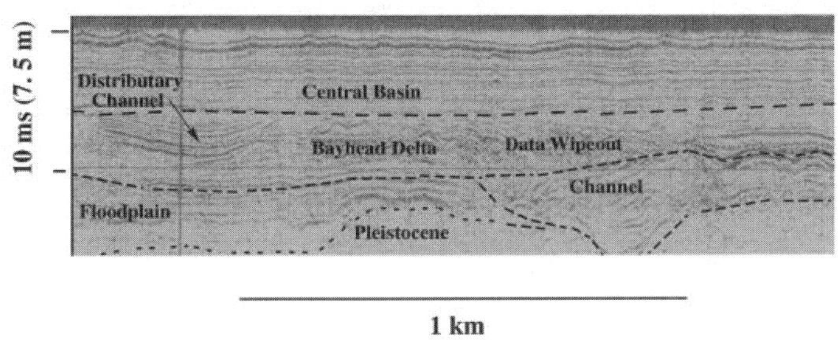

Figure 6. Representative seismic (boomer) profile illustrating seismic facies associated with the fluvial, bayhead delta, and central-basin environments.

Historical maps of the Trinity bayhead delta show active distributary channels and lobes to the west of the delta. This part of the delta has since been reworked by waves into a relatively straight shoreline (Fig. 2). Sediment cores collected just offshore of the shoreline sampled well-sorted sand with *Rangia* shells.

Interdistributary Bays and Lakes

Several small bays and lakes, including Lake Anahuac, occur in the upper part of Galveston Bay and along the north shore of East Bay and West Bay (Fig. 1). Sediment cores from these bays and lakes recovered organic-rich mud with *Rangia cuneata* shells and rare microfossils dominated by *Miliammina fusca* (Brann, 1969).

Central Basin

The diagnostic seismic facies of the central basin consists of horizontal, continuous, moderate- to low-amplitude reflections that onlap the valley margins (Figs. 5 and 6). Sediment cores that sampled the central-basin seismic facies recovered homogenous gray silty and sandy clay. Thin sand beds, generally less than 1 cm thick, occur along the margin of the bay. The macrofauna is dominated by *Crassotrea virginica* and *Mulinia cuneata*. Oyster reefs are mainly confined to the middle part of the bay where salinities range between 10 and 20 parts per thousand. The microfossil assemblage is dominated by the benthic foraminifera *Elphidium* spp. and *Ammonia beccarii*, with lesser numbers of *Trochammina* spp., *Elphidium delicatulum*, *Palmerinella gardenislandensis*, *Ammotium salsum*, and *Miliammina fusca* (Wantland, 1969).

Lower Estuary and Bolivar Tidal-Delta Complex

The 3-km-wide inlet between Bolivar Peninsula and Galveston Island is called Bolivar Roads (Fig. 1). Prior to dredging of the Galveston and Houston ship-channel entrance in the early 1900s, this was a natural inlet with prominent ebb-and flood-tidal deltas (Siringan and Anderson, 1993). Since the channel was dredged, the tidal deltas have shrunk to approximately one-half of their original size, and the inlet has been significantly modified.

Seismic profiles from the Bolivar tidal-delta complex show crosscutting and lateral accretion sets (Fig. 8). Reflections dip gently seaward within the ebb-tidal delta and gently landward within the flood-tidal delta (Siringan and Anderson, 1993).

A detailed sedimentological study of the modern Bolivar tidal-delta complex was conducted by Siringan and Anderson (1993). Sediment cores from the inlet sampled shell hash and minor sand beds. The inlet extends seaward and bayward into well-sorted sandy lobes, and these grade seaward and landward

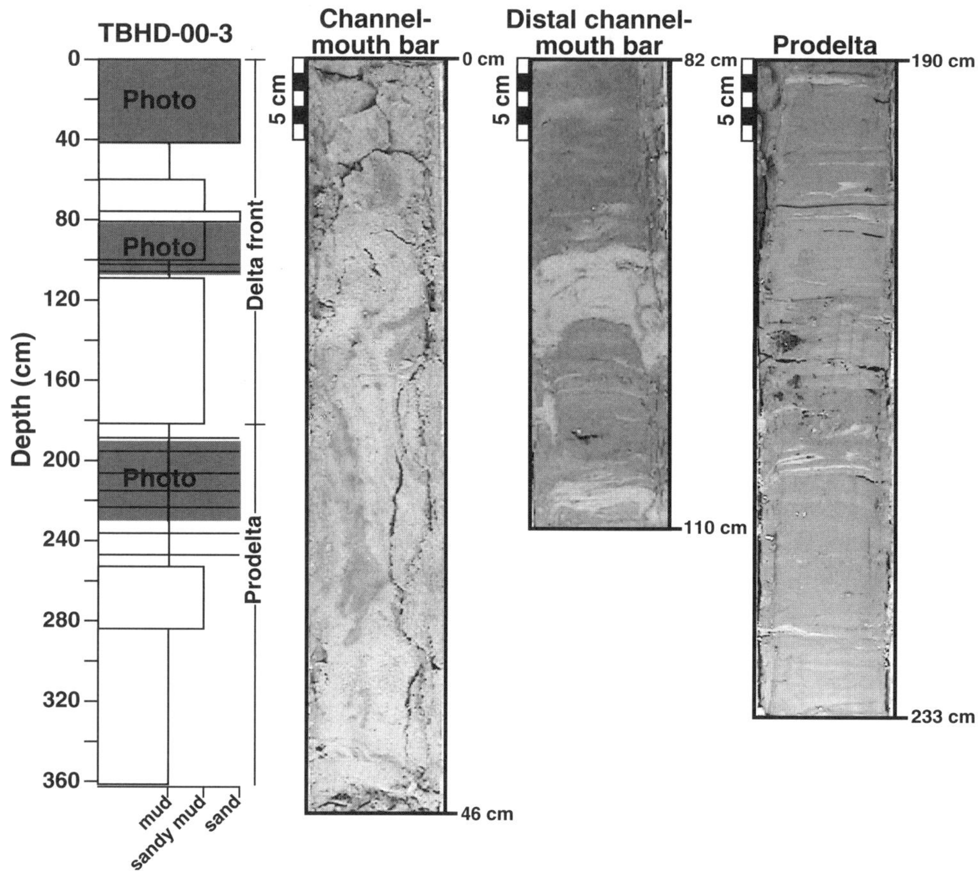

Figure 7. Generalized lithologic log for core TBHD-00-3 illustrating coarsening-upward bayhead delta sequence and core photographs illustrating bayhead delta lithofacies. Core locations are shown in Figure 2.

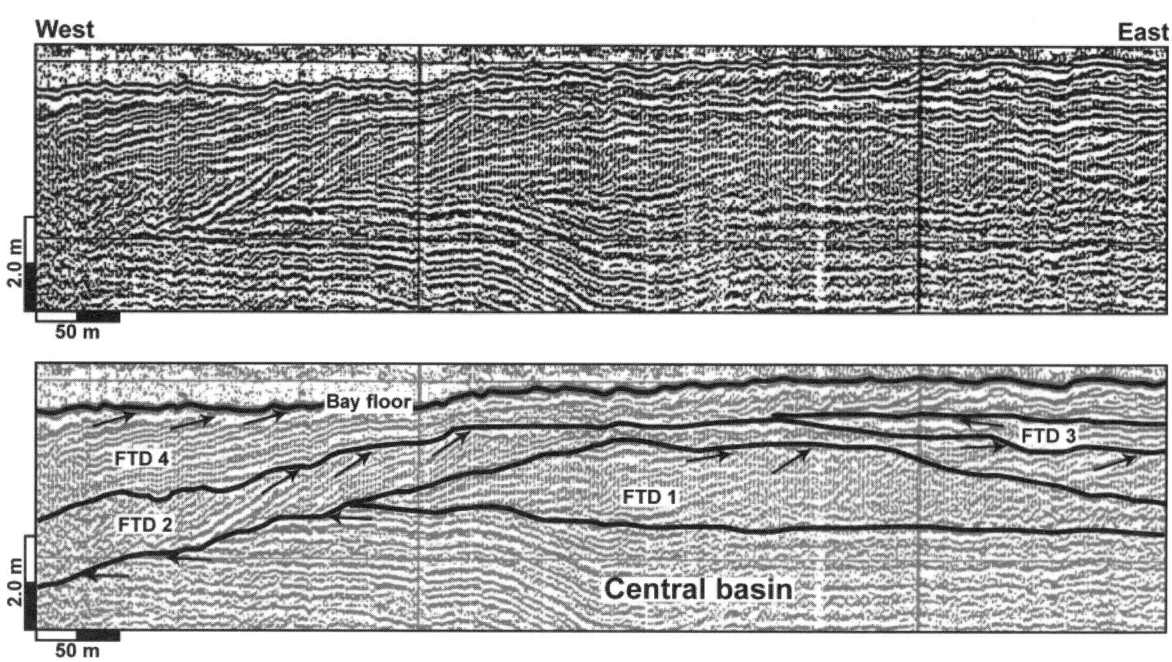

Figure 8. Representative seismic section illustrating seismic facies of the flood-tidal delta. FTD 1 through 4 designate different stages of inlet fill.

into tidal couplets composed of well-sorted sand and mud. The microfauna is more diverse than the mid-basin fauna, but it is still dominated by *Ammonia beccarii* and *Elphidium gunteri* (Wantland, 1969). Other, more diagnostic species include *Elphidium poeyanum*, *Elphidium matagordanum*, *Elphidium discoidale*, and several species of *Quinqueloculina*.

Coastal Barriers

Bolivar Peninsula. Bolivar Peninsula is a prograding spit shoreline characterized by prominent beach ridges that curve landward at the western end of the peninsula (Fig. 1). These ridges record westward accretion and infilling of the tidal inlet, which have gradually restricted the mouth of the estuary. The backside of the peninsula is dominated by washover and tidal-delta deposits that formed during the early phase of barrier evolution. Seismic profiles collected just north of the peninsula provide a record of the westward progradation of the peninsula (Fig. 9). These records also show a terraced erosion surface (tidal-ravinement surface) that reflects deepening of the inlet toward the west as it migrated over the Trinity River incised valley and its soft-sediment fill.

Sediment cores and sand pits through Bolivar Peninsula reveal barrier sands resting in sharp contact on back-barrier bay mud. A detailed study of the peninsula by Rodriguez et al. (2004) showed that it is a relatively young barrier, having formed after 2.5 ka.

Galveston Island. Galveston Island is a typical drumstick-shaped barrier island with prominent beach ridges that record the history of barrier progradation (Bernard et al., 1959; Rodriguez et al., 2004). The barrier consists of well-sorted quartz sand with thin interbeds of organic-rich clay, representing beach ridge and swale deposits, respectively. These proximal barrier deposits rest above and grade seaward into more distal, muddy, lower-shoreface deposits (Bernard et al., 1959). Radiocarbon ages indicate that the barrier prograded seaward between 5.5 and 1.2 ka (Bernard et al., 1959), followed by landward retreat after 1.2 ka (Rodriguez et al., 2004). The change from progradation to retrogradation occurred as the offshore sand supply was exhausted.

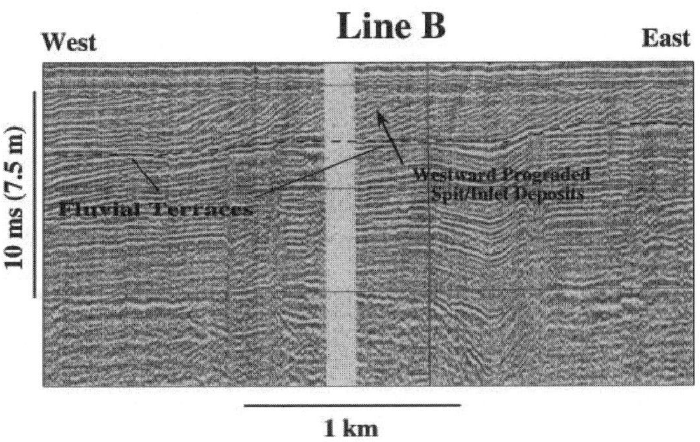

Figure 9. Portion of a seismic profile collected in East Bay, just north of Bolivar Peninsula, showing westward accreted flood-tidal delta and inlet deposits downlapping onto fluvial terraces.

Longitudinal Bays

Two longitudinal bays, East Bay and West Bay, exist on either side of Galveston Bay (Fig. 1). Seismic data from East Bay show acoustically laminated deposits onlapping fluvial terraces (Fig. 5). Sediment cores from the central and eastern part of the bay sampled oyster-bearing mud, which rests in sharp contact on Pleistocene deposits. Combined seismic data, sediment cores, and borings indicate that the estuary formed above old fluvial terraces and that sediments thicken toward the west over these terraces.

West Bay also rests on a terraced Pleistocene surface, and the thicknesses of the sediment fill and Galveston Island deposits decrease toward the west, away from the valley axis. Sediment cores from the eastern part of West Bay sampled estuarine mud resting directly on Pleistocene clay, except near the backside of the island, where washover sands occur. In the western half of the bay, bay mud rests on red clay that was deposited by the Brazos River when it flowed through what is now San Luis Pass.

Radiocarbon Chronology

Aten (1983) attempted to measure reservoir effects on radiocarbon ages for the upper Texas coast by dating charcoal and shell pairs that were believed to be contemporaneous. His work yielded a carbon reservoir for Galveston Bay of 225 ± 103 ^{14}C yr. Later, Ensor (1999) reexamined some of Aten's samples along with new samples from Trinity Bay in an effort to better constrain the carbon reservoir for the area. His results produced a reservoir correction of 540 ± 110 ^{14}C yr. We used Ensor's correction before making the conversion to calendar years.

DISCUSSION

Flooding History of Galveston Bay

Using drill cores acquired along the axis of Galveston Bay, Rehkemper (1969) subdivided strata into transgressive and regressive facies, based mainly on differences in sediment properties and foraminifera. Radiocarbon dates were used to constrain the age of these facies and to demonstrate that the overall change from transgressive to regressive facies occurred ~6000 yr ago, when the rate of sea-level rise decreased.

The detailed structure map of the Trinity–San Jacinto paleovalley (Fig. 4), in conjunction with the sea-level curves for the Gulf of Mexico (Toscano and Macintyre, 2003; Törnqvist et al., 2004; Milliken et al., this volume, Chapter 1) are used to help constrain the flooding history of the estuary. The valley has a terraced morphology. Thus, even with a relatively uniform rate of sea-level rise, flooding may have been episodic, where more rapid inundation took place as sea level reached the elevation of relatively flat fluvial terraces (Rodriguez et al., 2005).

High-resolution (boomer and chirp) seismic data reveal prominent surfaces that separate different seismic facies (Fig. 6). Long sediment cores through the valley fill sampled these surfaces, and they provide details about the sedimentary environments represented by the seismic facies and the extent of environmental change associated with these surfaces. With few exceptions, the surfaces separate more basinward facies above from more landward facies below, so they are flooding surfaces. What is most striking about these surfaces is the magnitude of environmental change indicated by the facies they bound. This is illustrated in Figure 10, which shows one of the more prominent flooding surfaces. Sediment core FR 1 sampled this interval and recovered greenish gray mud with oyster shells resting in sharp contact on organic-rich mud with thin sand layers and *Rangia cuneata* shells. These units are interpreted as central-basin deposits resting on delta-plain deposits. Distal bar and distributary mouth-bar deposits, which are the dominant facies in the modern bayhead delta (Fig. 6) (McEwen, 1969), are missing. A second example of abrupt changes in depositional environments is recorded in sediment cores TV99-2 and TV99-3 (Fig. 11). In these cores, oyster- and *Rangia*-bearing mud rests sharply on organic-rich clay at 730 cm in core TV99-3 and at 620 cm in core TV-99-2. These units are interpreted as central-basin and prodelta deposits resting on delta-plain deposits. In the modern estuary and prodelta, these facies are separated by 2–3 m water depth.

Given the magnitude of change associated with flooding surfaces, our next step was to acquire radiocarbon dates above and below these surfaces, preferably in more than one location, to constrain the timing and duration of flooding events. Figure 12 is a cross section along the axis of Galveston Bay that was compiled from the seismic data and cores. Cores TBHD-5-1 and TIV-00-1 actually sampled the deepest part of the valley, so the elevation of the bay line is lower at these sites relative to sites TV99-2 and TV99-3 in the upper part of Trinity Bay. The surfaces shown in Figure 12 were used to construct paleogeographic maps that depict the estuarine setting at various times, focusing on the main flooding events (Figs. 13A and 13B).

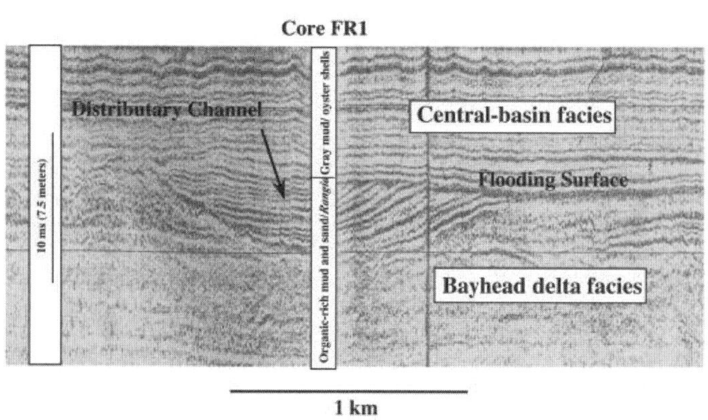

Figure 10. Representative seismic profile showing abrupt change in seismic facies, with central-basin facies above bayhead delta facies. Core FR sampled delta-plain deposits below, and they are in sharp contact with central-basin deposits at the level of the seismic unit boundary. Thus, this boundary represents a flooding surface.

The combined data indicate that initial flooding of the of the Trinity–San Jacinto valley to create what is now Galveston Bay took place just prior to ca. 9600 yr B.P. (Fig. 12). The flooding surface that records this event in the lower part of the bay (sites TGB-D and GB-00-1) has brackish upper-estuarine facies resting on fluvial deposits. This surface thus records initial flooding of the valley and is the bay line surface. This flooding event was apparently rapid, based on the relatively flat nature of the surface and the sharp contact between fluvial and upper-estuarine deposits in sediment cores GB-00-1 and TGB-D. The estuary that was formed as a result of this initial flooding was long and narrow, and its bayhead delta occupied the southern half of what is now Galveston Bay (Fig. 13A).

The organic-rich clay unit that overlies the bay line flooding surface is between 4 and 8 m thick and is confined to the deeper, narrow part of the valley. The unit has a laminated acoustic character, and sediment cores that sampled it contain alternating thin peat layers and clay near the base of the unit. Clay content increases and organic content decreases upward in the section, implying progressive flooding. Rare *Rangia flexuosa* shell fragments and arenaceous foraminifera (*Milliammina fusca*) occur in the upper part of the unit. The combined data indicate that this unit was deposited in an upper-estuarine setting, similar to modern interdistributary bays of the bayhead delta. Sandy distributary mouth-bar and distributary-channel facies are notably absent.

A prominent step in the bay line surface occurs at the location of core site SO2 (Fig. 11). This is at the approximate location where the Trinity and San Jacinto valleys merge (Fig. 4). It is also situated within the narrowest part of the bay where a prominent Pleistocene shoreline is known to exist. The step records a period of fluvial aggradation of ~4 m that occurred between ca. 9600 and ca. 8600 cal yr B.P. This is consistent with the changes in sea level that occurred during this time interval (Milliken et al., this volume, Chapter 1). Thus, fluvial aggradation within the narrow valley kept pace with the rate of sea-level rise during this time interval.

Circa 8200 cal yr B.P., the rate of creation of accommodation space exceeded the rate of fluvial aggradation, and the bay line shifted ~10 km up they valley to about the location of site FR (Fig. 12). At the same time, the bayhead delta shifted up the valley an equal distance. By the end of this flooding event, the upper part of the valley was occupied by brackish marsh and bayhead delta environments, and an open-estuary setting existed south of site TGB-2 (Figs. 12 and 13A). Following this flooding event, the bay line shifted slowly to the north, and the bayhead delta aggraded.

Another prominent flooding surface at ~–10 m below sea level can be traced throughout the bay using seismic records and cores (Fig. 12). This event is manifested in the upper estuary by central-basin mud resting sharply on delta-plain and interdistributary bay deposits (Fig. 10). The age of this surface

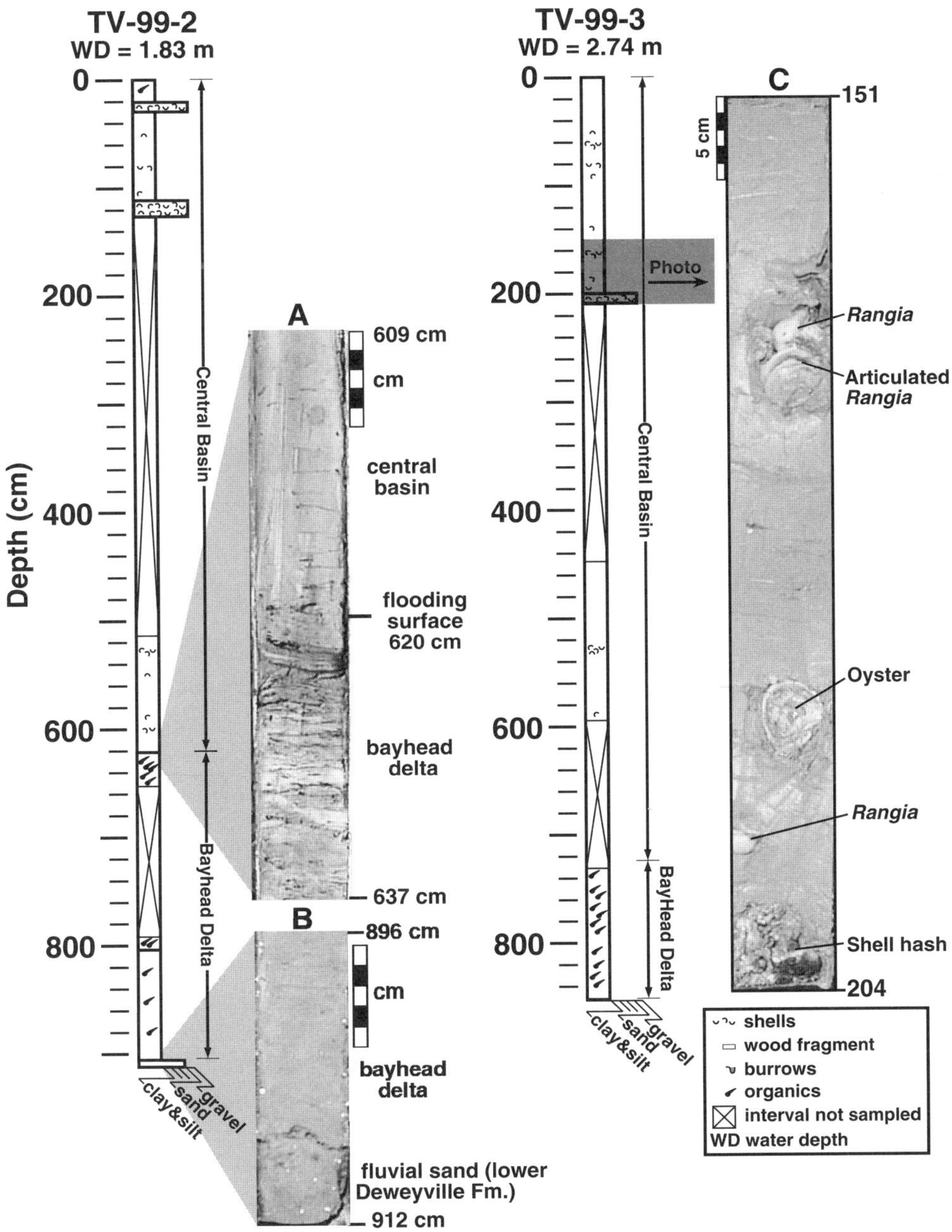

Figure 11. Lithologic logs for sediment cores TV-99-2 and TV-99-3 from Galveston Bay showing sharp facies changes in these cores. At the base of core TV-99-2, oxidized fine-grained fluvial sand is overlain by sediment that has abundant plant debris and lacks shells. The latter sediment type occurs today on the surface of the Trinity bayhead delta plain. Above 625 cm in core TV-99-2 and 730 cm in core TV-99-3, there is greenish gray mud with articulated *Rangia cuneata*, which is interpreted as a prodelta deposit. The contact between the delta-plain and prodelta deposits is a flooding surface that represents 2–4 m of sea-level rise, based on modern-day distributions of these facies.

is best constrained at site GB-00-1 in the lower part of the bay, where central-basin deposits rest on bayhead delta deposits. Two samples from just below the contact between bayhead delta and central-basin deposits yielded radiocarbon ages of 7740 and 7360 cal yr B.P. At site TV99-3 in the upper part of the estuary, an open-bay setting was established prior to 7370 cal yr B.P. Extrapolation of sedimentation rates at both sites yields an age between ca. 7700 and ca. 7400 cal yr B.P. for this flooding surface. Following this flooding episode, the upper bay–bayhead delta had shifted up the valley nearly 25 km at an unprecedented rate of 8.0 km/century (Fig. 13A). This was the most dramatic flooding event in the history of the estuary. At ca. 7100 cal. yr B.P., the gulf shoreline was located ~50 km offshore of its present location (Rodriguez et al., 2004). So, the bay was more than 100 km long after this flooding event.

The formation of Galveston Island and Bolivar Peninsula occurred over a period of several thousand years, beginning ca. 5500 yr B.P. in the case of Galveston Island and after ca. 2500 yr B.P. in the case of Bolivar Peninsula (Rodriguez et al., 2004). Prior to the formation of the Bolivar and Galveston barriers, a large tidal delta occupied the mouth of the bay, which at that time was ~25 km long (Rodriguez et al., 1998) (Fig. 13B). Cores TGB-D, GB-01-3, and TGB-2 sampled these flood tidal-delta deposits. Core TGB-D sampled up to 6 m of sand and

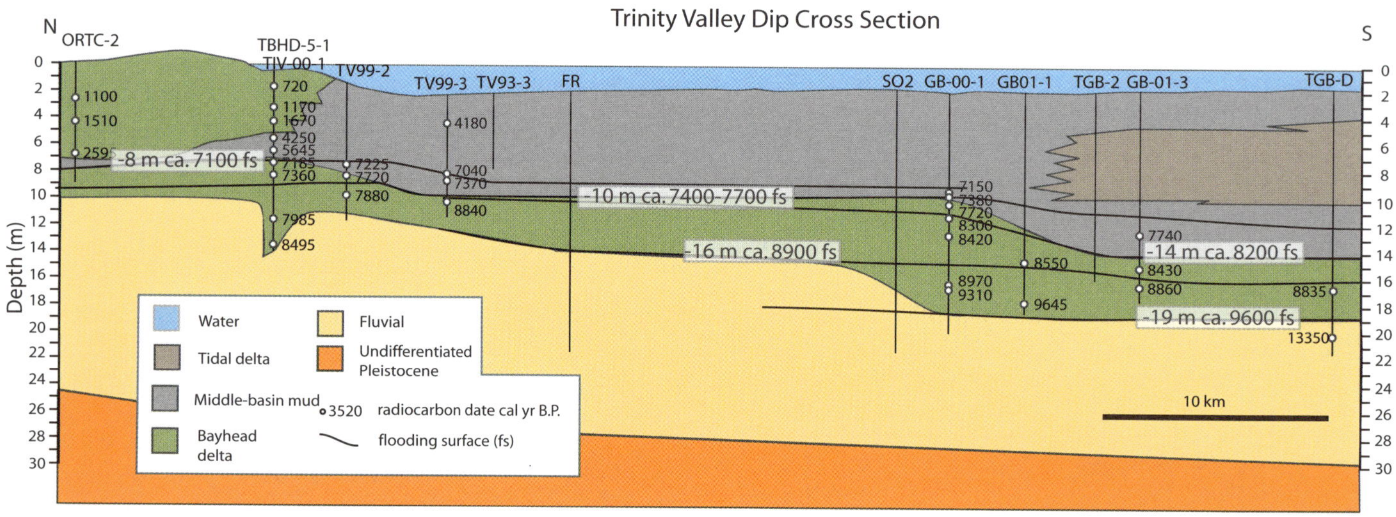

Figure 12. Axial section for Galveston Bay compiled from a seismic line and drill cores acquired along the valley. See Figure 3 for locations. Also shown are the main sedimentary facies, flooding surfaces, radiocarbon ages, and isochrons.

sand-mud couplets. The other cores sampled more distal tidal-delta facies interbedded with bay mud. No radiocarbon dates were obtained from this deposit, and seismic records indicate that it has an erosional base, so its age is not constrained.

As Bolivar Peninsula grew toward the west, the mouth of the estuary grew narrower, and by 2500 yr B.P., the inlet between the barriers was less than 10 km wide (Fig. 13B). As the estuary mouth grew narrower, the flood tidal delta shrank to its present size, which is a fraction of its size prior to 2500 yr B.P. (Rodriguez et al., 1998). The restriction of the inlet and associated change in tidal influx to the estuary occurred gradually, so the record of this change is hardly noticeable in middle-estuary cores that span this time interval. Progradation of the Trinity River bayhead delta into its present upper-estuary location occurred after ca. 2600 cal yr B.P., resulting in a delta package that is ~7 m thick at drill site ORTC-2 (Fig. 12).

Resolving Different Mechanisms for Bay Flooding Events

The combined data indicate that the Holocene evolution of the Galveston estuary complex was punctuated by abrupt flooding events. The questions we ultimately want to answer are (1) what caused these events, and (2) what is the probability that similar changes will occur in the next few centuries, given different scenarios for accelerated sea-level rise and reductions in sediment supply? To address these questions, we must first identify the forcing mechanisms responsible for the different flooding events.

Thomas and Anderson (1994) interpreted the flooding surfaces within the offshore Trinity-Sabine valley as having been caused by rapid increases in the rate of sea-level rise. This interpretation is consistent with the approximate depths of these surfaces and known intervals of rapid sea-level rise prior to ca. 11,000 yr B.P. (Bard et al., 1990). The initial flooding of the onshore portion of the valley at ca. 9600 yr B.P. occurred at a time when sea level was rising rapidly and perhaps episodically (Milliken et al., this volume, Chapter 1). The 8200 yr flooding event also occurred during a time of rapid sea-level rise. This flooding event may be correlative to the rapid sea-level event at ca. 8200 cal yr B.P. of Törnqvist et al. (2005). Within the study area, this flooding surface occurs at the approximate elevation of a prominent middle Deweyville terrace at −14 m (Fig. 5). Hence, this flooding event was amplified, at least in part, by flooding of relatively flat antecedent topography (Rodriguez et al., 2005). When the terrace was flooded, the estuary expanded rapidly, and there was a major shift in the locations of upper-estuary and central-basin environments (Fig. 13A). Inundation and erosion of extensive low-lying marshes and floodplains resulted in large amounts of organic material being delivered to the estuary and deposition of organic-rich upper-estuary facies. Flooding was also amplified by the increase in accommodation space for sedimentation brought about by expansion of the estuary. If the rate of sediment supply to the estuary remained constant, that sediment would have been dispersed over a larger area of the estuary. This is indicated by a dramatic decrease in sedimentation rates in the upper estuary at ca. 8500 cal yr B.P., from 4.6 mm/yr to 1.3 mm/yr. The 4.6 mm/yr rate approximately matches the rate of sea-level rise at this time (Milliken et al., this volume, Chapter 1). The 1.3 mm/yr rate is well below the rate of sea-level rise.

The most widespread flooding event occurred ca. 7700–7400 cal yr B.P. as the rate of sea-level rise was decreasing from an average rate of 4.2 mm/yr to 1.4 mm/yr (Milliken et al., this volume, Chapter 1). The ca. 7700–7400 yr flooding surface occurs at ~10 m below sea level, which corresponds to the elevation of another high Deweyville terrace (Fig. 5). So, antecedent

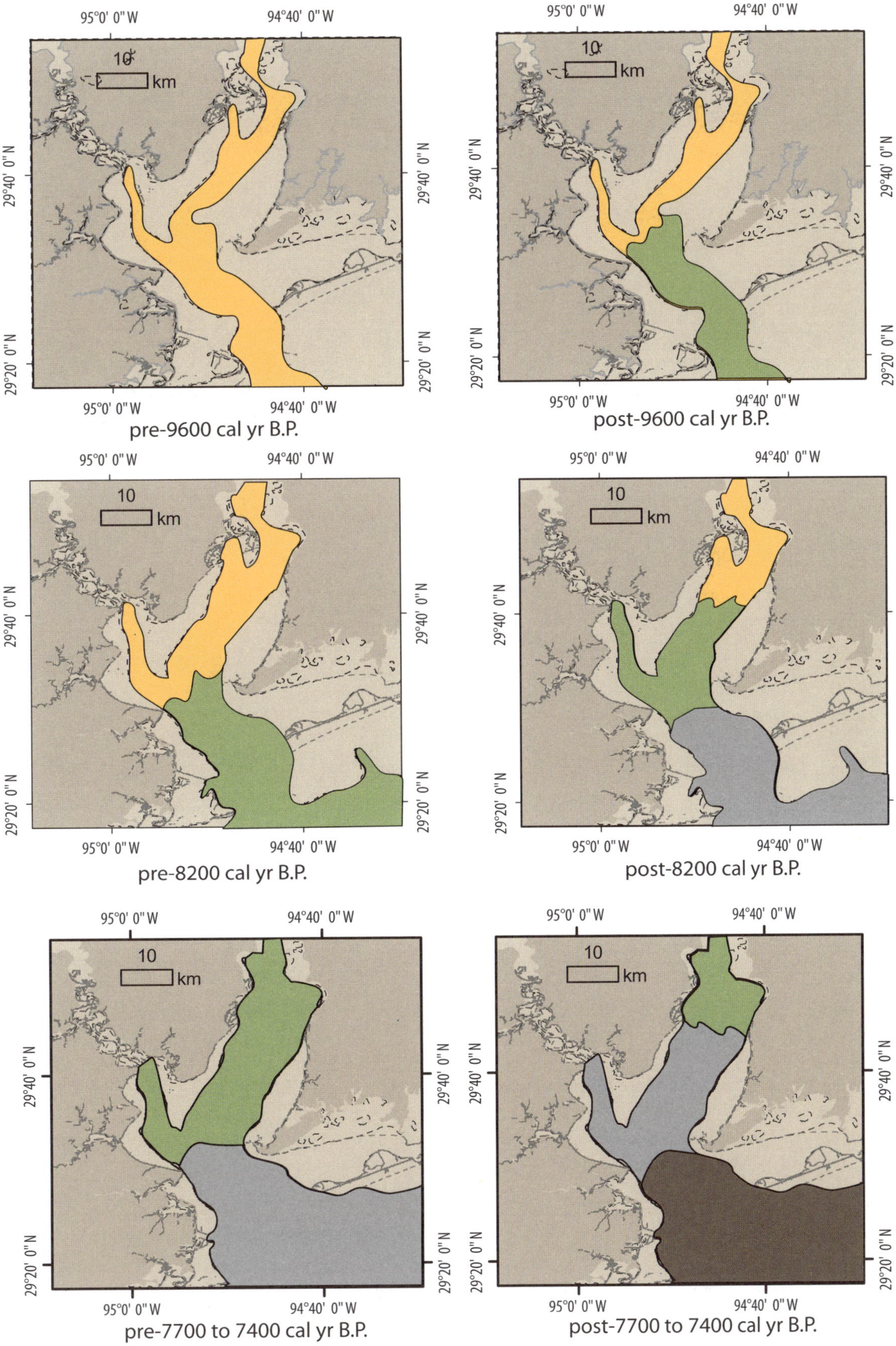

Figure 13 (*on this and following page*). Paleogeographic maps showing stages of evolution of Galveston estuary complex before and after major flooding events.

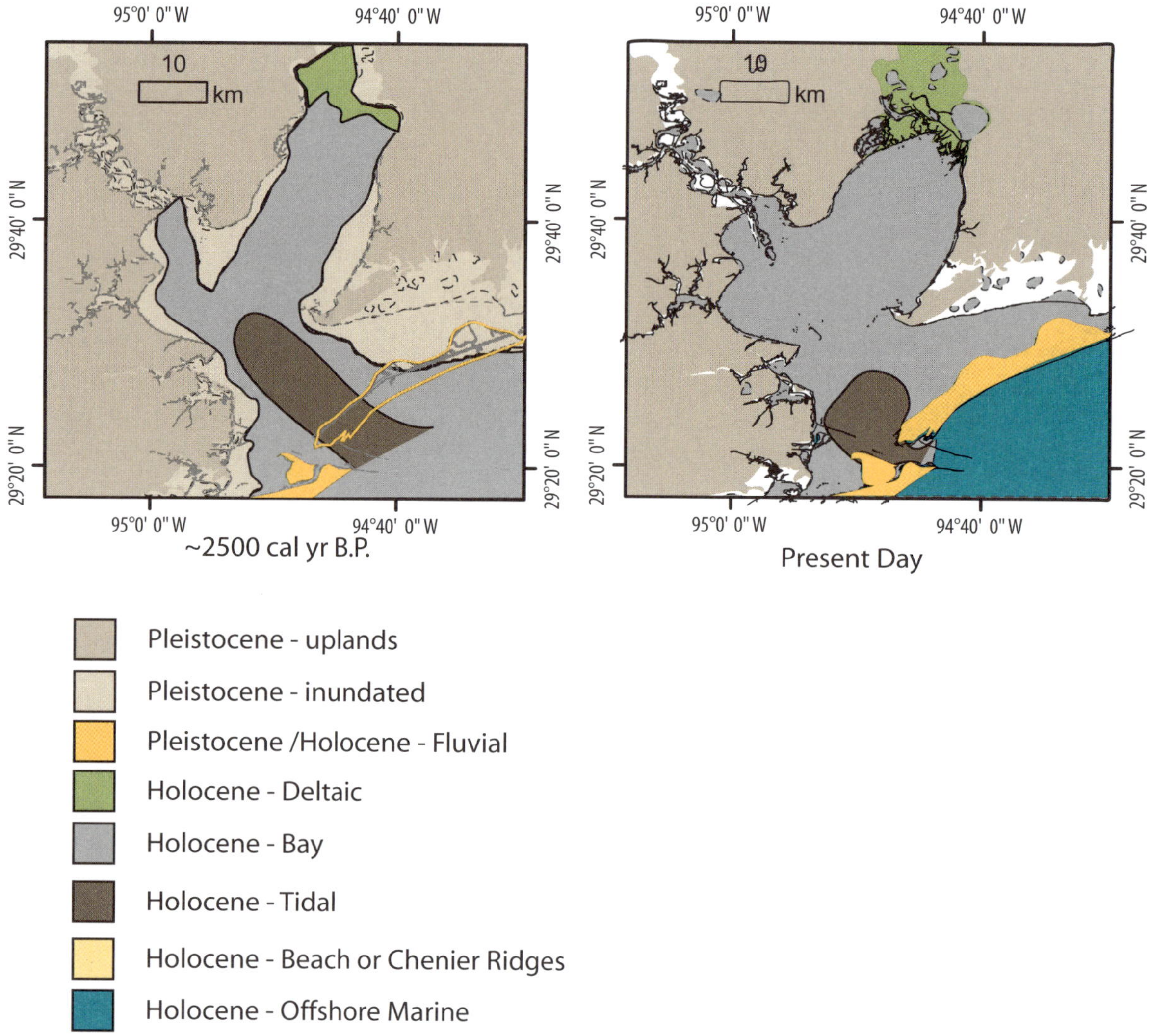

Figure 13 (*continued*).

topography contributed to the flooding of the estuary during this time interval. However, the extent and magnitude of flooding call for additional controls.

Major flooding events in Matagorda Bay to the west and Sabine Lake to the east also occurred around the ca. 7700–7400 cal yr B.P. time interval. In Matagorda Bay, the setting was transformed from an extensive bayhead delta that occupied most of the estuary to an open-estuary setting. Sedimentation rates in the estuary decreased significantly at this time (Maddox et al., this volume). In Sabine Lake, the bayhead delta shifted up the valley ~20 km, and the area of the estuary increased by ~30% between ca. 7300 and ca. 7100 cal. yr B.P. (Milliken et al., this volume, Chapter 4). This narrows the field of possible causes for this flooding event to a reduction in sediment supply to these estuaries.

While the Holocene climate history of east Texas is poorly known, it is generally accepted that the climate was in transition from cool and moist to warm and dry during the early Holocene (Bryant, 1977; Knox, 1983; Toomey et al., 1993; Humphrey and Ferring, 1994; Nordt et al., 1994, 2002; Bense, 1995). This suggests that climate change, and associated vegetation changes within the different drainage basins, led to a significant decrease in sediment supply to the east Texas estuaries (Sabine Lake, Galveston Bay, and Matagorda Bay).

The Trinity River bayhead delta started to prograde into the upper estuary ca. 2600 cal yr B.P., well after the rate of sea-level rise had decreased to between 0.4 and 0.6 mm/yr (Milliken et al., this volume, Chapter 1) (Fig. 13). The rate of growth of the delta increased dramatically after ca. 1600 cal yr B.P., which indicates an increase in sediment supply, perhaps related to climate change. This increase may also mark the beginning of agricultural activity by native Americans in the drainage basin of the river.

CONCLUSIONS

1. The Galveston estuary complex has had a history of rapid and dramatic changes in response to sea-level rise, flooding of the irregular topography of the ancestral Trinity River valley, and variations in sediment supply.
2. As the ancestral Trinity and San Jacinto River valleys were flooded, the estuary complex was transformed from a narrow and deep estuary, occupied mainly by upper-bay and bayhead delta environments, to the present wide, rounded, open-estuary setting.
3. Radiocarbon ages indicate that flooding events lasted only a few centuries and took place ca. 9600, ca. 8200, and between ca. 7700 and ca. 7400 cal yr B.P.
4. The ca. 9600 and ca. 8200 cal yr B.P. flooding events occurred when sea level was rising rapidly (average rate of 5.0 mm/yr), and perhaps episodically. Flooding of fluvial terraces enhanced the ca. 8200 and ca. 7700–7400 yr B.P. flooding events.
5. The ca. 7700–7400 cal yr B.P. flooding episode resulted in an ~30% increase in the area of the Galveston estuary complex in about three centuries. This flooding event occurred as the rate of sea-level rise was decreasing and at about the same time as flooding events in Matagorda Bay and Sabine Lake. These events may have resulted, at least in part, from decreased sediment supplies to these estuaries as the climate of east-central Texas was in transition from cool and moist to warm and dry.
6. A dramatic increase in the growth of the Trinity bayhead delta subsequent to ca. 1600 cal yr B.P. indicates an increase in the rate of sediment supply to the bay, probably due to a climate change that influenced the drainage basin and possibly to agricultural activity by native Americans.
7. Current rates of relative sea-level rise, due to both an increased rate of eustasy and subsidence, are approaching those rates associated with early flooding events in the Galveston estuary complex (4.2 mm/yr). This indicates that the Trinity bayhead delta is highly vulnerable to flooding in this century.

ACKNOWLEDGMENTS

We wish to thank David Krantz and Bill White for their thoughtful and helpful reviews of this manuscript. This research was funded by the National Science Foundation, grant EAR-0107650. We thank Craig Meyer, Scott Meyers, Chip Anderson, David Heroy, and Julia Wellner for their assistance in acquiring drill cores and seismic data used in this investigation.

REFERENCES CITED

Anderson, J.B., 2005, Diachronous development of late Quaternary shelf-margin deltas in the northwestern Gulf of Mexico: Implications for sequence stratigraphy and deep water reservoir occurrence, *in* Giosan, L., and Bhattacharya, J., eds., River Deltas—Concepts, Models and Examples: Society of Sedimentary Geology Special Publication 83, p. 257–278.

Aten, L.E., 1983, Indians of the Upper Texas Coast: New York, Academic Press, 370 p.

Bard, E., Hamelin, B., and Fairbanks, R.G., 1990, U/Th ages obtained by mass spectrometry in corals from Barbados: Sea level during the past 130,000 years: Nature, v. 346, p. 456–458, doi: 10.1038/346456a0.

Bard, E., Hamelin, B., Arnold, M., Montaggioni, L., Cabioch, G., Faure, G., and Rougerie, F., 1996, Deglacial sea level record from Tahiti corals and the timing of global meltwater discharge: Nature, v. 382, p. 241–244, doi: 10.1038/382241a0.

Bense, J.A., 1995, Archaeology of the Southeastern United States: London, Academic Press.

Bernard, H.A., Major, C.F., and Parrot, B.S., 1959, The Galveston barrier island and environs: A model for predicting reservoir occurrence and trend: Transactions of the Gulf Coast Association of Geological Societies, v. 9, p. 221–224.

Blum, M.D., Morton, R.A., and Durbin, J.M., 1995, "Deweyville" terraces and deposits of the Texas Gulf Coastal Plain: Transactions of the Gulf Coast Association of Geological Societies, v. 45, p. 53–60.

Brann, B.C., 1969, Microfossils of the Trinity River delta, *in* Lankford, R.R., and Rodgers, J.J.W., eds., Holocene Geology of Galveston Bay Area: Houston, Houston Geological Society, p. 118–156.

Bryant, V.M., 1977, A 16,000 year pollen record of vegetational change in central Texas: Palynology, v. 1, p. 143–156.

Ensor, H.B., 1999, Comparative review and discussion of *Rangia cuneata* dating methods on the upper Texas coast: Bulletin of the Texas Archeological Society, v. 70, p. 467–478.

Fairbanks, R.G., 1989, A 17,000-year glacio-eustatic sea level record: Influence of glacial melting on the Younger Dryas event and deep-ocean circulation: Nature, v. 342, p. 637–642, doi: 10.1038/342637a0.

Frazier, D.E., 1974, Depositional Episodes: Their Relationship to the Quaternary Stratigraphic Framework in the Northwest Portion of the Gulf Basin: Austin, Texas, Bureau of Economic Geology, Geologic Circular 74-1, 28 p.

Hogg, A.G., Higham, T.F.G., and Dahm, J., 1998, ^{14}C dating of modern marine and estuarine shellfish: Radiocarbon, v. 40, p. 975–984.

Humphrey, J.D., and Ferring, C.R., 1994, Stable isotopic evidence for latest Pleistocene and Holocene climatic change in north-central Texas: Quaternary Research, v. 41, p. 200–213, doi: 10.1006/qres.1994.1022.

Knox, J.C., 1983, Responses of river systems to Holocene climates, in late-Quaternary environments of the United States, *in* Wright, H.E., Jr., ed., The Holocene, Volume 2: Minneapolis, University of Minnesota Press, p. 26–41.

Le Blanc, R.J., 1949, Unpublished Shell Oil Company Report 128, 27 p.

Maddox, J., Anderson, J.B., Milliken, K., and Rodriguez, A.B., 2008, this volume, The Holocene evolution of Matagorda–Lavaca Bay complex, Texas, *in* Anderson, J.B., and Rodriguez, A.B., eds., Response of Upper Gulf Coast Estuaries to Holocene Climate Change and Sea-Level Rise: Geological Society of America Special Paper 443, doi: 10.1130/2008.2443(07).

McEwen, M.C., 1969, Sedimentary facies of the modern Trinity delta, *in* Lankford, R.R., and Rogers, J.J.W., eds., Holocene Geology of the Galveston Bay Area: Houston, Houston Geological Society, p. 53–77.

Milliken, K.T., Anderson, J.B., and Rodriguez, A.B., 2008, this volume (Chapter 1), A new composite Holocene sea-level curve for the northern Gulf of Mexico, *in* Anderson, J.B., and Rodriguez, A.B., eds., Response of Upper Gulf Coast Estuaries to Holocene Climate Change and Sea-Level Rise: Geological Society of America Special Paper 443, doi: 10.1130/2008.2443(01).

Milliken, K.T., Anderson, J.B., and Rodriguez, A.R., 2008b, this volume (Chapter 4), Record of dramatic Holocene environmental changes linked to eustasy and climate change in Calcasieu Lake, Louisiana, USA, *in* Anderson, J.B., and Rodriguez, A.B., eds., Response of Upper Gulf Coast Estuaries to Holocene Climate Change and Sea-Level Rise: Geological Society of America Special Paper 443, doi: 10.1130/2008.2443(04).

Morton, R.A., and McGowen, J.H., 1980, Modern Depositional Environments of the Texas Coast: Austin, University of Texas at Austin Bureau of Economic Geology, Guidebook 20, 167 p.

Morton, R.A., Blum, M.D., and White, W.A., 1996, Valley fills of incised coastal plain rivers: Transactions of the Gulf Coast Association of Geological Societies, v. 46, p. 321–331.

Morton, R.A., Bernier, J.C., and Barras, J.A., 2006, Evidence of regional subsidence and associated wetland loss induced by hydrocarbon production, Gulf Coast region, USA: Environmental Geology, v. 50, p. 261–274, doi: 10.1007/s00254-006-0207-3.

Nelson, H.F., and Bray, E.E., 1970, Stratigraphy and history of the Holocene sediments in the Sabine–High Island area, Gulf of Mexico, *in* Morgan, J.P., ed., Deltaic Sedimentation Modern and Ancient: Society for Economic Paleontologists and Mineralogists (SEPM) Special Publication 15, p. 48–77.

Nordt, L.C., Boutton, T.W., Hallmark, C.T., and Waters, M.R., 1994, Late Quaternary vegetation and climate changes in central Texas based on the isotopic composition of organic carbon: Quaternary Research, v. 41, p. 109–120, doi: 10.1006/qres.1994.1012.

Nordt, L.C., Boutton, T.W., Jacob, J.S., and Mandel, R.D., 2002, C_4 plant productivity and climate-CO_2 variations in south-central Texas during the late Quaternary: Quaternary Research, v. 58, p. 182–188, doi: 10.1006/qres.2002.2344.

Otvos, E.G., 2004, Holocene gulf levels: Recognition issues and an autocyclic mechanism for parasequence formation: Galveston Estuary, Texas: Journal of Sedimentary Research, v. 75, p. 608–620.

Paine, J.G., 1993, Subsidence of the Texas coast: Inferences from historical and late Pleistocene sea levels: Tectonophysics, v. 222, p. 445–458, doi: 10.1016/0040-1951(93)90363-O.

Paine, J.G., and Morton, R.A., 1986, Historical Shoreline Changes in Trinity, Galveston, West and East Bays, Texas Gulf Coast: University of Texas at Austin Bureau of Economic Geology Geological Circular 86-3, 58 p.

Phillips, J.D., Slattery, M.C., and Musselman, Z.A., 2004, Dam-to-delta sediment inputs and storage in the lower Trinity River, Texas: Geomorphology, v. 62, p. 17–34, doi: 10.1016/j.geomorph.2004.02.004.

Rehkemper, J.L., 1969, Sedimentology of Holocene estuarine deposits, Galveston Bay, *in* Lankford, R.R., and Rogers, J.J.W., eds., Holocene Geology of the Galveston Bay Area: Houston, Houston Geological Society, p. 12–52.

Rodriguez, A.B., Anderson, J.B., and Bradford, J., 1998, Holocene tidal deltas of the Trinity valley: Analogs for exploration and production: Transactions of the Gulf Coast Association of Geological Societies, v. 48, p. 373–380.

Rodriguez, A.B., Anderson, J.B., Siringan, F.P., and Taviani, M., 2004, Holocene evolution of the east Texas coast and inner continental shelf: Along-strike variability in coastal retreat rates: Journal of Sedimentary Research, v. 74, p. 405–422, doi: 10.1306/092403740405.

Rodriguez, A.B., Anderson, J.B., and Simms, A.R., 2005, Terrace inundation as an autocyclic mechanism for parasequence formation: Galveston Estuary, Texas, USA: Journal of Sedimentary Research, v. 75, p. 608–620, doi: 10.2110/jsr.2005.050.

Simms, A.R., Anderson, J.B., Taha, P.Z., and Rodriguez, A.B., 2006, Over-filled versus under-filled incised valleys: Examples from the Quaternary Gulf of Mexico, *in* Dalrymple, R., Leckie, D., and Tillman, R., eds., Incised Valleys in Time and Space: Society of Sedimentary Geology Special Publication 84, p. 117–139.

Siringan, F.P., and Anderson, J.B., 1993, Seismic facies, architecture, and evolution of the Bolivar Roads tidal inlet/delta complex, east Texas Gulf Coast: Journal of Sedimentary Geology, v. 63, p. 794–808.

Smyth, W.C., Anderson, J.B., and Thomas, M.A., 1988, Seismic facies analysis of entrenched valley-fill: A case study of the Galveston Bay area, Texas: Transactions of the Gulf Coast Association of Geological Societies, v. 38, p. 385–394.

Thomas, M.A., and Anderson, J.B., 1994, Sea-level controls on the facies architecture of the Trinity/Sabine incised-valley system, Texas continental shelf, *in* Dalrymple, R., Boyd, R., and Zaitlin, B.A., eds., Incised Valley Systems: Origin and Sedimentary Sequences: Society for Economic Paleontologists and Mineralogists (SEPM) Special Publication 51, p. 63–82.

Toomey, R.S., III, Blum, M.D., and Valastro, S., Jr., 1993, Late Quaternary climates and environments of the Edwards Plateau, Texas: Global and Planetary Change, v. 7, p. 299–320, doi: 10.1016/0921-8181(93)90003-7.

Törnqvist, T.E., Gonzalez, J.L., Newsom, L.A., van der Borg, K., de Jong, A.F.M., and Kurnik, C.W., 2004, Deciphering Holocene sea-level history on the U.S. Gulf Coast: A high-resolution record from the Mississippi Delta: Geological Society of America Bulletin, v. 116, no. 7–8, p. 1026–1039, doi: 10.1130/B2525478.1.

Törnqvist, T.E., Bick, S.J., Gonzalez, J.L., van der Borg, K., and de Jong, A.F.M., 2005, Tracking the sea-level signature of the 8.2 ka cooling event: New constraints from the Mississippi Delta: Geophysical Research Letters, v. 31, L23309, p. 1–4

Toscano, M.A., and Macintyre, I.G., 2003, Corrected western Atlantic sea-level curve for the last 11,000 years based on calibrated ^{14}C dates from *Acropora palmata* framework and intertidal mangrove peat: Coral Reefs, v. 22, p. 257–270, doi: 10.1007/s00338-003-0315-4.

Wantland, K.F., 1969, Distribution of modern brackish-water foraminifera in Trinity Bay, *in* Lankford, R.R., and Rogers, J.J.W., eds., Holocene Geology of the Galveston Bay Area: Houston, Houston Geological Society, p. 93–118.

White, W.A., and Morton, R.A., 1997, Wetland losses related to fault movement and hydrocarbon production, southeastern Texas coast: Journal of Coastal Research, v. 13, p. 1305–1320.

White, W.A., Morton, R.A., and Holmes, C.W., 2002, A comparison of factors controlling sedimentation rates and wetland loss in fluvial-deltaic systems, Texas Gulf Coast: Geomorphology, v. 44, p. 47–66, doi: 10.1016/S0169-555X(01)00140-4.

Williams, S.J., Prins, D.A., and Meisburger, E.P., 1979, Sediment Distribution, Sand Resources and Geologic Character of the Inner Continental Shelf Off Galveston County, Texas: U.S. Army Corps of Engineers Miscellaneous Report 79-4, 159 p.

Manuscript Accepted by the Society 10 March 2008

Printed in the USA

The Geological Society of America
Special Paper 443
2008

The Holocene evolution of the Matagorda and Lavaca estuary complex, Texas, USA

Jessie Maddox
John B. Anderson
K.T. Milliken
Department of Earth Science, Rice University, 6100 Main Street, MS 126, Houston, Texas 77005, USA

Antonio B. Rodriguez
Institute of Marine Sciences, University of North Carolina at Chapel Hill, 3431 Arendell Street, Morehead City, North Carolina 28557, USA

Timothy M. Dellapenna
Department of Marine Sciences, Texas A&M University at Galveston, 1001 Texas Clipper Road, Galveston, Texas 77554, USA

Liviu Giosan
Woods Hole Oceanographic Institution, Geology and Geophysics, MS#22, 360 Woods Hole Road, Woods Hole, Massachusetts 02543, USA

ABSTRACT

Rotary-drill cores from the Matagorda and Lavaca estuary complex sampled six distinct lithofacies, and high-resolution seismic data reveal four main seismic facies. The facies architecture of the bay indicates a history of Holocene flooding punctuated by major landward shifts in estuarine environments. Radiocarbon ages help to constrain the rate and timing of these flooding events and indicate significant reorganization of the estuarine environments at a decadal to century time scale.

The Holocene evolution of the Matagorda and Lavaca estuary complex began ca. 11,600 yr B.P. with initial flooding of the portion of the ancestral Lavaca River incised valley that is now occupied by lower Matagorda Bay. Early flooding of the lowstand valley was punctuated by landward shifts in the river mouth, followed by episodes of fluvial aggradation. The most pronounced phase of aggradation occurred between ca. 11,600 and 9600 yr B.P., when sea level was rising rapidly, and a step in the valley floor slowed the rate of flooding. By ca. 9500 cal yr B.P., the deep, narrow part of the Lavaca incised valley had been flooded and was occupied by a bayhead delta. The river mouth shifted landward within Lavaca Bay between 8500 and 8200 cal yr B.P., followed by another landward shift in the delta between 7900 and 7700 cal yr B.P. These changes were mainly caused by variations in the rate of sea-level rise. The most dramatic change in the bay setting occurred between 7300 and 6700 cal yr B.P., when the bayhead delta stepped landward at least 30 km, establishing

Maddox, J., Anderson, J.B., Milliken, K.T., Rodriguez, A.B., Dellapenna, T.M., and Giosan, L., 2008, The Holocene evolution of the Matagorda and Lavaca estuary complex, Texas, USA, *in* Anderson, J.B., and Rodriguez, A.B., eds., Response of Upper Gulf Coast Estuaries to Holocene Climate Change and Sea-Level Rise: Geological Society of America Special Paper 443, p. 105–119, doi: 10.1130/2008.2443(07). For permission to copy, contact editing@geosociety.org. ©2008 The Geological Society of America. All rights reserved.

most of modern-day Matagorda Bay. This flooding event is attributed to a combination of rapid sea-level rise and to a reduction in sediment supply to the estuary that was associated with a change in the regional climate from cool and moist to warm and dry conditions.

During the past 6700 yr, as the rate of sea-level rise decreased from an average rate of 4.2 mm/yr to an average rate of 1.4 mm/yr, the bayhead delta of the Lavaca River retreated slowly up the valley as the present-day outline of Lavaca Bay formed. The most significant change in the estuary setting during the past 6700 yr was the development of the present coastal barrier system, which led to gradual restriction of tidal exchange and associated changes in salinity.

Keywords: estuary, sea level, climate change, coastal.

INTRODUCTION

The objective of this investigation was to study how the Matagorda-Lavaca estuary complex evolved when sea level was rising faster than today and when climate variability may have significantly altered sediment input to the estuary. This information will allow better predictions of the responses of these estuaries to future changes in the rate of sea-level rise and sediment supply, both natural and anthropogenic.

In general, the evolution of an estuary is controlled by the size and shape of the valley that is being flooded and the balance between the rate of sea-level rise and sediment supply to the valley. The Holocene history of sea-level rise along the northern Gulf of Mexico is constrained to within a meter or two (Toscano and Macintyre, 2003; Törnqvist et al., 2004; Milliken et al., this volume), and the geomorphology of the Lavaca incised valley is well established (Wright, 1980).

Here, we will demonstrate that the Matagorda and Lavaca estuary complex evolved in a highly episodic manner, where episodes involved flooding and major reorganization of the estuary setting. Radiocarbon dates are used to constrain the rate and timing of these events. Using the existing sea-level history for the northern Gulf of Mexico, a detailed geomorphic map of the Lavaca incised valley, information about neotectonic features in the area, and published information concerning the Holocene climate change for the region, we address the possible causes of flooding events.

STUDY AREA

The Matagorda and Lavaca estuary complex consists of Matagorda Bay and five auxiliary bays: Lavaca, Keller, Carancahua, Cox, and Tres Palacios Bays (Fig. 1). Collectively, they encompass an area of ~1014 km^2. Water depths average 2 m in Lavaca Bay and deepen to 4 m in Matagorda Bay. With a mean diurnal tidal range of 43 cm, Matagorda Bay is classified as a microtidal, wave-dominated estuary (Dalrymple et al., 1992).

Six prominent rivers and streams provide freshwater and sediment input into the estuary complex, the largest of which are the Lavaca and Navidad Rivers. Together, these rivers drain an area of ~6004 km^2. The headwaters of both rivers are located 120 km inland from the mouth of the river at an elevation of ~156 m, which characterizes them as small, coastal-plain rivers. The drainage basin of the rivers and streams that flow into Matagorda Bay and Lavaca Bay includes mostly unconsolidated Miocene, Pliocene, and Pleistocene deposits, and soils are primarily clay-rich (Fig. 2). The drainage basin is situated within a climate transition zone separating more humid conditions to the east from more arid conditions to the west (Bryant, 1977; Toomey et al., 1993; Nordt et al., 1994, 2002; Hall and Valastro, 1995). Thus, the area may be especially sensitive to variations in river and stream discharge and changes in vegetation, which control sediment supply. The estuary complex resides within the subhumid climatic zone, and precipitation averages 107 cm/yr. Monthly mean temperatures range from 10 °C (50 °F) during the winter to 32 °C (90 °F) during the summer. Prevailing winds are out of the southeast, although persistent winds from the northwest occur as fronts pass through the area during winter months.

METHODS

A complete stratigraphic record of estuary evolution exists only within the central confines of the Lavaca incised valley. Thus, in order to acquire the most accurate records of coastal change and valley flooding, it was imperative to collect data along the axis of the valley. Using the Pleistocene structural map of Wright (1980) as a guide, 110 km of seismic data were collected, including one line up the axis of the Lavaca paleo–fluvial valley and four strike lines, two each in Lavaca Bay and Matagorda Bay (Fig. 1). An Applied Acoustics Engineering 200 boomer system with a center frequency of 1.5 kHz provided the source for the survey. It transmitted 100 Joules every 500 ms, and data were recorded at a 150 ms scan rate. In the field, a bandwidth filter, consisting of a low pass of 2500 Hz and a high pass of 120 Hz, was applied to the data. Data were acquired digitally using a Delph seismic acquisition system. The digital data were reprocessed with a high-pass bandwidth filter of 300 Hz, and horizontal stacking was applied. In addition to uniboom data, chirp sonar

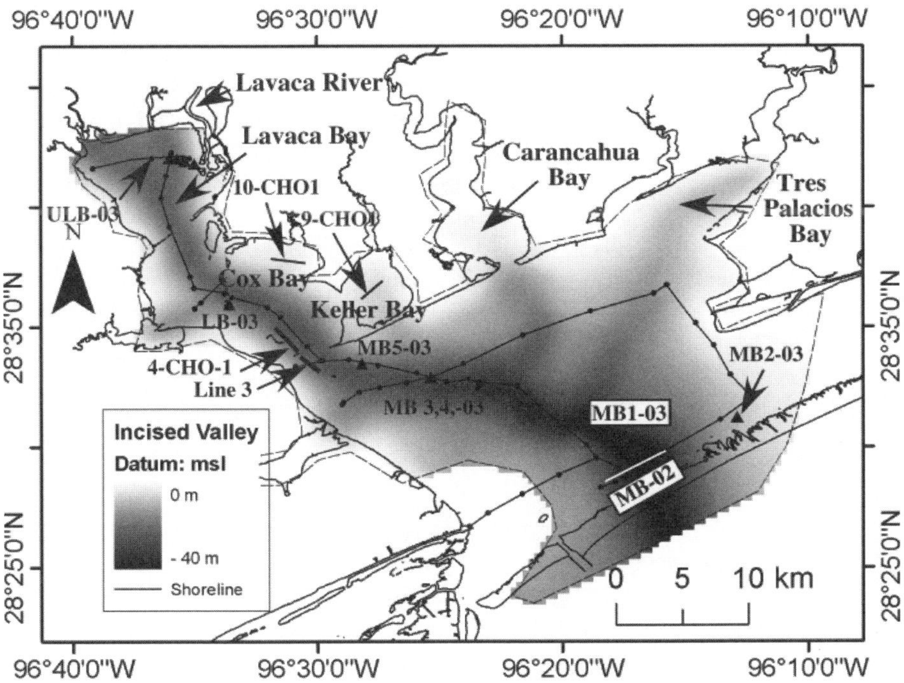

Figure 1. Digital elevation map of the Pleistocene exposure surface beneath Matagorda and Lavaca Bays (modified from Wright, 1980). Also shown are seismic track lines, core locations, and the locations of the seismic sections, which are illustrated with bold lines.

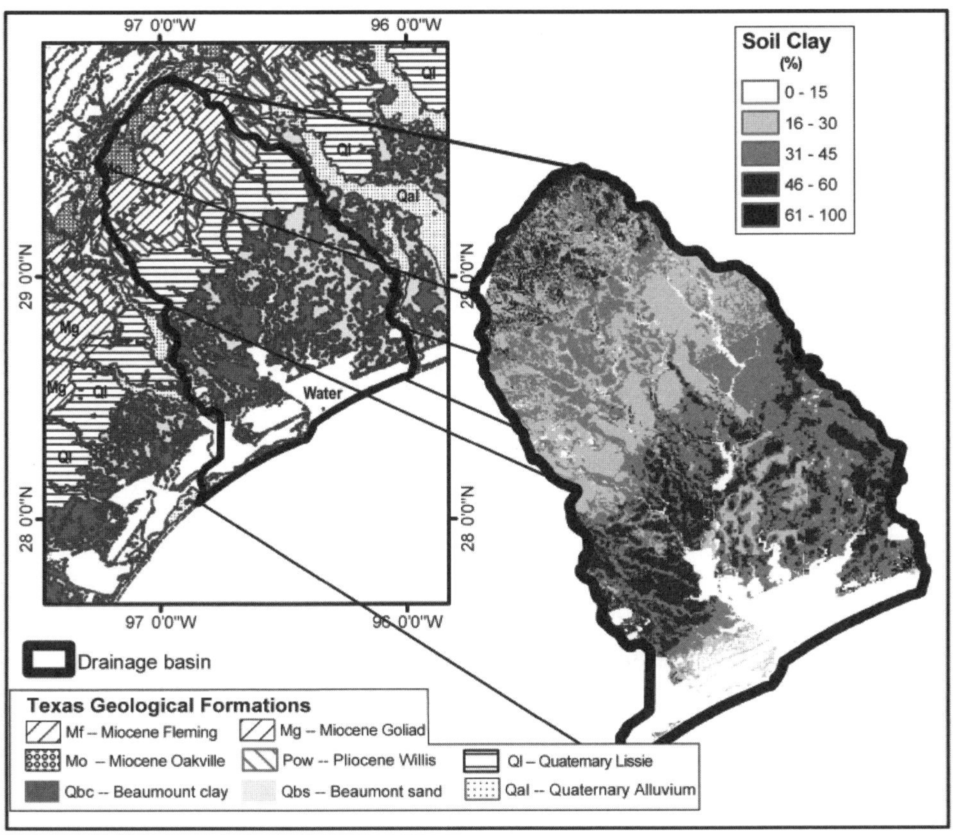

Figure 2. Geological and soils map of the Lavaca-Matagorda drainage basin (Stoeser et al., 2005; USDA, 2007). The drainage area encompasses Miocene- to Quaternary-age deposits. The clay percentage derived from the USDA soil database shows that the deposits in the upper part of the drainage basin have a lower clay content than the deposits in the lower portion of the drainage basin.

data were collected within Lavaca Bay using an Edgetech 216 Full Spectrum Subbottom Chirp tow fish with a frequency range of 2–12 kHz. Data were digitally acquired using Delph seismic software. Navigational control was provided by a global positioning system (GPS) that is accurate to within 10 m.

Four rotary-drill cores were collected within the central part of the Lavaca incised valley. The core locations were spaced at approximately equal distance from the northern extent of Lavaca Bay to the southern extent of Matagorda Bay (Fig. 1). A short core (1.5 m) on the eastern side of Matagorda Bay (MB2-03, Fig. 1) was also collected in an attempt to sample the Pleistocene-Holocene contact. Coring was accomplished using a hydraulic rotary-drill rig aboard the R/V *Trinity*. During the summer of 2004, an additional six vibracores were collected. Five of the cores ranged between 4 and 6 m in length and sampled the modern Lavaca River delta. Another vibracore (MB5-03) was collected from the northern part of Matagorda Bay in an attempt to acquire shells to radiocarbon date the youngest flooding surface in the bay (Fig. 1).

Sediment cores were split in the laboratory, and the archive halves were stored in D-tubes in a refrigerated core locker. The sample halves were photographed and described for sediment texture, composition, color, structures, fauna, organics, and carbonate concretions. To verify visual estimates of grain size, ~2 cm^3 samples were each disaggregated in 250 mL of water and then measured by a Malvern Mastersizer 2000. Additional samples were wet sieved through a 500 μm sieve for macrofaunal (primarily mollusks) identification. Discrete samples from two cores were scanned with an X-ray fluorescence whole-core scanner at Woods Hole Oceanographic Institution. Cores were correlated to seismic data using an average velocity of 1500 m/s. This correlation permitted independent validation of flooding and/or regressive surfaces interpreted in seismic data.

Radiocarbon (accelerator mass spectrometry [AMS]) analysis was used to establish a chronostratigraphic framework for estuarine deposits and to evaluate the timing and rates of environmental change within the Matagorda and Lavaca estuary complex. Twenty carbonate samples, taken primarily from articulated mollusks, and two organic carbon samples were sent to Woods Hole National Ocean Sciences Accelerator Mass Spectrometry Facility (NOSAMS), and radiocarbon analyses were conducted following procedures documented in *NOSAMS General Statement of ^{14}C Procedures* (1999) (Table 1). The reported dates were calibrated using Calib 4.0 (Stuiver and Braziunas, 1993; Stuiver and Reimer, 1993; Stuiver et al., 1998).

RESULTS

Valley Morphology

Through the use of 58 jet-down holes in Lavaca Bay and 12 seismic-reflection traverses in Matagorda Bay, Byrne (1975) contoured the Pleistocene exposure surface within these bays. He interpreted three separate fluvial valleys underlying Matagorda Bay: Keller, Carancahua, and Lavaca-Navidad, where the latter two are the largest. Working with a grid of high-resolution sparker profiles collected by the U.S. Geological Survey and the Texas Bureau of Economic Geology, Wright (1980) constructed a second structural map depicting the different valleys coalescing near the middle of present-day Matagorda Bay and

TABLE 1. RADIOCARBON AGES AND DESCRIPTION OF MATERIAL SAMPLED

Sample name	Material	Lab code	Age (^{14}C yr B.P.)	Age error	Age (cal yr B.P.)†	Age error	Depth from sea level (m)
ULB-03_6.59m	*Nuculana acuta*‡	OS-48161	7160	45	7490	200	8.09
ULB-03_9.57m	Carbonate nodule	OS-48159	15,550	70	18,320	370	11.07
LB-03_6.33m	*Nuculana concentrica*‡	OS-48160	3550	35	3180	260	7.24
LB-03_9.25m	*Rangia flexuosa*	OS-48162	6000	50	6240	250	10.16
LB-03_11.50m	*Nuculana concentrica*‡	OS-48163	7680	50	7940	240	12.41
LB-03_13.74m	*Rangia flexuosa*‡	OS-48164	8260	50	8560	240	14.65
LB-03_15.14m	*Rangia flexuosa*‡	OS-48165	9020	50	9430	310	16.05
LB-03_18.86m	Wood (fragments)	OS-48380	8610	45	9600	80	19.77
ULB1_1.15m	*Macoma*‡	OS-49877	2600	75	2040	300	2.65
ULB1_1.77m	*Rangia flexuosa*‡	OS-49878	2010	85	1360	280	3.27
LB2_0.53m	*Rangia flexuosa*‡	OS-49879	1030	60	485	200	1.14
LB2_1.52m	*Chione cancellata*‡	OS-49880	1610	75	1000	260	2.13
MB5_2.15m	*Abra aequalis*	OS-49881	1390	65	810	180	5.05
MB4_2.67m	*Abra aequalis*	OS-49882	3240	95	2800	370	6.02
MB3_10.30m	Shell fragments	OS-49883	7190	100	7530	270	13.65
MB1_3.69m	*Mulinia lateralis*	OS-49884	4820	90	4910	390	6.14
MB1_5.33m	*Abra aequalis*	OS-49886	4970	45	5070	260	7.78
MB1_9.99m	*Abra aequalis*‡	OS-49927	6420	50	6700	290	12.44
MB1_11.57m	*Abra aequalis*‡	OS-49928	7000	65	7370	220	14.02
MB1_13.31m	*Rangia flexuosa*	OS-49929	7560	60	7830	220	15.76
MB1_16.38m	Shell fragments	OS-49930	7850	70	8160	230	18.83
LB1_1.19m	Wood (whole)	OS-49769	430	35	480	50	1.50

Notes: Samples were processed at the National Science Foundation (NSF) accelerator mass spectrometry (AMS) laboratory at the Woods Hole Oceanographic Institution.
†See text for details on age calibration.
‡Articulated bivalve.

extending offshore as one large valley. The seismic data and cores acquired during this study support Wright's (1980) interpretation. Thus, a slightly modified version of Wright's (1980) map was chosen as the Pleistocene structural map to be used for this study (Fig. 1).

Like other incised valleys of the northern Gulf Coast region, the Lavaca valley is characterized by a deep and narrow incision near (average 5 km wide) the center of the valley. Adjacent to this incision, the valley flanks are broad, shallow, and bounded by the modern bay shoreline (Fig. 1). Smaller tributary valleys that are now occupied by Cox, Keller, Carancahua, and Tres Palacios Bays have similar cross-valley profiles (Fig. 3). Wave erosion along the bayshore has modified the shape of the valley, especially along the northern shore of Matagorda Bay, but the overall shape of the bay still outlines the incised-valley confluence and its tributaries (Fig. 1).

Previous work in Galveston Bay has revealed a number of relatively flat fluvial terraces that are believed to have been formed during the overall fall in sea level between ca. 120,000 and ca. 18,000 yr B.P. (Rodriguez et al., 2005). Rodriguez and his colleagues argued that the flooding history of the bay was punctuated by rapid flooding events that occurred when sea level reached the elevation of these terraces. Two terraces on the Pleistocene surface occur within the western and eastern regions of Matagorda Bay. These terraces are located at approximately −19 to −17 m and −12 m to −9 m water depth.

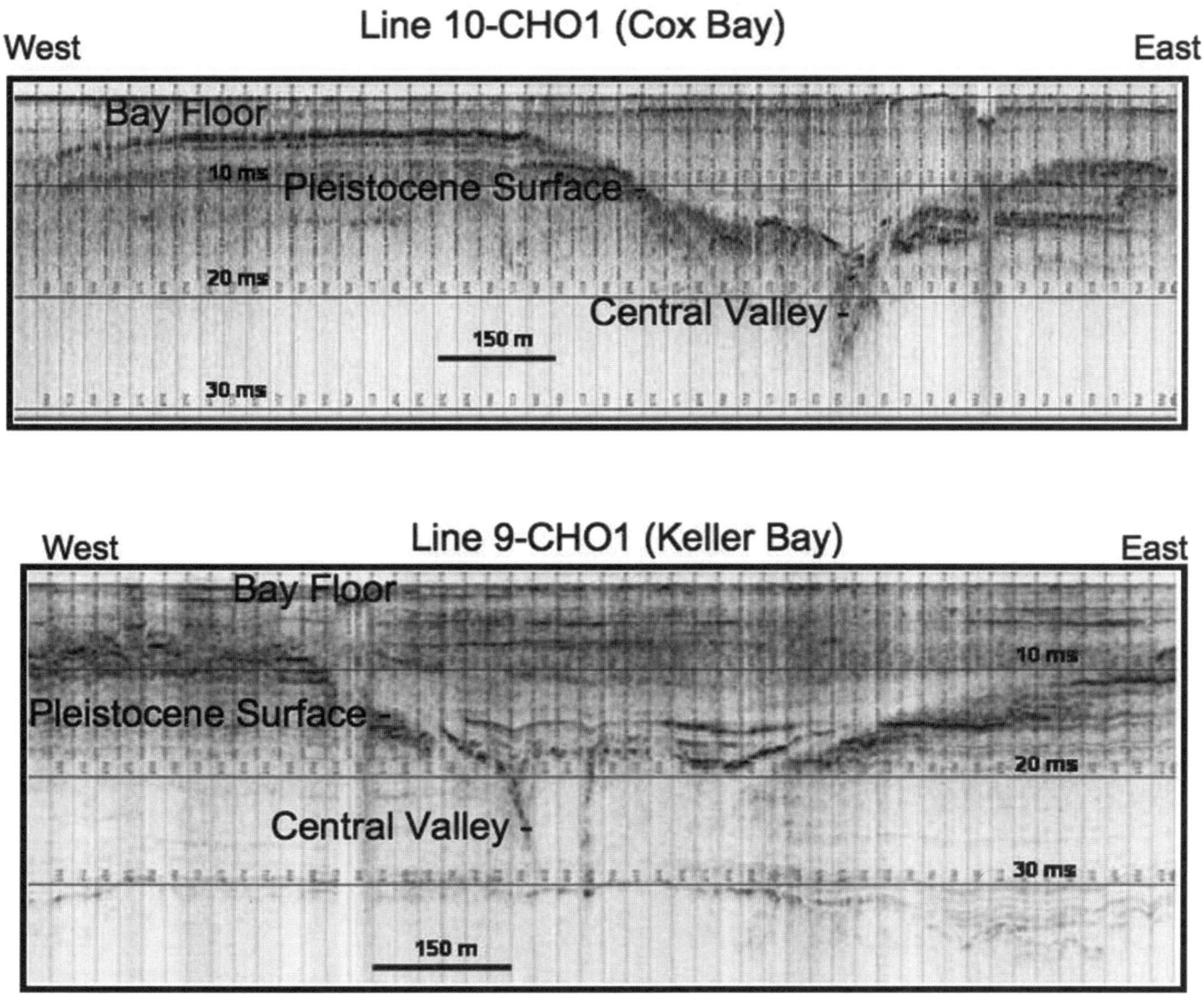

Figure 3. Profiles 10-CHO1 and 9-CHO1, crossing Cox Bay and Keller Bay, respectively, show the morphology of the incised valleys that occur beneath these bays.

Neotectonic Influence on Bay Evolution

Within the study area, subsurface Oligocene growth faults of the Vicksburg and Frio fault zones have been mapped beneath Matagorda Bay (Watkins et al., 1996). Yildiz (2003) also identified Miocene growth faults in southern Matagorda Bay related to the Clemente-Tomas fault zone. Recent research by White et al. (2002) documented the surface expression of active faults crosscutting northeast Matagorda Peninsula, just east of the study area. Thus, it was important to examine the possible influence of neotectonic activity on estuary evolution.

For the most part, seismic data from the bay complex do not image the Pleistocene-Holocene contact well enough to see offsets due to faulting. However, there is ~10 m of relief on the Pleistocene surface in the southern part of Matagorda Bay, between core sites MB1 03 and MB3 03 (Fig. 1). This relief may well be due to growth faults at this location. There are no obvious offsets in the Holocene part of the section.

Offshore seismic data image the Pleistocene-Holocene contact well and shows no offsets in the Pleistocene-Holocene contact, although faults do occur within the Pleistocene section (Eckles et al., 2004). This implies that long-term fault movement does occur and/or that fault activity during the Holocene was localized and episodic. Regardless of the reason, fault-induced subsidence does not appear to have significantly affected the Holocene evolution of Matagorda Bay.

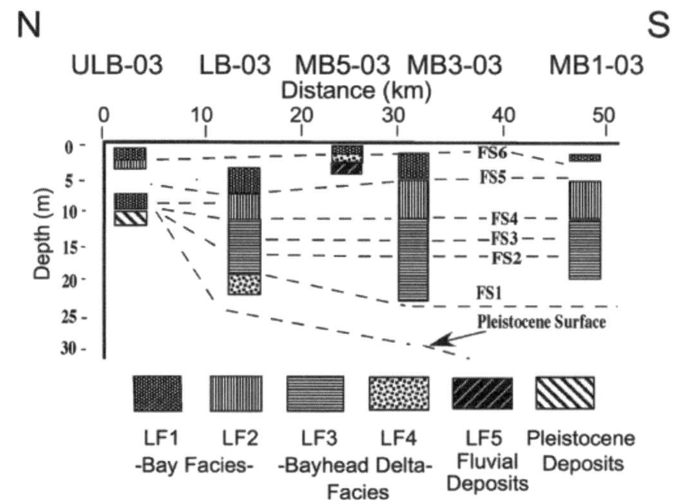

Figure 4. Axial cross section for Matagorda and Lavaca Bays based on lithofacies from sediment cores. Flooding surfaces FS1 through FS6 are also shown. See text for lithofacies descriptions.

Lithofacies

Six distinct lithofacies (LF) occur within the Matagorda and Lavaca estuary complex. The down-core occurrence of these facies in each core is presented in Figure 4.

Dark Gray Mud (LF1)

This lithofacies is a dark gray (5Y/4/1), homogeneous mud that is typically composed of ~75% silt, 15% clay, and 10% sand; however, sand contents as high as 42% occur in some intervals. Shell content varies from sparse to common and includes *Nuculana concentrica*, *Mulinia lateralis*, *Abra aequalis*, *Anadara transversa*, and *Anadara floridana*. Organic material is not present. This is the youngest facies encountered in sediment cores and is interpreted to be an open-bay deposit. Where sand content is elevated, it is interpreted as a bay-margin deposit. This facies is equivalent to the upper gray estuarine-mud facies of Wilkinson and Basse (1978).

Gray-Brown Mud (LF2)

Lithofacies 2 is a homogeneous mud that ranges in color from dark grayish brown (10YR/4/2) to brown (10YR/4/3). It is usually composed of ~65% silt, 23% clay, and 12% sand; however, sand content is as high as 63% in intervals interpreted to be bay-margin deposits. Shell content varies from sparse to common, and dominant species include *Mulinia lateralis*, *Abra aequalis*, *Anadara transversa*, and *Rangia flexuosa*. Organic material is absent. This is interpreted as an open-bay deposit and occurs stratigraphically below LF1. This facies is equivalent to the red-brown estuarine mud facies of Wilkinson and Basse (1978).

Olive Gray Mud (LF3)

This lithofacies is a homogeneous, olive gray (5Y/4/2) mud. Grain size is composed of ~67% silt, 13% clay, and 20% sand. Shell content is sparse to absent, but when present, it consists of fragments of *Rangia flexuosa* and *Abra aequalis*. Pyrite is sparse to common. Vertical burrows composed of very-fine sand occur within this facies in some locations. This lithofacies is interpreted as a distal deltaic deposit and occurs below LF2 within Matagorda Bay and middle Lavaca Bay.

Gray-Brown Fine Sand (LF4)

This lithofacies is a grayish brown (10YR/4/2), very-fine to fine sand consisting of ~9% silt, 2% clay, and 89% sand. Shells are common and consist of fragments of *Rangia flexuosa*, *Rangia cuneata*, and *Macoma mitchelli*. Organic matter in the form of wood fragments and leaf material is also common. Very dark gray (2 5Y/3/1), homogeneous mud with sparse shell fragments and organic matter occurs as thick beds within the sand. These deposits are interpreted as proximal deltaic and interdistributary bay deposits.

Mottled Sandy Mud and Sand (LF5)

This lithofacies is composed of well-sorted fine sand and tan (7.5YR/4/4) sandy mud mottled with gray (2.5Y/4/1) mud. Shell material is absent, but millimeter-sized, reworked carbonate nodules are abundant. This lithofacies is interpreted as an alluvial plain and associated crevasse splay deposit.

Oxidized Mud

This lithofacies is composed of very stiff sandy clay. It is greenish gray (10GY/6/1), mottled with olive yellow (2.5Y/6/8) sandy clay with intervals of burnt orange and purple, indicative of an oxidizing environment. Calcareous nodules are common, while shell content is absent. These are oxidized Pleistocene deposits.

X-Ray Fluorescence Results

Discrete samples from two cores, LB-03 from Lavaca Bay and MB3-03 from Matagorda Bay, were examined using an X-ray fluorescence whole-core scanner. The objective was to determine if this method would show down-core geochemical variations that might be used to determine the causes of observed lithofacies changes, especially as they might relate to changes in sediment source with time. The X-ray fluorescence logs for cores LB-03 and MB-03 show spikes and abrupt transitions in the concentrations of Ti, Si, K, Fe, Ca, Sr, Rb, and Zr, which, in the majority of cases, are associated with flooding surfaces (Fig. 5).

Changes in the ratio of Ca/Ti indicate that some of these surfaces are associated with changes in biogenic carbonate content between adjacent lithofacies units, whereas spikes in Ca/Ti appear to be linked to shell-rich lag deposits typical of some flooding surfaces. Similarly, changes in Zr/Rb and Si/Ti ratios, which are indicative of grain size, suggest that some flooding surfaces may be associated with sharp textural changes. However, high-resolution continuous scanning of fresh cores is needed for a systematic study of compositional changes associated with the flooding history of the estuary complex.

Seismic Facies and Bounding Surfaces

Three distinct seismic facies were identified above the Pleistocene-Holocene unconformity (Fig. 6). The uppermost seismic facies consists of horizontal, parallel, continuous high-amplitude reflections and low-amplitude reflections and corresponds to dark gray and gray-brown open-bay and bay-margin lithofacies in Lavaca and Matagorda Bays. It is referred to as the central-basin seismic facies (Fig. 6A). The middle seismic facies consists of subparallel, medium-amplitude, horizontal and oblique reflections and includes small channels. It corresponds to lithofacies LF3 and LF4, which are interpreted as bayhead delta deposits. The lower seismic facies is composed of high-amplitude, wavy to chaotic reflections and includes larger, deeper channels than those of the overlying bayhead delta facies. This facies includes fluvial deposits (LF5) that rest above and below the Pleistocene-Holocene contact.

Seismic profiles from both Matagorda and Lavaca Bays imaged sharp surfaces that bound different seismic facies. These surfaces separate facies that indicate a shift from more landward to more basinward environments, and so they are inferred to be flooding surfaces (Figs. 4 and 6). The uppermost surface (FS6) is imaged only in chirp profiles (Fig. 6B).

Chronostratigraphy

Twenty-two radiocarbon dates were obtained from cores within the Matagorda and Lavaca estuary complex (Table 1). Two of these dates were taken from wood samples. The remaining 20 dates were obtained from mollusk shells.

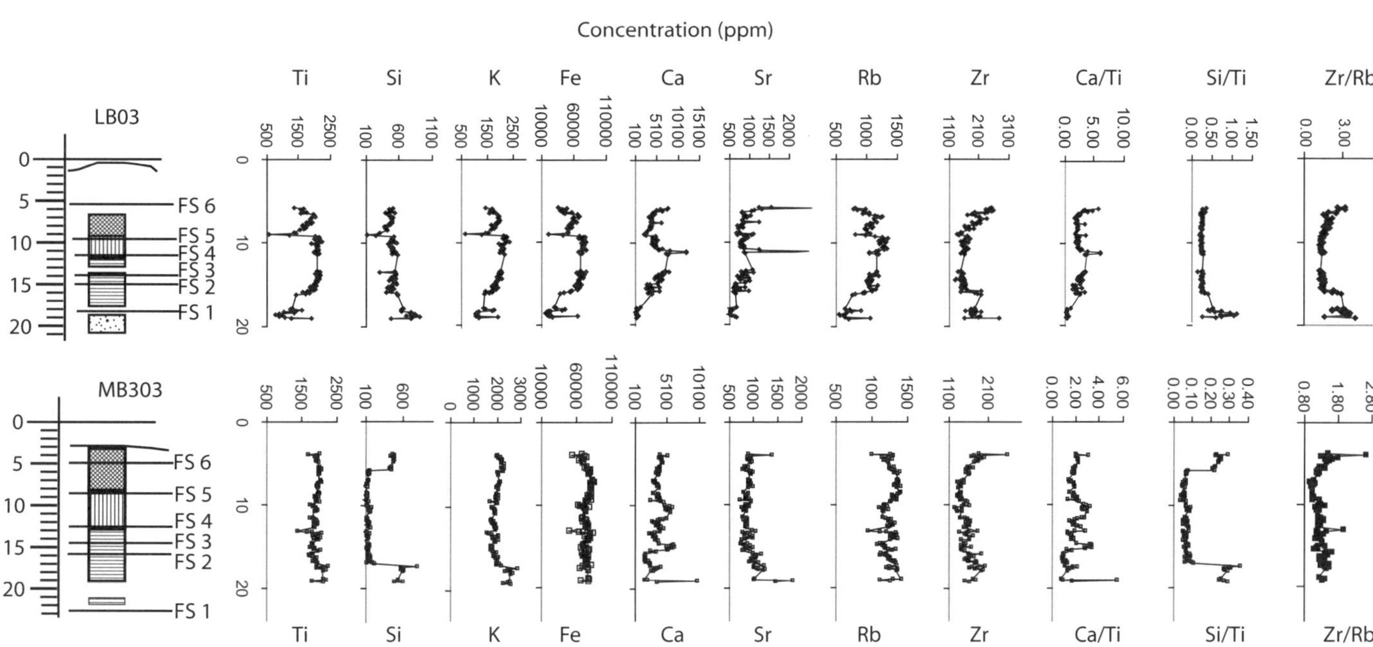

Figure 5. X-ray fluorescence profiles for sediment cores LB-03 and MB3-03. Flooding surfaces FS1 through FS6 are also shown.

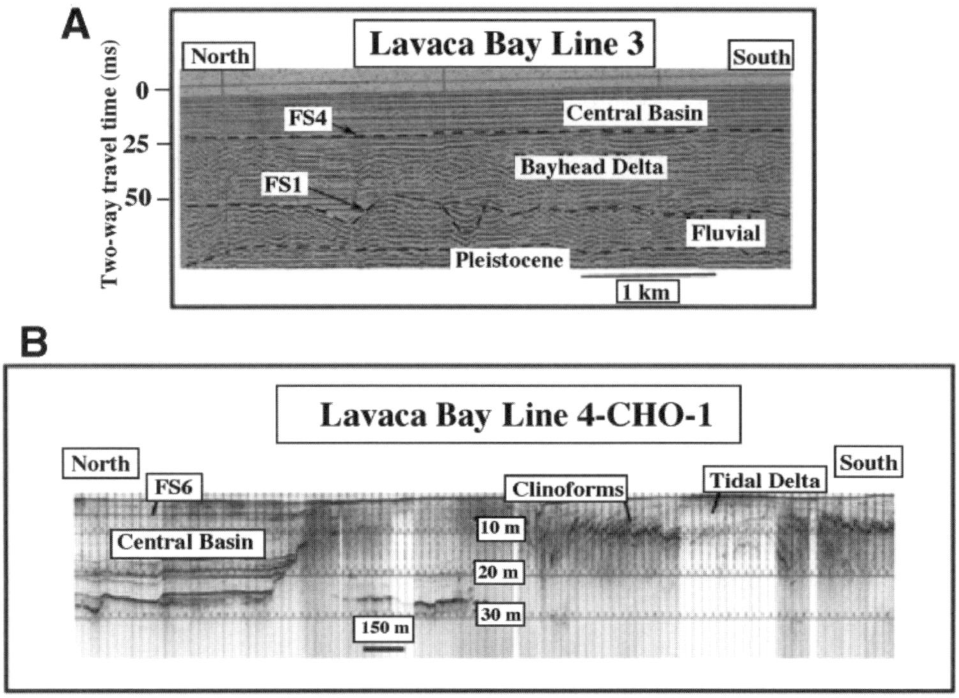

Figure 6. Lavaca Bay boomer line 3 (A) and chirp line 4-CHO-1 (B) are used to illustrate seismic facies imaged by these different seismic sources. FS—flooding surface.

Aten (1983) attempted to empirically quantify the carbon reservoir effect in the study area by dating charcoal and shell pairs collected within a shell midden that he thought to be contemporaneous, yielding a carbon reservoir correction of 945 ± 295. However, fire hearths were often built on top of old shell middens, so this material may not have been contemporaneous (Aten, 1983; Ricklis, 1999; Kibler, 1999). Additionally, at the time of Aten's research, it was not standard procedure to correct for $^{12}C/^{13}C$ stable isotopic fractionation (Kibler, 1999; Ensor, 1999). This resulted in apparent radiocarbon ages that were 300–400 yr younger than true ages (Howard et al., 1991).

In an attempt to further constrain the radiocarbon reservoir for Matagorda and Lavaca Bays, an articulated mollusk (*Rangia flexuosa*) and a wood fragment in core LB-03 were dated. The mollusk shell was situated just above the wood fragment at 18.6 m depth in the core. Both are believed to have remained in situ and in proper stratigraphic order; however, ages of the samples are reversed. This reversal is believed to be due to the carbon reservoir effect of the mollusk. We found that a carbon reservoir effect of 500 ^{14}C yr is needed to realign the ages. This, combined with the fact that the Lavaca River does not drain any old carbonates, leads to the interpretation that the true carbon reservoir effect in Matagorda Bay is close to the 540 ± 110 ^{14}C yr reservoir for Trinity Bay (Ensor, 1999).

All radiocarbon ages reported here are given in calendar yr B.P. and range from 9600 to 480 cal yr B.P. Mean age uncertainties are ±250 cal yr. Dates are consistent with depth in the sediment column in all cores except ULB1-03, in which two dates are out of sequence.

DISCUSSION

Stratigraphic Relationships

Combined seismic facies and lithofacies were used to identify and map sedimentary units that accumulated during Holocene flooding of the estuary complex. These units indicate changes from fluvial to bayhead delta to open-bay environments (Figs. 4 and 6A). The main depositional units and flooding surfaces that bound them are illustrated using an axial section constructed from these combined data (Fig. 7). Seismic line MB-02, a strike line from the southern part of Matagorda Bay, illustrates the changes that occurred in the lower portion of the bay as it was flooded (Fig. 8).

Where seismic data were of sufficient quality, the Pleistocene-Holocene surface was recognized by the presence of onlapping reflectors above it and terminating reflectors below it (Fig. 8). Where seismic data quality is poor or data were unavailable, Wright's (1980) Pleistocene-Holocene structural map was used to constrain the location of this surface.

The surface separating fluvial deposits from bayhead delta deposits is the bay line. It is a prominent surface that separates high-amplitude, chaotic reflectors below from low-amplitude, subparallel, horizontal and oblique reflections above (Figs. 6A and 8). The bay line is a diachronous surface that represents the initial drowning of the valley at a given location.

Typical of other bays we have studied, the bay line is a surface of several meters relief when viewed in cross-valley profiles, which represents drowned fluvial topography (Figs. 6A and 8).

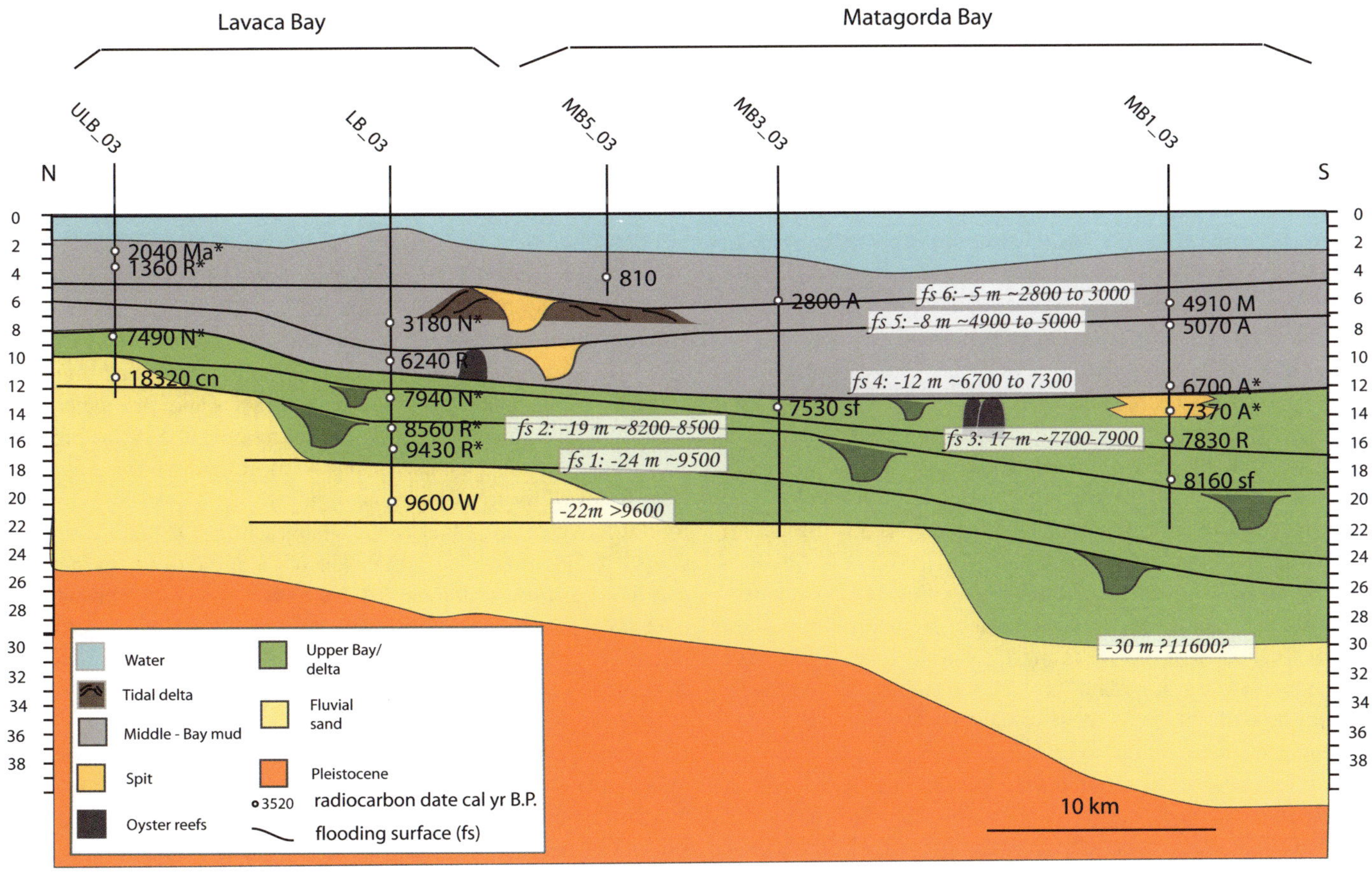

Figure 7. Chronostratigraphic profile along the axis of the Lavaca incised valley showing flooding surfaces (fs) and radiocarbon dates used to constrain the ages of these surfaces. Note the steps in the bay line flooding surface, which separates fluvial (yellow) from bayhead delta (green) deposits.

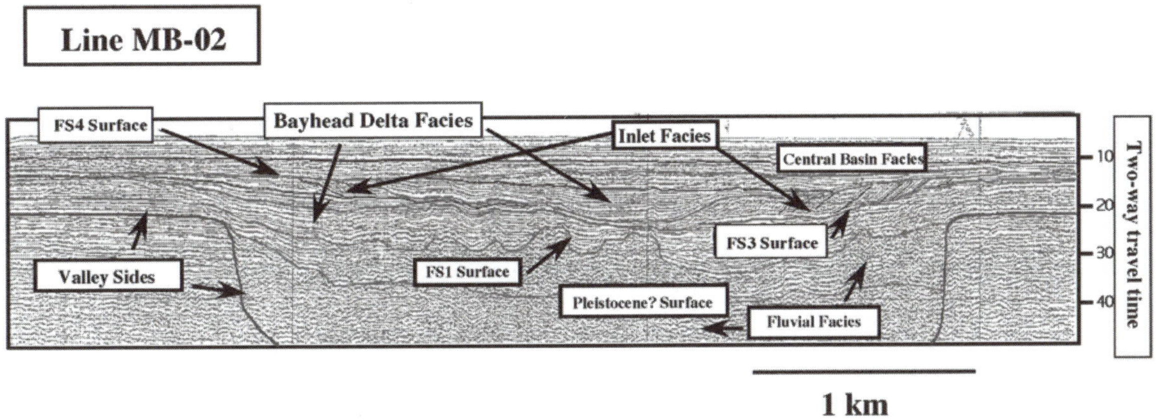

Figure 8. Strike-oriented seismic profile MB-02 across the Lavaca incised valley in southern Matagorda Bay showing seismic facies and flooding surfaces. Note the lateral accretion surfaces on either side of the incised valley, which are bounded by flooding surfaces FS3 and FS4. See Figure 1 for line location.

Prominent steps in the bay line surface occur along the axis of the valley. These steps represent prolonged periods of fluvial aggradation. Their position within the valley was likely controlled by the antecedent topography of the valley, although this cannot be demonstrated due to poor resolution of the Pleistocene surface.

The bay line was sampled at only one location in Matagorda Bay (site MB3-03), but there was a lack of material for radiocarbon dating of the surface. Thus, the age of the surface is estimated from its elevation using the composite sea-level curve of Milliken et al. (this volume). The bay line surface was sampled in Lavaca Bay at sites LB 03 and ULB 03 (Fig. 7). This is the oldest flooding surface that we are able to date, and it is designated the FS1 surface. Its age is ca. 9500 cal yr B.P.

The next youngest flooding event (FS2) is recognized in seismic records as a landward shift of the bay line of ~4 km. This flooding surface is not manifest as a lithofacies change in cores. However, the X-ray fluorescence logs for cores LB-03 and MB-03 both show geochemical excursions just below the FS2 surface on seismic records, which likely mark the actual surface (Fig. 5).

In lower Matagorda Bay, the FS3 flooding surface is easily recognized in seismic data because it truncates underlying reflectors near the center of the valley, and it is a downlap surface for clinoforms that prograded into the valley from the east and west (Fig. 8). These clinoforms reflect spit accretion across the valley following the FS3 flooding event. These spits are analogous to the ones that exist today along the northern shore of Matagorda Bay (Fig. 1). In upper Matagorda Bay, the FS3 surface occurs within bayhead delta facies, so it is not easily recognized in either core or seismic records.

Flooding surface FS4 separates bayhead delta deposits below from central-basin facies above and was sampled in cores LB-03, MB3-03, and MB1-03 (Fig. 7). In the seismic data, this surface is characterized by low-amplitude, parallel, horizontal reflections onlapping low-amplitude, subparallel, horizontal and oblique reflections (Fig. 6A). In the southern part of the bay, at the location of seismic line MB-02, this surface corresponds to the tops of clinoforms (Fig. 8). Flooding surface FS4 marks a major landward shift in the bayhead delta and establishment of open-estuary conditions throughout much of present-day Matagorda Bay.

Flooding surface FS5 is marked by a subtle change in seismic facies at a depth between −9 and −7 m below the bay floor. This surface is also marked by the abandonment of a deep tidal channel at the confluence of Lavaca and Matagorda Bays and by the termination of oyster reef growth in the upper part of Matagorda Bay (Fig. 7).

Sediment cores LB-03 and MB3-03 sampled a lithologic boundary at the flooding surface that separates the dark gray mud lithofacies above from gray-brown mud lithofacies below. *Rangia flexuosa* occur in the lower unit, which implies slightly more brackish conditions for this facies relative to the upper unit.

Chirp profiles throughout Lavaca Bay show a prominent reflection at approximately −5 m below the bay floor (Fig. 6B, line 4-CHO-1). This is the FS6 flooding surface. Chirp profiles and the axial seismic profile from the upper part of Matagorda Bay and lower part of Lavaca Bay cross a feature that has positive relief and clinoforms that dip both north and south (Fig. 6B). Hence, we interpret this feature as a tidal delta, similar to the one that occurs in the upper part of Corpus Christi Bay (Simms et al., this volume). This interpretation is consistent with the fact that the bayhead delta had shifted to the north within Lavaca Bay prior to the time this feature formed. Line 4-CHO-1 is used to illustrate the manner in which the FS6 surface truncates the top of a delta that occurs in the lower part of Lavaca Bay and in the northern part of Matagorda Bay, north of core site MB3-03 (Fig. 3). Vibracore MB5-03 sampled the FS6 surface, which consists of bay mud resting on sandy mud at this location.

It is important to emphasize that the placement of flooding surfaces is based on changes in seismic facies and lithofacies that record marine inundation of the valley, where each surface represents more-seaward facies stacked on more-landward facies. The abruptness of these surfaces, both in sediment cores and in seismic records, implies that rapid changes in the estuarine setting occurred. However, the ultimate test of the rapid flooding hypothesis requires robust radiocarbon dating of the surfaces throughout the estuary.

Estuary Evolution

Radiocarbon dating was focused on flooding surfaces and on acquiring enough dates from bayhead delta and central-basin facies to determine average rates of sedimentation for these depositional environments. They assume a radiocarbon reservoir of 500 yr. These data, combined with lithofacies and seismic facies data, were used to construct paleogeographic maps that depict the bay setting before and after different flooding events (Fig. 9).

In response to lowered sea level during the Last Glacial Maximum, the Lavaca River incised ~40 m into Pleistocene fluvial-deltaic sediments along the modern coastline of the Gulf of Mexico (Fig. 1). At this time, sediments bypassed the area now occupied by Matagorda Bay and were deposited on the continental shelf and slope (Eckles et al., 2004). During the postglacial sea-level rise, the river began to aggrade and deposited lowstand and transgressive fluvial sands within its valley.

The balance between sediment supply and sea-level rise controls the bay line location. Times of relatively high sediment supply are manifest by fluvial aggradation, and times of relatively low supply are manifest by landward shifts in the bay line surface. This results in steps in a landward-stepping (backstepping) bay line surface (Fig. 7). The most dramatic step in the bay line is located about halfway between drill sites MB1-03 and MB3-03. No radiocarbon dates were acquired from this part of the stratigraphic section, but the depths of the bay line at −30 m and −22 m coincide with ages of ca. 11,600–9600 cal yr B.P., based on the composite sea-level curve for the northern Gulf of Mexico (Milliken et al., this volume). During this time interval, sea level was rising rapidly and perhaps episodically; the fastest rates were on the order of 18 mm/yr (Milliken et al.,

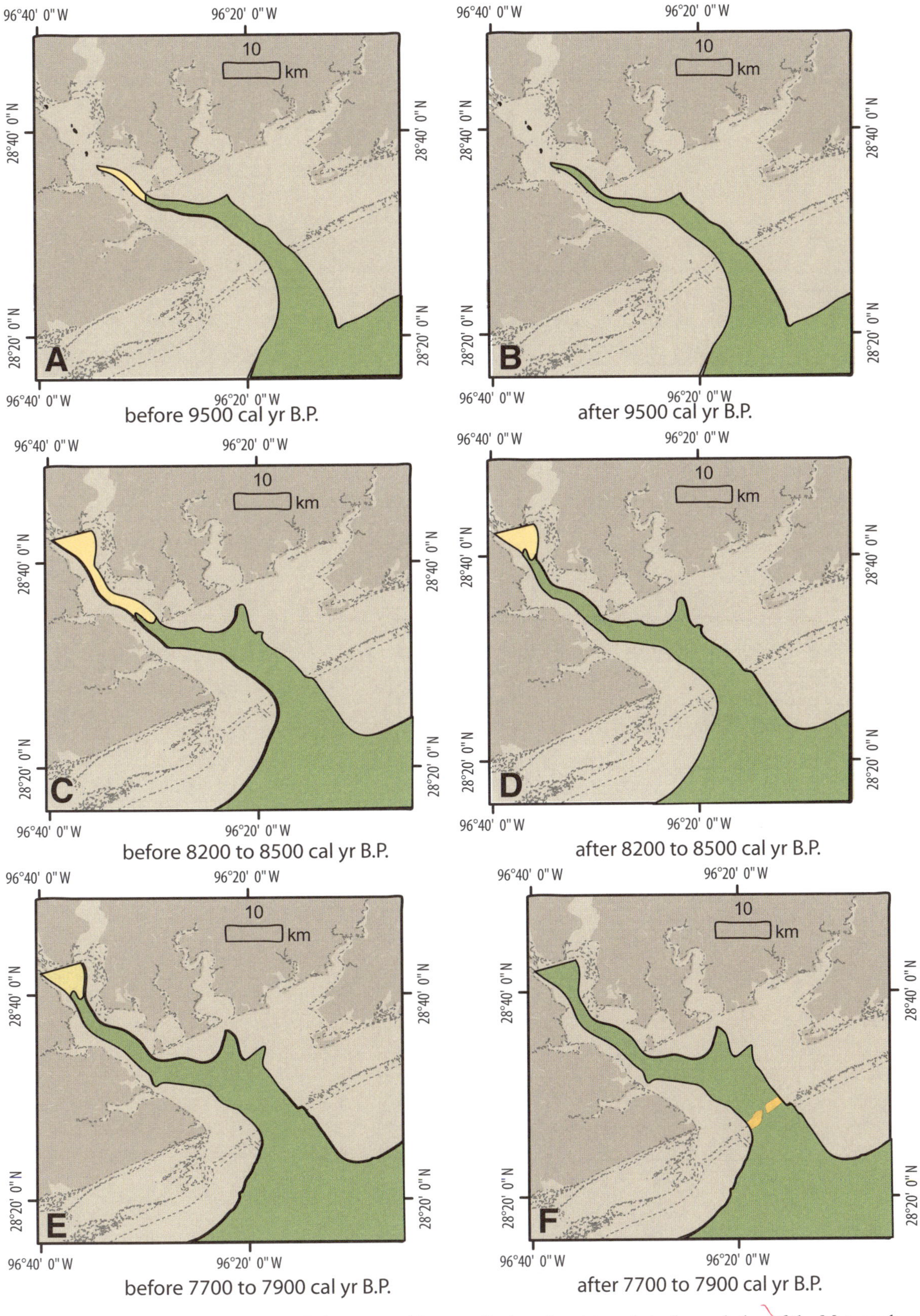

Figure 9 (*on this and following page*). Paleogeographic maps for key time intervals in the evolution of the Matagorda-Lavaca estuary complex.

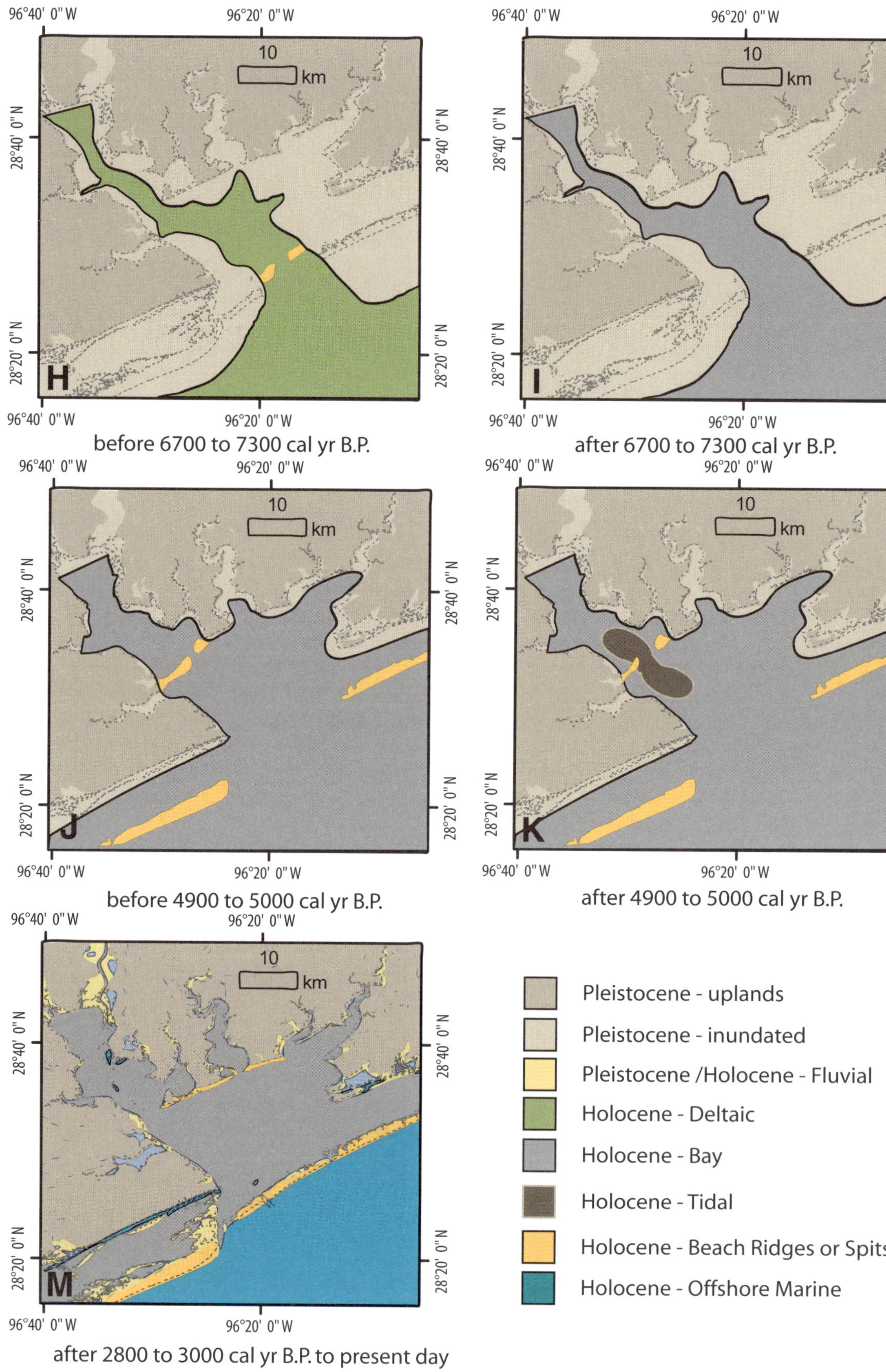

Figure 9 (*continued*).

this volume). However, because sedimentation was confined to the narrow part of the valley, sediment accumulation kept pace with sea-level rise until ca. 9600 yr B.P. Following this initial flooding episode, a bayhead delta existed in the narrow part of the valley.

At ca. 9500 cal yr B.P., sea level reached the elevation of the lowest fluvial terrace within the valley, resulting in an episode of rapid flooding when the bay line stepped landward ~15 km. Following this flooding event, an extensive bayhead delta occupied all of present-day Matagorda Bay (Fig. 9A). During the next 1000 yr, the location of the bay line shifted to the south ~10 km, followed by a flooding event that shifted the bay line to the north. This flooding event (FS2) culminated ca. 8200 cal yr B.P. and occurred during another phase of rapid sea-level rise (Törnqvist et al., 2005; Milliken et al., this volume).

Between 7900 and 7700 cal yr B.P., the final stages of flooding of the deep, narrow part of the valley took place in the upper part of the valley (FS3). This flooding event also corresponds to an apparent episode of rapid (~18 mm/yr) sea-level rise (Milliken et al., this volume). At the end of this flooding event, the area of Lavaca Bay increased by ~20%. During this time interval, prominent spits formed just landward of the current shoreline location (Fig. 9A). These spits formed in a bay setting and were probably similar to those that exist today along the northern shore of Matagorda Bay (Fig. 1) and Corpus Christi Bay (Simms et al., this volume).

The spits that were imaged in seismic profile MB-02 (Fig. 8) prograded until ca. 6700 cal yr B.P., when flooding once again forced bay and coastal environments to step back and emplaced estuarine deposits on top of deltaic deposits (FS4, Figs. 6A and 7). This was the most dramatic flooding event in the history of the Matagorda and Lavaca estuary complex. Within a few centuries, most of the area now occupied by Matagorda and Lavaca Bays was transformed from a bayhead delta to an open estuary that was several meters deep (Fig. 9B). The elevation of the FS4 flooding surface (−12 m) corresponds to another step in the antecedent topography between −12 m and −9 m. However, the magnitude of this flooding event can only partly be attributed to flooding of fluvial terraces.

Flooding surface FS4 formed at a time (ca. 7300–6700 cal yr B.P.) when the rate of sea-level rise in the northern Gulf of Mexico was decreasing from an average rate of 4.2 mm/yr to 1.4 mm/yr (Milliken et al., this volume). However, this event occurred when the climate in southeast Texas was transitioning from cool and moist to a warm and dry, which was associated with a change from forests to grasslands (Bryant, 1977; Toomey et al., 1993; Nordt et al., 1994, 2002; Humphrey and Ferring, 1994). This change likely had a twofold effect on sediment supply to the estuary. The strong root network of the remaining trees and shrubs acted to keep soils in place while the decreasing effective moisture diminished the erosive capacity of streams. Thus, we interpret the dramatic flooding and reorganization of estuarine environments between 7300 and 6700 cal yr B.P. as having been caused mainly by a decrease in sediment supply to the estuary.

Following the 7300–6700 cal yr B.P. (FS4) flooding event, the bay was influenced by sediment input from the ancestral Colorado River, which began prograding into the bay at this time (Wilkinson, 1975; Wilkinson and Basse, 1978). This resulted in higher sedimentation rates in Matagorda Bay relative to Lavaca Bay.

The evolution of Matagorda Peninsula undoubtedly influenced the evolution of the Matagorda and Lavaca estuary complex. However, the peninsula's evolution has been poorly constrained by radiocarbon age dating (Wilkinson, 1975; Wilkinson and Basse, 1978). Wilkinson and Basse (1978) obtained five radiocarbon dates from back-barrier bay deposits beneath modern barrier sands that range in age from 6670 ± 100 yr to 930 ± 60 yr (uncorrected ages). Thus, the evolution of the modern barrier appears to have coincided with the landward shift of bay environments after ca. 6700 cal yr B.P. This was also the time when oyster reefs shifted from the southern part of Matagorda Bay to the upper part of the bay (Fig. 7).

During the past 6700 yr, Matagorda and Lavaca Bays grew wider as the shallow margins of the valley were slowly flooded. It was during this time that the modern estuarine complex began to take on its present-day appearance (Fig. 9B). Wave reworking of the surrounding Pleistocene bluffs provided a source of sand for spit accretion, which has subsequently given Matagorda Bay its rectangular shape.

The most notable changes that have taken place within the bay complex since ca. 6700 cal yr B.P. are marked by a subtle change in seismic facies at −7 to −8 m (FS5) and by an upward change from dark gray mud to gray-brown mud lithofacies in cores LB-03 and MB3-03. The age of this surface is constrained by radiocarbon dates at core site MB1-03 to between ca. 5000 and ca. 4900 cal yr B.P. (Fig. 7). This flooding event was followed by the formation of spits and a tidal delta at the mouth of Lavaca Bay (Fig. 9B). By this time, the modern barrier complex had significantly evolved (Fig. 9B).

The youngest flooding surface (FS6) occurs at an average depth of −5 m below the bay floor. This flooding event submerged the spits and tidal delta that previously existed in the narrows between Matagorda Bay and Lavaca Bay. The age of the youngest surface is poorly constrained, but, based on extrapolation of sedimentation rates at site MB3-03, the age of FS6 is ca. 2800–3000 cal yr B.P. By this time, the estuary and barriers had pretty much evolved to their present configuration.

CONCLUSIONS

Using a combination of high-resolution seismic data and core data, six distinct lithofacies and four unique seismic facies were identified within the Matagorda and Lavaca estuary complex. The seismic facies and lithofacies together include Holocene estuarine, bayhead delta, and fluvial deposits resting on Pleistocene fluvial deposits. Flooding surfaces that reflect significant landward shifts of fluvial and estuarine environments bound these facies. Radiocarbon age constraints for these surfaces indicate dramatic changes in the estuarine complex at decadal to century time scales and lead to the following conclusions.

1. A back-stepping bay line surface records initial flooding of the Lavaca incised valley. Flooding events occurred when the river was not able to aggrade and keep pace with rapidly rising sea level.
2. By ca. 9500 cal yr B.P., most of the deep, narrow portion of the incised valley had been flooded, and a bayhead delta occupied the valley. Flooding events at ca. 8500–8200 cal yr B.P. and at 7900–7700 cal yr B.P. occurred during rapid, and possibly episodic, sea-level rise. These flooding events resulted in expansions in estuary area, which resulted in the inability of the rivers to deliver sediment to the valley fast enough to keep pace with increased rates of sea-level rise.
3. Between 7300 and 6700 cal yr B.P., widespread flooding occurred within the Lavaca valley. Prior to this time, the valley was occupied by an extensive bayhead delta. Following this flooding event, the bayhead delta was located north of its current location, and the valley had been transformed into an open-estuary setting. This flooding event took place at a time when the rate of sea-level rise was decreasing and when the climate was transitioning from cool and moist to warm and dry, which was associated with a change in vegetation cover. This dramatic change in the estuary setting is believed to have resulted from decreased sediment supply to the estuary.
4. Since ca. 6700 cal yr B.P., only minor changes in the estuary setting have occurred, most notably, the development of the modern barrier system and associated changes in tidal circulation and salinity.

ACKNOWLEDGMENTS

The authors would like to thank Bruce Wilkinson and one anonymous reviewer for helpful comments and suggestions that greatly improved this manuscript. This research was funded by the National Science Foundation grant EAR-0107650. We thank Chip Anderson for assistance in acquiring drill cores.

REFERENCES CITED

Aten, L.E., 1983, Indians of the Upper Texas Coast: New York, Academic Press, 370 p.

Bryant, V.M., 1977, A 16,000 year pollen record of vegetational change in central Texas: Palynology, v. 1, p. 143–156.

Byrne, J.R., 1975, Holocene Depositional History of Lavaca Bay, Central Texas Gulf Coast [Ph.D. thesis]: Austin, Texas, University of Texas at Austin, 150 p.

Dalrymple, R.W., Zaitlin, B.A., and Boyd, R., 1992, Estuarine facies models: Conceptual basis and stratigraphic implications: Journal of Sedimentary Petrology, v. 62, no. 6, p. 1130–1146.

Eckles, B.J., Fassell, M.L., and Anderson, J.B., 2004, Late Quaternary evolution of the wave-storm-dominated central Texas shelf, *in* Anderson, J.B., and Fillon, R., eds., Late Quaternary Stratigraphic Evolution of the Northern Gulf of Mexico Margin: Society for Economic Paleontologists and Mineralogists (SEPM) Special Publication 79, p. 271–287.

Ensor, H.B., 1999, Comparative review and discussion of *Rangia cuneata* dating methods on the upper Texas coast: Bulletin of the Texas Archeological Society, v. 70, p. 467–478.

Hall, S.A., and Valastro, S., Jr., 1995, Grassland vegetation in the southern Great Plains during the Last Glacial Maximum: Quaternary Research, v. 44, p. 237–245, doi: 10.1006/qres.1995.1068.

Howard, M.A., Bailey, G.L., Bousman, C.B., Gardner, K.M., and Fields, R.C., 1991, National Register Eligibility Testing at the Spanish Moss Site (41GV10) and 41GV53, Galveston County, Texas: Austin, Prewitt and Associates Reports of Investigations No. 77.

Humphrey, J.D., and Ferring, C.R., 1994, Stable isotopic evidence for latest Pleistocene and Holocene climatic change in north-central Texas: Quaternary Research, v. 41, p. 200–213, doi: 10.1006/qres.1994.1022.

Kibler, K.W., 1999, Radiocarbon dating of *Rangia cuneata*: Correction factors and calibrations for the Galveston Bay area: Bulletin of the Texas Archeological Society, v. 70, p. 457–466.

Milliken, K.T., Anderson, J.B., and Rodriguez, A.B., 2008, this volume, A new composite Holocene sea-level curve for the northern Gulf of Mexico, *in* Anderson, J.B., and Rodriguez, A.B., eds., Response of Upper Gulf Coast Estuaries to Holocene Climate Change and Sea-Level Rise: Geological Society of America Special Paper 443, doi: 10.1130/2008.2443(01).

Nordt, L.C., Boutton, T.W., Hallmark, C.T., and Waters, M.R., 1994, Late Quaternary vegetation and climate changes in central Texas based on the isotopic composition of organic carbon: Quaternary Research, v. 41, p. 109–120, doi: 10.1006/qres.1994.1012.

Nordt, L.C., Boutton, T.W., Jacob, J.S., and Mandel, R.D., 2002, C_4 plant productivity and climate-CO_2 variations in south-central Texas during the late Quaternary: Quaternary Research, v. 58, p. 182–188, doi: 10.1006/qres.2002.2344.

Ricklis, R.A., 1999, Atmospheric calibration of radiocarbon ages on shallow-water estuarine shells from Texas coast sites and the problem of questionable shell-charcoal pairing: Bulletin of the Texas Archeological Society, v. 70, p. 479–488.

Rodriguez, A.B., Anderson, J.B., and Simms, A.R., 2005, Terrace inundation as an autocyclic mechanism for parasequence formation: Galveston estuary, Texas, USA: Journal of Sedimentary Research, v. 75, p. 608–620, doi: 10.2110/jsr.2005.050.

Simms, A.R., Anderson, J.B., Rodriguez, A.B., and Taviani, M., 2008, this volume, Mechanisms controlling environmental change within an estuary: Corpus Christi Bay, Texas, USA, *in* Anderson, J.B., and Rodriguez, A.B., eds., Response of Upper Gulf Coast Estuaries to Holocene Climate Change and Sea-Level Rise: Geological Society of America Special Paper 443, doi: 10.1130/2008.2443(08).

Stoeser, D.B., Shock, N., Green, G.R., Dumonceaux, G.M., and Heran, W.D., 2005, Geological Map Database of Texas, Version 1: Denver, Colorado, U.S. Geological Survey Open-File Data Series DS 170, http://pubs.usgs.gov/ds/2005/170 (last accessed 2 May 2008).

Stuiver, M., and Braziunas, T.F., 1993, Modeling atmospheric ^{14}C influences and ^{14}C ages of marine samples back to 10,000 BC: Radiocarbon, v. 35, p. 137–189.

Stuiver, M., and Reimer, P.J., 1993, Extended ^{14}C database and revised CALIB radiocarbon calibration program: Radiocarbon, v. 35, p. 215–230.

Stuiver, M., Reimer, P.J., Bard, E., Beck, J.W., Burr, G.S., Hughen, K.A., Kromer, B., McCormac, F.G., van der Plicht, J., and Spurk, M., 1998, INTCAL98: Radiocarbon age calibration 24,000–0 cal B.P.: Radiocarbon, v. 40, p. 1041–1083.

Toomey, R.S., III, Blum, M.D., and Valastro, S., Jr., 1993, Late Quaternary climates and environments of the Edwards Plateau, Texas: Global and Planetary Change, v. 7, p. 299–320, doi: 10.1016/0921-8181(93)90003-7.

Törnqvist, T.E., Gonzalez, J.L., Newsom, L.A., van der Borg, K., de Jong, A.F.M., and Kurnik, C.W., 2004, Deciphering Holocene sea-level history on the U.S. Gulf Coast: A high-resolution record from the Mississippi Delta: Geological Society of America Bulletin, v. 116, no. 7–8, p. 1026–1039, doi: 10.1130/B2525478.1.

Törnqvist, T.E., Bick, S.J., Gonzalez, J.L., van der Borg, K., and de Jong, A.F.M., 2005, Tracking the sea-level signature of the 8.2 ka cooling event: New constraints from the Mississippi Delta: Geophysical Research Letters, v. 31, L23309, p. 1–4.

Toscano, M.A., and Macintyre, I.G., 2003, Corrected western Atlantic sea-level curve for the last 11,000 years based on calibrated ^{14}C dates from *Acropora palmata* framework and intertidal mangrove peat: Coral Reefs, v. 22, p. 257–270, doi: 10.1007/s00338-003-0315-4.

U.S. Department of Agriculture, Natural Resources Conservation Service, 2007, Soil Data Mart: http://SoilDataMart.nrcs.usda.gov/ (Fort Worth, Texas, SSURGO [Soil Survey Geographic database, last accessed 2 May 2008]).

Watkins, J.S., Bryant, W.R., and Buffler, R.T., 1996, The Gulf of Mexico structural and stratigraphic structural synthesis project: Introduction and overview: American Association of Petroleum Geologists (AAPG) Bulletin, v. 80, no. 9, p. 1517.

White, W.A., Morton, R.A., and Holmes, C.W., 2002, A comparison of factors controlling sedimentation rates and wetland loss in fluvial-deltaic systems, Texas Gulf coast: Geomorphology, v. 44, p. 47–66, doi: 10.1016/S0169-555X(01)00140-4.

Wilkinson, B.H., 1975, Matagorda Island, Texas: The evolution of a Gulf Coast barrier complex: Geological Society of America Bulletin, v. 86, p. 959–967, doi: 10.1130/0016-7606(1975)86<959:MITTEO>2.0.CO;2.

Wilkinson, B.H., and Basse, R.A., 1978, Late Holocene history of the central Texas coast from Galveston Island to Pass Cavallo: Geological Society of America Bulletin, v. 89, p. 1592–1600, doi: 10.1130/0016-7606(1978)89<1592:LHHOTC>2.0.CO;2.

Wright, S.S., 1980, Seismic Stratigraphy and Depositional History of Holocene Sediments on the Central Texas Gulf Coast [M.A. thesis]: Austin, Texas, University of Texas at Austin, 123 p.

Yildiz, S., 2003, Sequence Stratigraphy and Depositional Systems of Late Cenozoic Sediments, Matagorda Bay, Gulf of Mexico [M.S. thesis]: Austin, Texas, University of Texas at Austin, 157 p.

Manuscript Accepted by the Society 10 March 2008

Printed in the USA

The Geological Society of America
Special Paper 443
2008

Mechanisms controlling environmental change within an estuary: Corpus Christi Bay, Texas, USA

Alexander R. Simms
Boone Pickens School of Geology, Oklahoma State University, 105 NRC, Stillwater, Oklahoma 74074, USA

John B. Anderson
Department of Earth Science, Rice University, 6100 Main Street, MS 126, Houston, Texas 77005, USA

Antonio B. Rodriguez
Institute of Marine Sciences, University of North Carolina at Chapel Hill, 3431 Arendell Street, Morehead City, North Carolina 28557, USA

Marco Taviani
Instituto di Scienze Marine, Consiglio Nazionale delle Ricerche, via Gobetti 101, Bologna, 40129, Italy

ABSTRACT

Over 400 km of high-resolution seismic data and 53 sediment cores up to 30 m in length were collected from Corpus Christi Bay along the central Texas coast in order to study the impact of sea-level and climate change on coastal environments over the last 10 k.y. Although coastal environments experienced a general landward migration as relative sea level rose over the last 10 k.y., this retreat was punctuated by three, possibly four, major flooding events. These flooding events are marked by abrupt changes in lithologic and seismic facies interpreted to represent rapid landward shifts of bay environments. Such changes include the back stepping of bayhead deltas, tidal deltas, oyster reefs, and other bay environments. Flooding events occurred at 8.0, 4.8, 2.6 ka, and possibly at 9.6 ka, and lasted only a few hundred years. The 9.6 ka flooding surface represents the initial drowning of the ancestral Nueces River valley and may or may not have been rapid in nature. The flooding surface that formed around 8.0 ka is interpreted to record either an increase in the rate of relative sea-level rise or the flooding of relict fluvial terraces that formed by the Nueces River during the stepped fall in sea level since marine isotope stage (MIS) 5e (120 ka). The 4.8 ka flooding event is thought to have formed as a result of either a climatic change during the mid-Holocene, characterized by warmer and drier conditions compared to present, and/or the flooding of another fluvial terrace. The most recent flooding event (2.6 ka) is thought to have resulted from a decrease in sediment delivery to the bay associated with a return to more mesic conditions similar to those of the present climatic regime.

Keywords: Holocene, Gulf of Mexico, sea level, incised valley, coastal evolution.

Simms, A.R., Anderson, J.B., Rodriguez, A.B., and Taviani, M., 2008, Mechanisms controlling environmental change within an estuary: Corpus Christi Bay, Texas, USA, *in* Anderson, J.B., and Rodriguez, A.B., eds., Response of Upper Gulf Coast Estuaries to Holocene Climate Change and Sea-Level Rise: Geological Society of America Special Paper 443, p. 121–146, doi: 10.1130/2008.2443(08). For permission to copy, contact editing@geosociety.org. ©2008 The Geological Society of America. All rights reserved.

INTRODUCTION

With the recent attention given to global warming, many studies have focused on the incipient increase in the rate of sea-level rise and its potential impact on coastal systems. However, coastal systems are dynamic and sensitive not only to changes in the rate of sea-level rise but also to other factors, such as climate change and the flooding of relict topography. The purpose of this study is to document the changes that occurred within the Corpus Christi Bay system over the last 10 k.y., a time period that is known to have experienced varying rates of sea-level rise and climatic changes. From this work, we document the effects of these and other agents of coastal change, such as the flooding of relict topography.

Corpus Christi Bay was chosen as a study site due to its location within the flooded ancestral valley (incised valley) of the Nueces River (Brown et al., 1976; Wright, 1980; Shideler, 1986; Durbin et al., 1997; Morton and Kindinger, 1998; Simms et al., 2006b). Incised valleys often preserve the most complete records of coastal change due to their high preservation potential in light of transgressive ravinement (Belknap and Kraft, 1981, 1985). In addition, most coastal studies have focused on systems that lie in humid or subhumid climate zones (Belknap and Kraft, 1985; Bratton et al., 2003; Baeteman et al., 2002), while this study documents the changes that have occurred within a semi-arid climatic setting.

STUDY AREA

Corpus Christi Bay lies along the central Texas coast of the northwestern Gulf of Mexico (Fig. 1). This portion of the Gulf of Mexico represents a passive margin with a very gentle coastal plain. The average coastal-plain gradient is 0.28 m/km, and the average continental-shelf gradient is 3 m/km, calculated from the location of the 200 m contour. The subsidence rates along the central Texas coast are very low, averaging less than 0.05 mm/yr (Paine, 1993).

Corpus Christi Bay has an average water depth of between 3 and 4 m (Fig. 1B). Nueces Bay, a shallower (average 1 m depth) tributary bay of Corpus Christi Bay, occupies the landward-most reaches of the flooded Nueces River. It is separated from Corpus Christi Bay by Rincon and Indian Points. Corpus Christi Bay is connected to the Gulf of Mexico through two ephemeral inlets: Aransas Pass and the historically active Corpus Christi Pass (Morton and McGowen, 1980; Fig. 1). The bay lies within a subhumid to semiarid climate zone (Thornthwaite, 1948) and receives 68–89 cm/yr of precipitation at the present coastline, with a decrease in precipitation with greater distances from the coast (Carr, 1967; White et al., 1983). Evaporation within the area exceeds precipitation by 16–20 cm/yr (White et al., 1983). Winds are predominately from the southeast (Lohse, 1956), although winter storms bring winds from the north (Morton and McGowen, 1980; Shideler, 1984). Hurricanes have undoubtedly played an important role in the evolution of the coast, especially in the lower bay area (Snedden et al., 1988). They are an important agent in the delivery of sand to the estuarine system via overwash and breaching of the barrier-island system, as well as the flooding of rivers and streams flowing into the bay. Over historic times, the probability of their occurrence is ~7% in any one year (Simpson and Lawrence, 1971; White et al., 1983), but the past frequency of hurricanes throughout the Holocene is unknown. The current astronomical tidal range of the bay is <0.3 m (Marmer, 1954; Shideler, 1984).

The Nueces River is the principal supplier of freshwater and sediment (Yeager et al., 2006), delivering 6.3×10^8 m^3 yr^{-1} of freshwater (Henley and Rauschuber, 1981; Mannino and Montagna, 1996) and 750,000 tons yr^{-1} of sediment to the bay (pre-dam; Shepard, 1953). The climate along the length of the Nueces River is similar to that of Corpus Christi Bay—generally a semi-arid climate with slightly drier conditions in its upper reaches, which receive just over 50 cm/yr of precipitation (Carr, 1967).

Sea Level

Sea level along the northwestern Gulf Coast has risen between 90 and 120 m over the last 20 k.y. (Fig. 2; Shepard and Suess, 1956; Curray, 1960; McFarlan, 1961; Nelson and Bray, 1970; Törnqvist et al., 2004a; Simms et al., 2007). Two camps have risen within the coastal and shelf community as to whether the postglacial rise was continuous (Shepard and Suess, 1956; McFarlan, 1961; Coleman and Smith, 1964; Törnqvist et al., 2004a) or episodic, with rapid increases in the rate of sea-level rise around 15, 11, 9, 8.2, 7, 2.5, and 1.2 ka (Rehkemper, 1969; Nelson and Bray, 1970; Penland et al., 1988; Thomas and Anderson, 1994; Törnqvist et al., 2004b). Recent work in the Trinity incised valley suggests that some of these sea-level events might have been the result of other mechanisms, such as the flooding of flat antecedent topography (Rodriguez et al., 2005). Morton et al. (2000) and Blum et al. (2001, 2002) have made arguments for a mid-Holocene highstand and fall in sea level since 6.5 ka.

Coral-based sea-level records indicate two rapid increases in the rate of sea-level rise occurring at 14.5 ka (meltwater pulse [MWP] 1-A) and 11.3 ka (MWP 1-B; Fairbanks, 1989; Chappell and Polach, 1991; Bard et al., 1996). Evidence for a much smaller climatic event with a possible small "eustatic" rise in sea level around 8.2 ka has been presented by Alley et al. (1997), and within the Gulf of Mexico, by Törnqvist et al. (2004b). Given the uncertainty in dating, the timing of the first two Gulf of Mexico events (15 ka and 11 ka) could correspond to MWP 1-A and 1-B, but, with the exception of the 8.2 ka event, the latter five events have no known "eustatic" equivalents. However, the resolution of these coral-based "eustatic" records is only 5 m (Lighty et al., 1982), and lower-amplitude events could have occurred.

Work from northwest Europe (Morner, 1980; Streif, 2004), the Atlantic Coast of the United States (Fletcher et al., 1993), and the Gulf Coast (Suku and Pizzuto, 1995; Donoghue and White, 1995; Walker et al., 1995; Froede, 2002; Stapor and Stone, 2004) suggests the presence of higher-order oscillations in sea level

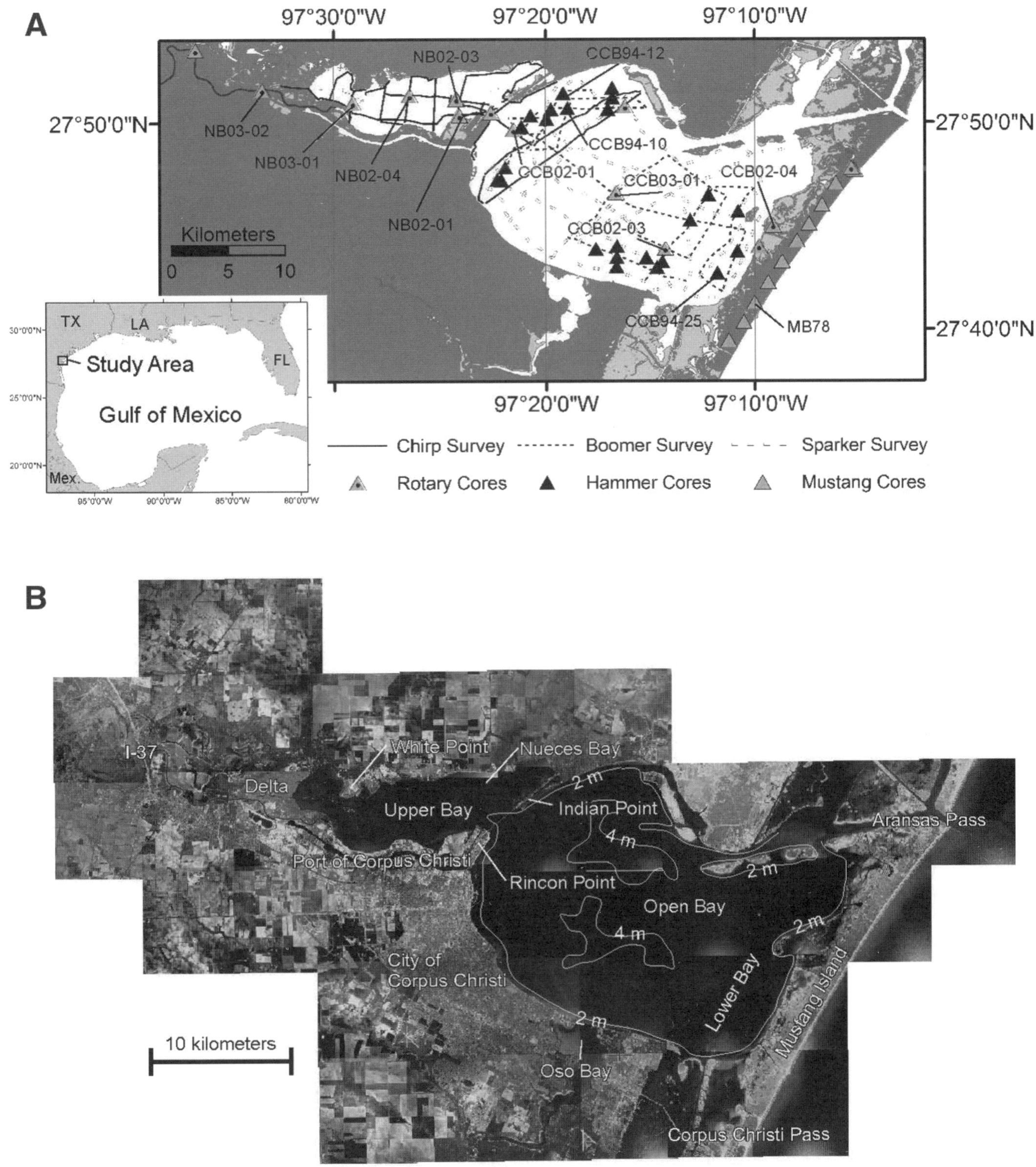

Figure 1. (A) Study area and data map. White areas represent water, dark gray areas represent land, and light gray areas represent intertidal areas. (B) Aerial photograph of Corpus Christi Bay showing the locations of major geographic features and bathymetric contours for the bay (data courtesy of the Texas General Land Office).

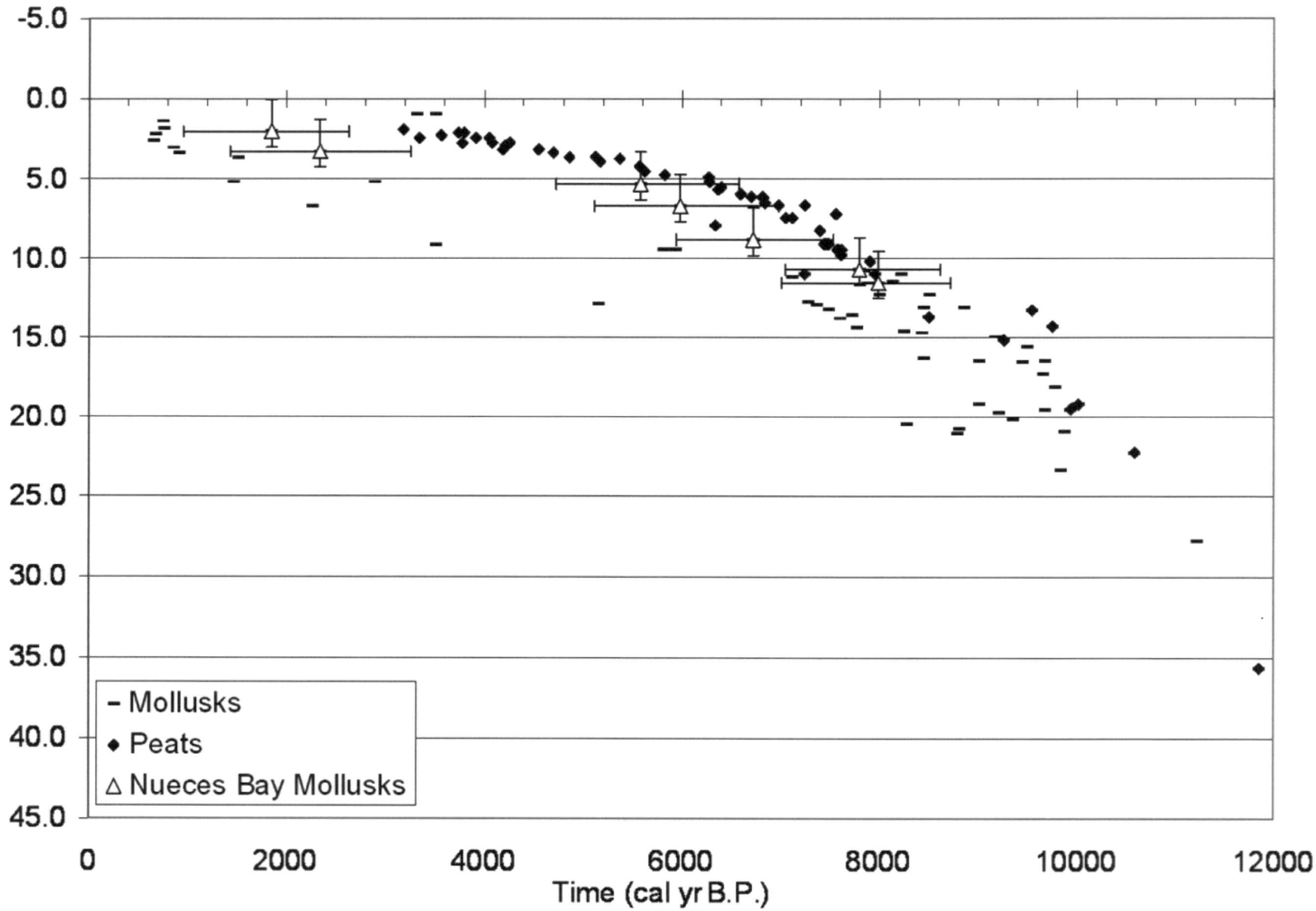

Figure 2. Sea-level indices from across the coastal areas of the northwestern Gulf of Mexico. Data come from Shepard and Moore (1955), Curray (1960), Rehkemper (1969), Nelson and Bray (1970), Rodriguez et al. (2004), Törnqvist et al. (2004a), and this study. Estuarine deposits were obtained from either wood fragments or mollusk shells. Peats were interpreted by each of the respective authors to represent basal peats. Nueces Bay dates were obtained from mollusk shells (this study).

with amplitudes less than 1 m over the last 6 k.y. We were unable to test the validity of these higher-amplitude oscillations within the study area since oscillations of this magnitude are below the resolution of this data set.

In addition, other effects such as glacio-hydro-isostasy and subsidence will cause differences between the relative sea-level history of the Gulf of Mexico and the study area and sea-level records from other locations globally (Lambeck and Chappell, 2001; Simms et al., 2007). Given the Gulf of Mexico's intermediate-field location (Peltier et al., 1978), the amplitudes but not direction of sea-level history could be different from the "eustatic" sea-level curve. For the purposes of this study, we use a compilation of relative sea-level curves based on several data sets from the Gulf of Mexico to bracket the sea-level history experienced by the Corpus Christi Bay system (Curray, 1960; Nelson and Bray, 1970; Rodriguez et al., 2004; Törnqvist et al., 2004a). Each of these sea-level curves suggest a continual slow rise in relative sea level since ca. 8 ka and decreases in the rate of relative sea-level rise at 5.5 and 3.2 ka (Fig. 2). This behavior has also been documented along the Atlantic Coast of the United States (van de Plassche et al., 1989; van de Plassche, 1990).

METHODS AND DATA

Data

Over 400 km of high-resolution seismic data and 53 sediment cores were used in this study (Fig. 1). In addition, over 90 core descriptions from the U.S. Army Corp of Engineers and the Texas Department of Transportation were used to help supplement this data set. Three seismic data sets were used, each from a different source: a minisparker survey conducted by the U.S. Geological Survey in 1978, a boomer survey conducted by Rice University in 1994, and another boomer survey conducted by the U.S. Geological Survey in 1996. The only processing of seismic lines was simple band-pass filters and automatic gain control. Three methods of coring were used, 25 hammer cores up to 4 m in length taken on the R/V *Lone Star*, 11 rotary cores up to 22 m in length, and 8 vibracores up to 5 m in length taken on the R/V *Trinity*. Nine push cores up to 30 m in length taken by the U.S. Geological Survey on Mustang Island were also used (Simms et al., 2006a). In addition to lithological data, 10 of the 11 rotary core sites were logged using a Mount Sopris 2PGA-1000 gamma logger. Gamma logs helped to constrain the lithology of sections where core sampling was incomplete or limited. This was most helpful in delineating the exact thicknesses of very sandy intervals with limited core recovery. Lithological data with radiocarbon ages were tied to seismic data to construct paleogeographic maps of the bay for the past 10 k.y.

Lithology

Grain-size analyses using laser diffractometry (Malvern Sizemaster 2000) were conducted at approximately every 25 cm within the cores. A refractive index of 1.544 was used following the "pipette" method of Sperazza et al. (2004). In addition to grain-size analyses, the samples were washed, and the sand-sized fraction (>63 um) was analyzed. Coarse-fraction analysis (relative percentage of sand, organic material, forams, and shell material) was developed for distinguishing marginal-marine and shallow-marine environments along the central Texas shelf and coast by Shepard and Moore (1954, 1955) and found to be of great use by Rehkemper (1969) in environmental interpretation within Galveston Bay in east Texas. A modified version of this method was used to help distinguish the 13 lithological units described in this study. Over 500 samples of the >63 μm fraction from 25 cores were randomly selected and sorted (without knowledge of their origin). In addition to the relative importance of major constituents within the >63 μm component, sorting and lithology were also used to help subdivide the samples. The macrofauna (mainly mollusks) and sedimentary structures (identified as homogeneous, or structureless, mottled, irregular bedding, or regular bedding; Moore and Scruton, 1957) within each core were also examined to help define and interpret the lithological facies.

Seismic Velocity

Seismic velocities were calculated based on the assumption that the most drastic seismic facies change came at the Pleistocene-Holocene contact or a distinct mud-sand contact. Six such contacts in cores and seismic profiles were found, resulting in calculated velocities of 1636 m/s, 1559 m/s, 1500 m/s, 1557 m/s, 1548 m/s, and 1680 m/s. The mean of these values results in an average seismic velocity of 1580 m/s. This number falls within the values obtained in a check-shot survey conducted on similar sediments offshore (Abdulah, 1995) and to other seismic studies of late Pleistocene–Holocene sediments of the Gulf Coast and shelf (Shideler, 1986; Sydow and Roberts, 1994; Anderson et al., 2004; Bartek et al., 2004). Although variations are expected from the averaged value, a remarkably good fit was found between lithologic and seismic facies boundaries.

Chronostratigraphy

Over 50 radiocarbon dates were obtained from mollusk and wood fragments (Table 1). When possible, only articulated mollusks were used. The samples were sent to the National Ocean Sciences Accelerator Mass Spectrometry Facility (NOSAMS) and Beta Analytic for dating and corrected at each laboratory for the ratio of $^{13}C/^{12}C$. Dates were remarkably consistent with depth in all of the cores taken within Corpus Christi Bay. Some inconsistencies in the age with respect to depth were found in cores obtained from the lower bay environment.

Carbon reservoir effects on ^{14}C dating are known to occur within the estuaries of the Gulf Coast (Aten, 1983) and within fluvial and marine systems around the world (Raymond and Bauer, 2001). To test for a possible carbon reservoir within the deposits from Corpus Christi Bay, two dates were obtained from an articulated barnacle growing on a wood fragment found within a muddy unit in core NB02-04 at a depth of 15.0 m. The barnacle

TABLE 1. RADIOCARBON DATES COLLECTED FOR THIS STUDY

Core	Depth (m)	Depth (m bsl)	Species	Uncorrected age (yr)	Error (±yr)	Reservoir (yr)	Reservoir error (yr)	Calibrated age (yr B.P.)	2σ (yr)
Corpus Christi Bay									
CCB02-01	3.47	8.04	*Nuculana concentrica*†	2310	40	760	380	1500	830
CCB02-01	4.78	9.35	*Abra ovalis*†	2530	45	760	380	1740	970
CCB02-01	6.22	10.79	*Nuculana concentrica*‡	3290	35	760	380	2610	950
CCB02-01	8.12	12.69	*Mulinia lateralis*†	4550	40	760	380	4200	1090
CCB02-01	9.47	14.04	*Nuculana concentrica*†	6270	40	760	380	6310	940
CCB02-01	10.94	15.51	*Nuculana concentrica*	7520	130	760	380	7620	870
CCB02-01	12.6	17.17	*Mulinia lateralis*	7750	50	760	380	7850	820
CCB02-01	15.62	20.19	*Mulinia lateralis*	8750	40	760	380	8900	890
CCB02-01	17.01	21.58	*Brachidontes exustus*	9360	55	760	380	9650	1040
CCB02-01	20.29	24.86	Wood fragment	5690	30	0	0	6470	80
CCB02-03	4.7	10.18	*Mulinia lateralis*	4480	40	760	380	4110	1160
CCB02-03	8.23	13.71	*Mulinia lateralis*	4280	50	760	380	3850	990
CCB02-03	12.22	17.7	*Nuculana concentrica*	1730	35	760	380	930	930
CCB02-04	5.6	7.73	*Anomalocardia cuneimeris*	2130	30	0	0	2110	150
CCB02-04	7.01	9.14	*Anomalocardia cuneimeris*	2130	35	0	0	2110	150
CCB02-04	11.82	13.95	*Mulinia lateralis*	4620	40	0	0	5420	70
CCB02-04	14.85	16.98	*Anomalocardia cuneimeris*	2060	35	0	0	2020	95
CCB03-01	3.52	7.85	*M. lateralus* or *R. flexuosa*‡	2390	30	760	380	1585	850
CCB03-01	6.66	10.99	*Mulinia lateralis*†	4470	35	760	380	4100	1160
CCB03-01	7.64	11.97	*Rangia flexuosa*‡	5010	30	760	380	4800	1070
CCB03-01	12.71	17.04	Unknown fragment	6520	55	760	380	6590	840
CCB03-01	13.48	17.81	*M. campechiensis texana*	6780	45	760	380	6870	860
CCB03-01	14.73	19.06	*M. campechiensis texana*	7500	45	760	380	7600	800
CCB03-01	15.44	19.77	*Nuculana concentrica*†	7890	60	760	380	7980	970
CCB03-01	16.29	20.62	*Abra aequalis*†	8050	40	760	380	8140	860
CCB03-01	17.25	21.58	*Nuculana concentrica*†	8470	40	760	380	8590	880
CCB03-01	20.09	24.42	*Nuculana concentrica?*	9240	65	760	380	9480	1000
Nueces Bay									
NB02-01V	0.48	2.92	*Mulinia lateralis*†	730	45	760	380	Modern	
NB02-01 V	1.66	4.1	*Crassostrea virginica*†	3260	60	760	380	2570	950
NB02-01	5.18	7.62	*Nuculana concentrica*	4710	55	760	380	4400	1050
NB02-01	7.31	9.75	*Nuculana concentrica*	4880	60	760	380	4630	990
NB02-01	10.28	12.72	*Nuculana concentrica*†	7060	60	760	380	7140	820
NB02-01	12.07	14.51	Barnacle†	8000	65	760	380	8090	910
NB02-03	5.4	6.62	*Nuculana concentrica*†	4420	55	3660	380	4030	1030
NB02-03	7.13	8.35	*Abra ovalis*‡	6420	60	5660	380	6480	840
NB02-04	15.03	16.25	Barnacle†	8430	45				
NB02-04	15.04	16.25	Wood	7670	45	0	0	8484	100
NB03-01	1.2	2.02	*Mulinia lateralis*†	2630	45	1870	380	1850	900
NB03-01	2.45	3.27	*Mulinia lateralis*‡	3060	45	2300	380	2340	920
NB03-01	4.5	5.32	*Mulinia lateralis*‡	5640	60	4880	380	5580	1000
NB03-01	5.9	6.72	*Macoma mitchelli*†	5980	60	5220	380	5980	1000
NB03-01	8	8.82	*Nuculana acuta*	6640	50	5880	380	6720	830
NB03-01	9.86	10.68	*Rangia flexuosa*†	7700	70	6940	380	7800	800
NB03-01	10.71	11.53	*Nuculana concentrica*†	7900	100	7140	380	7990	980
NB03-02	0.99	0.99	Plant/Wood	30	–33.2	0	0	30	33
NB03-02	3.67	3.67	Plant/Wood	35	–60.5	0	0	35	60
NB03-02	3.96	3.96	*Macoma mitchelli*‡	5160	40	760	380	4990	1010
NB03-02	4.77	4.77	*Macoma constricta*†	2990	30	760	380	2250	910
Mustang Island									
MB78	7.5	6.5	*Anadara* sp.‡	5440	40	0	0	6240	65
MB78	13.5	12.5	*Donax* sp.‡	4500	35	0	0	5170	130
MB78	16.7	15.7	*Mulinia lateralis*‡	5700	30	0	0	6480	80
MB78	20.8	19.8	*Donax* sp.‡	5730	30	0	0	6520	110

†Articulated.
‡Valve.

and wood fragment were dated separately. The date obtained from the barnacle was 760 yr older than that obtained from the wood fragment. The difference in age between the barnacle and wood fragment is consistent with a radiocarbon reservoir—the mollusk yielded an older age in comparison with the plant fragment. Although variability in the radiocarbon reservoir probably exists over time and with species, this value was assumed to represent the reservoir correction for mollusks within Corpus Christi Bay. To account for variations in the reservoir, we used a reservoir uncertainty of 50% when calibrating the ages. A reservoir value of 760 yr falls within the range of other reservoir corrections found in other estuaries of the Gulf Coast (Aten, 1983) and brings near-surface dates obtained in this study to modern values.

This radiocarbon reservoir was applied to all dates obtained from mollusks that are known to inhabit estuaries. Dates obtained from whole valves of *Donax* sp. were assumed not to be affected by a radiocarbon reservoir due to their habitation of the high-energy surf zone adjacent to the open ocean, which is thought to have a rapid exchange of ^{14}C with the atmosphere (Stuiver and Braziunas, 1993). After correcting for the reservoir effect where appropriate, carbon dates were calibrated to calendar years using the terrestrial Northern Hemisphere curve of the Calib 4.4 program (Stuiver and Braziunas, 1993; Stuiver et al., 1998). All dates discussed in this paper are calibrated, corrected for ^{13}C content, corrected for reservoir effects, and given in calendar yr B.P. When dates from other sources are given, the same procedure was followed.

RESULTS AND INTERPRETATION

Lithology

Apart from Pleistocene deposits, four broad types of facies were distinguished within the Holocene deposits: mud facies, sand facies, shell hash facies, and oyster facies. Of these four facies, mud is the most variable and widespread. All mud facies contain less than 20% sand and record values of >20 counts per second (cps) in gamma logs. All the sand facies consist of greater than 20% sand, most around 50% sand, and a gamma log value <20 cps. A brief summary of each lithofacies is given in Table 2.

Mud Facies (MF)

MF1. Mud facies 1 is a light gray structureless mud with an average sand content of 10% (Fig. 3A). Plant fragments are not present in appreciable amounts. The sand component is extremely well-sorted, quartz-rich (>95%), and fine to very fine in size. The facies contains sparse shells, and forams constitute, on average, a larger portion of the coarse fraction than shell material. Of the few shells present, the populations are dominantly *Nuculana acuta*, *N. concentrica*, and *Mulinia lateralis*. This facies occurs within cores from depths down to 20 m below sea level (bsl) from two general areas: middle Nueces Bay updip from Rincon and Indian Point and middle Corpus Christi Bay. Samples from the middle of Corpus Christi Bay contained an occasional *Anomia simplex*, *Ostrea equestris*, *Mercenaria campechiensis texana*, *Abra aequalis*, or *Dentalium texasianum*, while the samples from Nueces Bay contained an occasional juvenile *Rangia flexuosa* and *Macoma mitchelli*. Ostracodes and possible fecal pellets were also present within the facies within Nueces Bay. Mud facies 1 is interpreted to represent an open-bay deposit and reflects deposition that is well removed from any significant sources of sand (i.e., bayhead delta or bay margins). The open-bay environment in Corpus Christi Bay is more saline than within Nueces Bay, indicating an open-bay environment located distal from the bayhead delta within Corpus Christi Bay and restricted or interdistributary bay located behind the main distributaries of the active bayhead delta in Nueces Bay.

MF2. Mud facies 2 is generally darker gray and contains more sand and shell material than MF1 (Fig. 3B). Mud facies 2 lacks sedimentary structures. Of its coarse fraction, shells are the most abundant constituent. Forams are less abundant than sand grains. The mineralogic fraction is an extremely well-sorted, quartz-rich (>95%), fine to very fine sand. The shells within this facies are often concentrated in discrete layers, presumably storm deposits. The most abundant shells within this facies are *Nuculana acuta*, *N. concentrica*, *Mulinia lateralis*, *Mercenaria campechiensis texana*, *Abra aequalis*, *Chione cancellata*, and *Dentalium texasianum*. This facies is present in the upper 3 m of almost every core collected in Corpus Christi Bay. It is never found below a depth of 5 m. MF2 is interpreted to be deposits that accumulated in an open-bay setting analogous to the environmental conditions within the present Corpus Christi Bay.

MF3. Mud facies 3 consists of a laminated gray mud with sparse to no shells (Fig. 3E). Mineralogic sand is the most abundant coarse fraction component (>90%). Forams, shells, and plant fragments are present but in very small quantities. Generally, the mineralogic sand component is well-sorted, fine to very fine in size, and quartz-rich. Shells include juvenile *Rangia flexuosa* and *Macoma mitchelli*. This lithofacies is interpreted to represent a prodelta environment.

MF4. Mud facies 4 consists of an irregular (nonplanar, discontinuous, or wavy) bedded gray mud (Fig. 3C). Mineralogic sand is the dominant coarse fraction, and forams and shells constitute only a minor component. The mineralogic sand fraction is generally poorly sorted and is mostly fine-grained; however, medium lithic grains also occur. Organic material is present within this facies, often in discrete irregular layers. Shells include juvenile *Rangia flexuosa*, *Mulinia lateralis*, and an occasional oyster (*Crassostrea virginica* or *Ostrea equestris*). This lithofacies is interpreted to represent upper-bay deposits.

Sand Facies (SF)

SF1. Sand facies 1 is actually a sandy mud. A large proportion of the sand component is found within sand-filled burrows or root casts (Fig. 3D). The sand fraction is moderately sorted and composed dominantly of quartz. Shells and forams are rare to absent. Plant material is an abundant constituent of this facies.

TABLE 2. SUMMARY OF LITHOFACIES DESCRIBED IN THIS STUDY, THEIR EQUIVALENT SEISMIC FACIES, AND ENVIRONMENTAL INTERPRETATION

Lithofacies	Brief description	Seismic facies	Interpreted depositional environment
MF1	Structureless gray mud, <10% sand, few shells	F1, F4	Open bay
MF2	Sandy gray mud, shell layers	F1	Open bay
MF3	Well-laminated gray mud, very little shell material	F2	Prodelta
MF4	Irregularly bedded gray sandy mud	F2	Upper bay
SF1	Sandy organic-rich mud	F2	Delta plain
SF2	Poorly sorted, medium sand, lithic-rich	F2, F6	Mouth bar
SF3	Muddy sand/sandy mud, mixed brackish-marine fauna	F5	Lower bay
SF4	Clean, very-well-sorted quartz sand	NA	Barrier island
SF5	Structureless, light gray sandy mud	F2, F5	Delta front
SF6	Interbedded fine sand and mud	F5	Distal flood-tidal delta
ShF1	Well-sorted shell hash, dominantly oyster fragments	Cuspate seismic features	Interbay tidal inlet
ShF2	Sandy shell hash, mixed marine/brackish fauna	NA	Tidal inlet
OF1	Muddy oyster shells	F3	Oyster reef

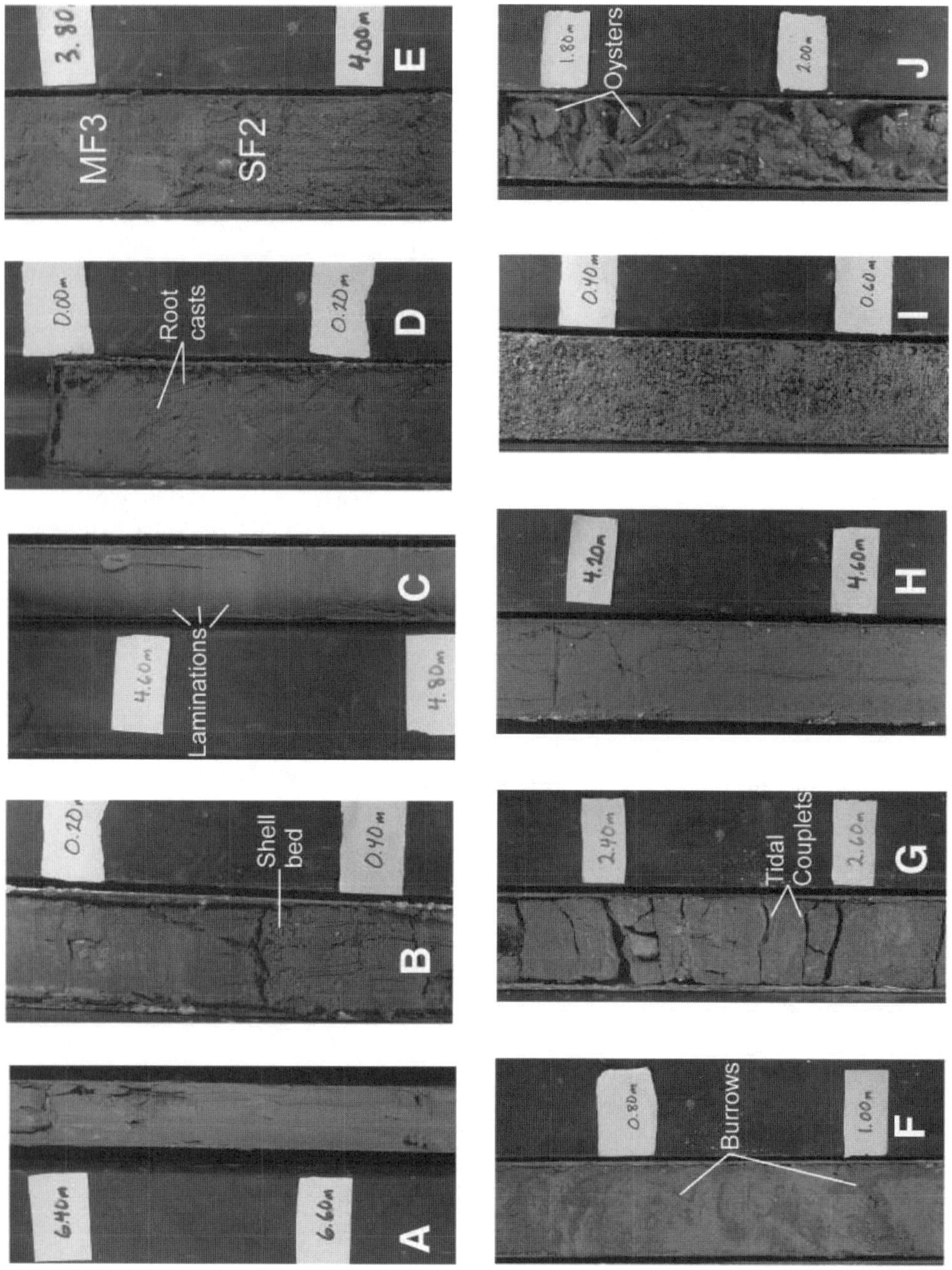

Figure 3. Examples of selected lithofacies from cores taken within Corpus Christi Bay. (A) MF1 from core CCB03-01, (B) MF2 from core CCB94-10, (C) MF4 from core NB03-01, (D) SF1 from core NB03-02, (E) SF2 below MF3 from core NB03-02, (F) SF3 from core CCB94-25, (G) SF5 from core CCB94-3, (H) SF6 from core NB02-01VC, (I) ShF1 from core NB02-03VC, and (J) OF1 from core NB02-03 VC. The vertical scale for each photograph is 10 cm. MF—mud facies; SF—sand facies; Shf—shell-hash facies.

SF1 is interpreted to represent delta-plain deposits of the Nueces bayhead delta.

SF2. Sand facies 2 is a structureless, muddy, poorly sorted, medium sand (Fig. 3E). The mineralogic sand component is relatively lithic-rich (<75% quartz). It has abundant to no shell and plant material. Occasionally, this facies has abundant oyster-reef fauna (*Crassostrea virginica*, *Ostrea equestris*, *Brachidontes* sp.). Other fauna include *Macoma mitchelli*, *Tagelus plebeius*, *Rangia flexuosa*, and *Turbonilla* sp. It is interpreted to represent a mouth bar within a bayhead delta.

SF3. Sand facies 3 is a muddy sand/sandy mud. The mineralogic sand component is fine, well-sorted, and quartz-rich (>95%). It has few to abundant shells. The fauna, being mixed in origin with both brackish- and open-marine species, is the most diagnostic characteristic. *Anomalocardia auberiaus*, *Polymesoda maritima*, *Crassinella lunulata*, *Lucina amiantus*, *Lucina multilineata*, *Donax* sp., *Cerithium lutosum*, *Anadara transversa*, *Ostrea equestris*, *Mulinia lateralis*, and *Chione cancellata* are found within this facies. Occasionally, the facies shows irregular bedding, mottling, or burrows (Fig. 3F). In longer cores, it lacks sedimentary structures. It is interpreted to represent lower-bay (washover) deposits.

SF4. Sand facies 4 is a clean, quartz-rich, very well-sorted, fine to very fine sand. It has a uniform grain size of 2.6–2.7 phi and contains less than 10% mud. No sedimentary structures were preserved within the cores that sampled SF4. Shell and organic material are occasionally found. When present, fauna are dominantly *Donax* sp. and *Anomalocardia cuneimeris*. This facies is interpreted as undifferentiated barrier-island deposits. Distinction between barrier-flat, beach, and eolian deposits was attempted (Simms et al., 2006a) but not important to this study.

SF5. Sand facies 5 is similar to sand facies 1. However, sand facies 5 is generally lighter gray in color and contains less sand and more shell material (Fig. 3H). In addition, root casts and burrows are absent, and no other sedimentary structures are found. The fauna within this facies include *Turbonilla* sp., *Nuculana acuta*, *Nuculana concentrica*, and *Dentalium texasianum*. This facies is interpreted to be delta-front deposits based on these criteria and its association with clinoforms in seismic profiles.

SF6. Sand facies 6 is an interbedded fine sand and mud (tidal couplets; Fig. 3G). The sand component is fine to very fine in size. Very few shells are found directly within this facies; however, this facies is always found in association with SF3 and its accompanying fauna. This deposit is interpreted to be a distal flood-tidal delta facies.

Shell-Hash Facies (ShF)

ShF1. This shell-hash facies is a structureless well-sorted shell hash (Fig. 3I). It is composed almost exclusively of millimeter-sized oyster fragments (*Crassostrea virginica* and possibly *Ostrea equestris*) and associated mussels (*Brachidontes* sp.). This facies is only found in the top of a core near Rincon and Indian Points (Fig. 1), which separate Corpus Christi and Nueces Bays. It is interpreted to be a modern interbay tidal-inlet deposit. Wind tides rather than tidal circulation are thought to have been the dominant process operating in this environment due to the relatively low tidal range and depth of Nueces Bay and the strong winds that persist within the bay, which is aligned parallel to the dominant wind direction (Shideler, 1984).

ShF2. Shell-hash facies 2 is a structureless, poorly sorted sandy shell hash. Occasional whole valves can be found. The fauna is more diverse than ShF1, containing both estuarine and shallow-marine fauna. Fragments of *Nuculana acuta*, *Nuculana concentrica*, *Lucina multilineata*, *Lucina amiantus*, and *Chione cancellata* are all present. This deposit is interpreted to be a proximal tidal-inlet facies.

Oyster Facies (OF)

OF1. Oyster facies 1 consists of articulated oysters (mostly *Crassostrea virginica* with an occasional *Ostrea equestris*) within a muddy matrix (Fig. 3J). Other fauna include *Brachidontes* sp., *Chione cancellata*, *Nuculana acuta*, and *Anomia simplex*.

Pleistocene

Pl. Undifferentiated Pleistocene deposits were sampled in a few locations. Generally, they consist of a very stiff gray or tan mottled mud with calcareous nodules and a gamma log value >60 cps. A mollusk shell taken from the stiff mud unit beneath Mustang Island yielded a radiocarbon dead result (Simms et al., 2006a). Beneath Mustang Island, a sand unit was also interpreted to be Pleistocene in age. It is thought to represent alluvium associated with buried fluvial "Deweyville" terrace deposits based on its more lithic and coarse-grained nature, yellow color, and stratigraphic position (Simms et al., 2006a).

Seismic Facies

Six seismic facies were found above the interpreted Pleistocene-Holocene contact. These are illustrated in Figure 4.

F1

Seismic facies 1 consists of horizontal, parallel, medium-amplitude continuous reflectors (Fig. 4). Seismic facies 1 often contains large mounds of seismic facies 3. In terms of lithology, seismic facies 1 corresponds with MF1 and MF2. F1 is interpreted as open-bay deposits.

F2

Seismic facies 2 is composed of nearly parallel, high-amplitude, nearly continuous reflectors (Fig. 4). F2 has more wavy and irregular reflectors relative to F1 and often contains smaller mounds of F3. When sampled, this seismic facies corresponds with lithological facies SF1, SF2, SF5, MF3, MF4, and, in two cases, MF1. F2 is interpreted to represent an upper-bay environment.

F3

Seismic facies 3 is composed of mounded, high-amplitude reflectors (Fig. 4). Continuous reflectors on either side onlap these mounds. This seismic facies was sampled by core NB02-01 VC, which sampled lithological facies OF1. In addition, cores sampling the surrounding seismic facies F1, F2, and F4, when near mounds of F3, often contained fragments of oyster shells (*Crassostrea virginica* and *Ostrea equestris*). F3 is interpreted as oyster reefs.

F4

Seismic facies 4 is nearly acoustically transparent (Fig. 4). Faint continuous, parallel reflectors occasionally occur. In terms of lithology, F4 corresponds with MF1. In addition, samples of MF1 within F4 always contain very little sand (<5%). This seismic facies is interpreted to represent open-bay deposits.

F5

Seismic facies 5 is composed of sigmoid-oblique to oblique-parallel reflectors (Fig. 4). Cores that sampled F5 contain lithological facies SF3 or SF5. This seismic facies is interpreted to represent deltaic deposits (either bayhead or tidal deltas).

F6

Seismic facies 6 is chaotic in nature (Fig. 4). The cores that sampled F6 contain lithological facies SF2. This seismic unit is

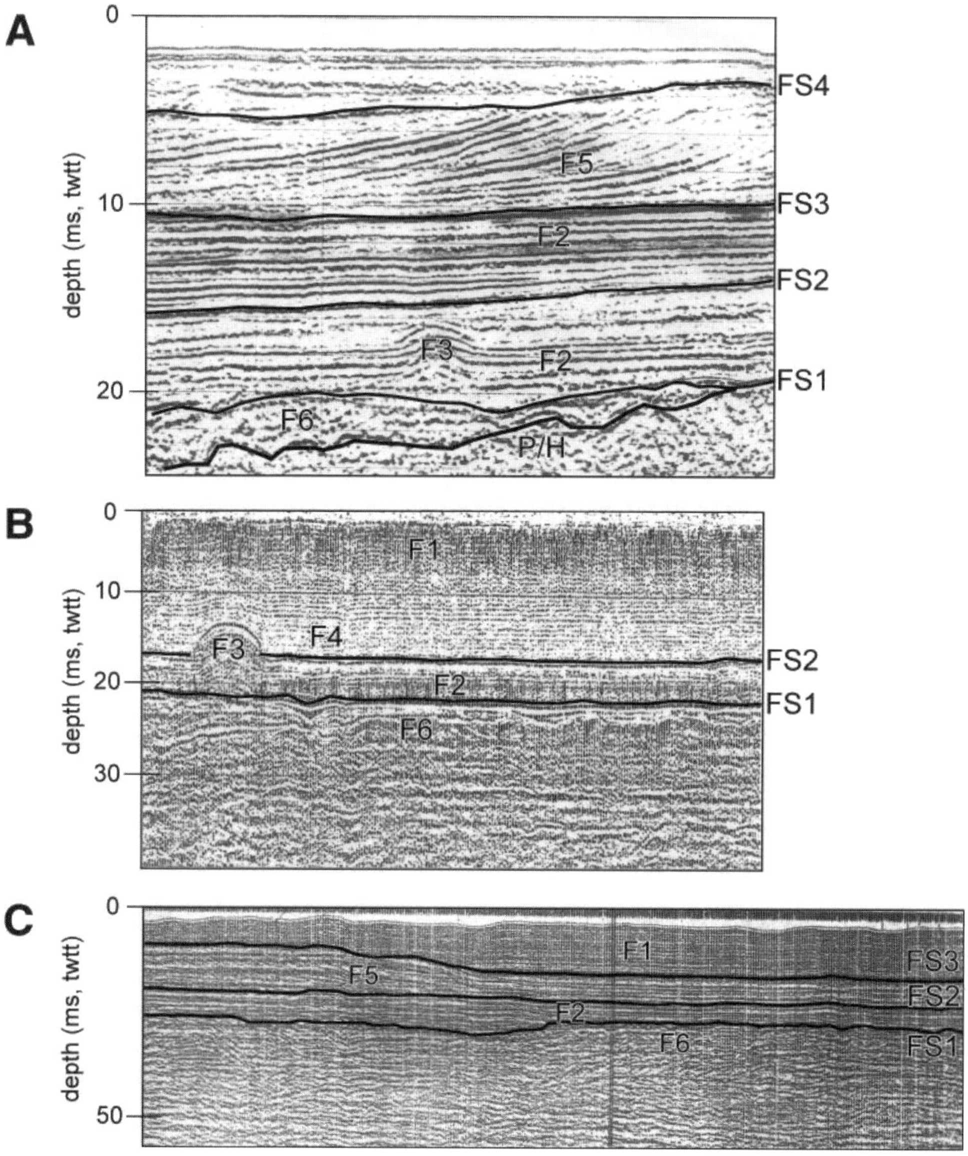

Figure 4. Illustration of the six major seismic facies and four major flooding surfaces recognized in this study: (A) selected facies for the U.S. Geological Survey boomer data, (B) selected facies for Rice University boomer data, and (C) selected facies for sparker data. twtt—two-way traveltime.

interpreted to represent fluvial-deltaic deposits that rest directly above the Pleistocene-Holocene contact. The U.S. Geological Survey Boomer data did not image this contact well when fluvial deposits rested above it. This is most likely due to the low impedance contrast between fluvial terrace deposits formed during the late Pleistocene fall in sea level and fluvial deposits formed during the Holocene rise in sea level. However, the lower-frequency boomer and sparker data did a better job imaging the Pleistocene-Holocene contact.

DISCUSSION

Bay Evolution

The general succession of facies within the deposits beneath Corpus Christi Bay illustrates an overall transgression. This general vertical succession of facies is punctuated by landward shifts in facies bounded by discrete flooding surfaces (sensu Anderson et al., 2001; transgressive overlap boundaries (TOBs) of Fletcher et al., 1993). Three, possibly four (FS1–FS4), major flooding surfaces occur. Flooding surfaces were observed in both seismic profiles and cores by sharp contacts between seismic and lithologic facies (Figs. 4 and 5). Radiocarbon ages from material above and below the flooding surfaces suggest that they formed in as little as 200 yr. Paleogeographic maps were constructed based on the facies found between flooding surfaces.

Pleistocene-Holocene Structure Map

The first task undertaken in this study was the identification and mapping of the Pleistocene-Holocene contact (sequence boundary) beneath the deposits of Corpus Christi Bay. This irregular surface is the antecedent topography that was flooded during the transgression and represents the incised valley of the Nueces River that formed during the fall in sea level that culminated during the Last Glacial Maximum at 20 ka (Wright, 1980; Eckles et al., 2004; Simms et al., 2006b). In seismic profiles, this boundary was best delineated with sparker data. It was identified by reflectors that are erosionally truncated below and onlap above the surface (Mitchum et al., 1977; Fig. 6). In cores and seismic profiles, this surface represents the Pleistocene-Holocene boundary. This boundary was picked in Corp of Engineer boring descriptions as the contact between a soft or very soft gray mud or silty sand and a tan or gray stiff mud with calcareous nodules. Anything in the descriptions with pocket penetrometer values less than 1 (all values fell between 0.0 and 0.25) was assumed to be Holocene, while values measuring greater than 1 (all values were >2.0) were assumed to be Pleistocene. The contacts that were picked using this method matched remarkably well with the seismic interpretations and those from nearby rotary-drill cores. The map created using these methods is illustrated in Figure 7.

One prominent feature of the Pleistocene-Holocene surface is the steps observed in seismic profiles and cores (Fig. 6). The steps observed in the Pleistocene Holocene contact are interpreted to represent fluvial terraces that formed during the fall in sea level (120–20 ka). The same terraces have been mapped and dated using thermoluminescence dating in the lower portions of the Nueces River valley landward of the study area (Fig. 6B; Durbin et al., 1997).

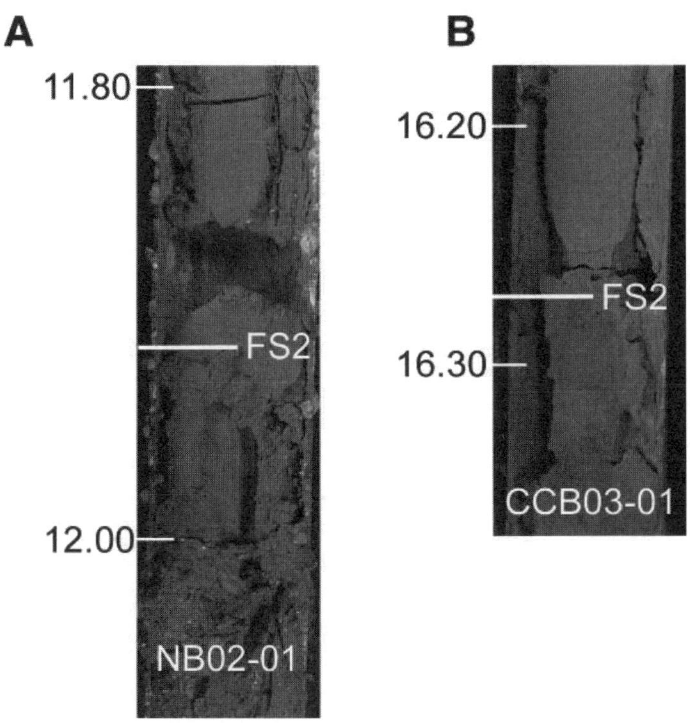

Figure 5. Core photographs of sharp lithological contacts between (A) facies SF2 and SF1 in core NB02-01 and (B) facies MF4 and MF1 in core CCB03-01, associated with the FS2 flooding surface. MF—mud facies; SF—sand facies.

Initial Drowning (FS1)

Seismic facies F6 lies directly above the Pleistocene surface and is interpreted as fluvial deposits of the Nueces River. A Texas Department of Transportation boring that sampled this unit was described by Wright (1980) as fining-upward coarse gravels to sand. These deposits lie above a large unconformity (our Pleistocene-Holocene structure map) interpreted by Wright (1980) to be the sequence boundary. The oldest bay deposits sampled in this study consist of lithofacies MF4 and correspond to seismic facies F2. These deposits lie above the fluvial coarse gravels and sands described by Wright (1980) and above a unit of SF2 interpreted to represent a mouth-bar environment. Two radiocarbon dates obtained from the bay unit resulted in ages of 9480 ± 1000 yr B.P. within CCB03-01 and 9650 ± 1040 yr B.P. within CCB02-01 (Figs. 8A and 8B). Seismic profiles indicate that the boundary between the underlying fluvial (F6) and upper bay (F2) deposits is sharp.

Flooding Surface 2 (FS2)

Based on radiocarbon dates in cores NB02-01, CCB02-01, and CCB03-01, FS2 formed around 8.0 ka (Fig. 8B). At this point, a large portion of central Corpus Christi Bay experienced a change from an upper-bay (seismic facies F2) to open-bay environment (F1 or F4), and, in places, these open-bay facies directly onlap Pleistocene strata (Fig. 6B). At the location of core CCB03-01, the lithofacies changes from MF4 to MF1 (Figs. 5 and 8A), marking a transition from an upper- to open-bay environment. Within cores NB02-01 and CCB02-01, the lithofacies changes from bayhead delta to upper bay (Figs. 5 and 8A). Figure 5 illustrates the sharp lithologic contacts that mark this flooding event. This change in lithology is also manifest in gamma-ray logs (Fig. 9). Prior to the formation of FS2, most oyster reefs (seismic facies F3) existed seaward of core CCB03-01. During this flooding event, they shifted landward to a location between core sites CCB03-01 and CCB02-01. These seismic and lithological changes represent back stepping of the upper-bay environment by 15 km in less than 200 yr (Figs. 10A and 10B).

Bayhead Delta Progradation

Around 6.8–7.0 ka, after FS2 but before FS3 formation, the bayhead delta of the Nueces River stopped retrograding and prograded into what is now Nueces Bay. In seismic profiles, clinoforms dipping both seaward and landward characterize this feature (Fig. 11). It has been interpreted as both a bayhead delta (Morton and Kindinger, 1998) and an interbay tidal-delta complex (Simms et al., 2003). Seismically, it has the attributes of a tidal-delta complex in that the clinoforms dip both landward and seaward. Several *Lucina amiantus* and *Crassinella lunulata* were sampled in other nearby shallow cores (i.e., hammer cores CCB94-8 and CCB94-9) from more seaward portions of the feature, suggesting proximity to an inlet or some other marine influence. Also, a fluvial feeder was not identified. However, long cores through the delta sampled plant fragments, along with restricted and more freshwater-influenced fauna (*Rangia*

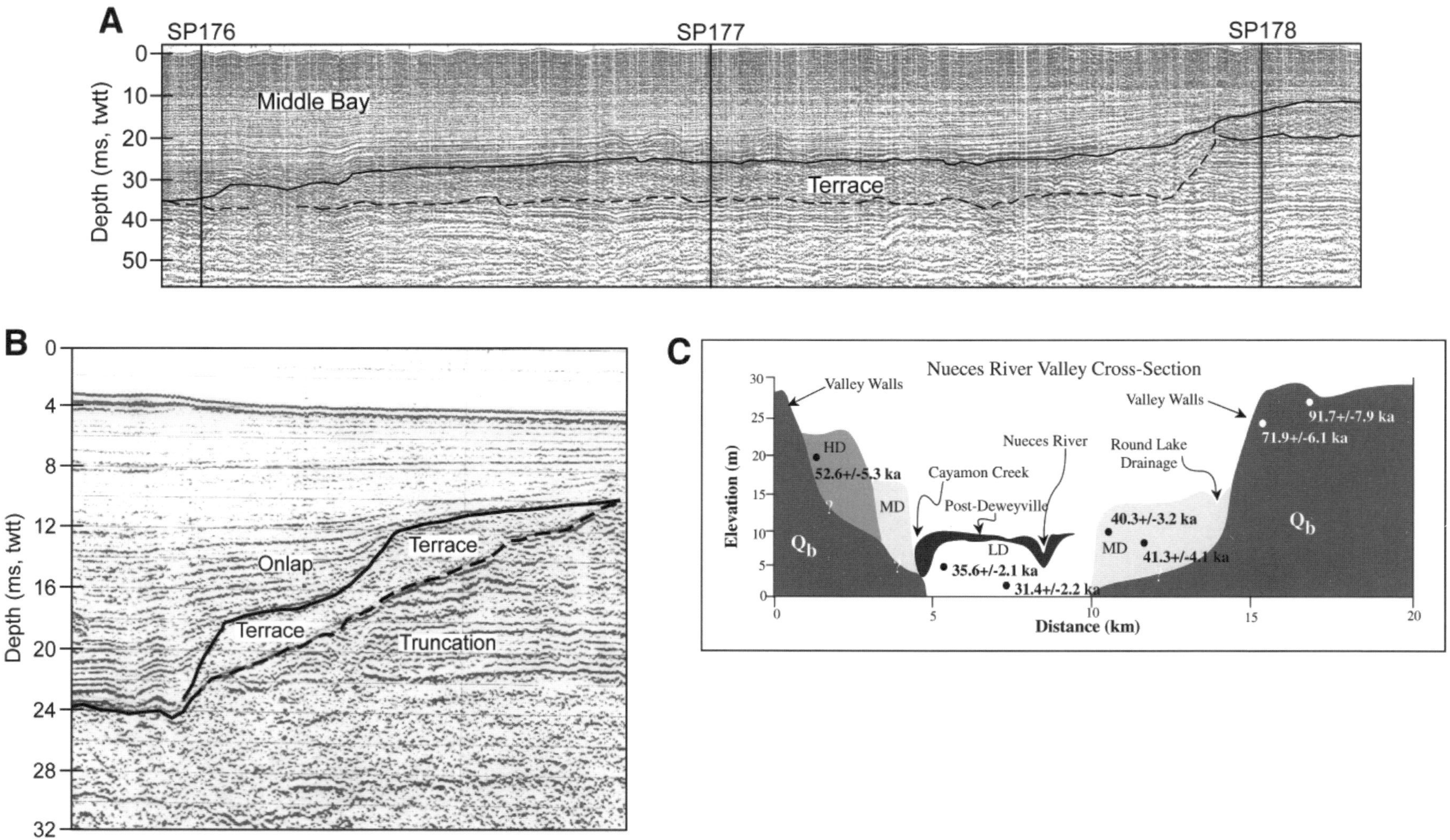

Figure 6. (A) Seismic line illustrating truncation below, onlap above, and distinct steps in the interpreted Pleistocene-Holocene contact. These steps are thought to be fluvial terraces formed during the last fall in sea level. (B) Fluvial terrace as illustrated in lower-resolution sparker profile. (C) Cross section of the Nueces River valley ~15 km updip from Corpus Christi Bay, illustrating fluvial terraces mapped on shore (Durbin et al., 1997). LD—Lower Deweyville; HD—High Deweyville; MD—Middle Deweyville; Qb—Quaternary Beaumont Formation; twtt—two-way traveltime.

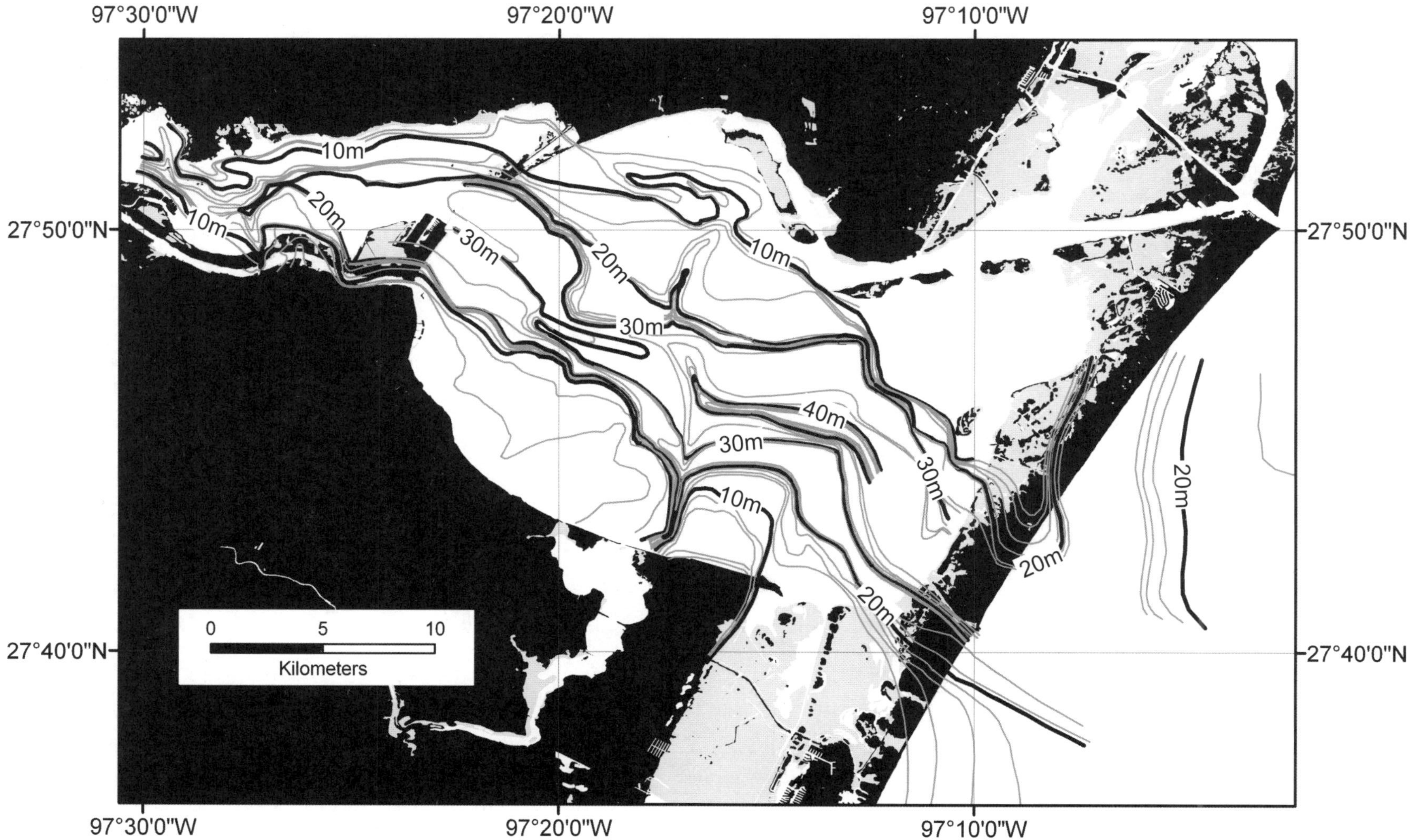

Figure 7. Elevation (meters below sea level) of the Pleistocene-Holocene contact beneath Corpus Christi Bay. Contour interval is 2 m.

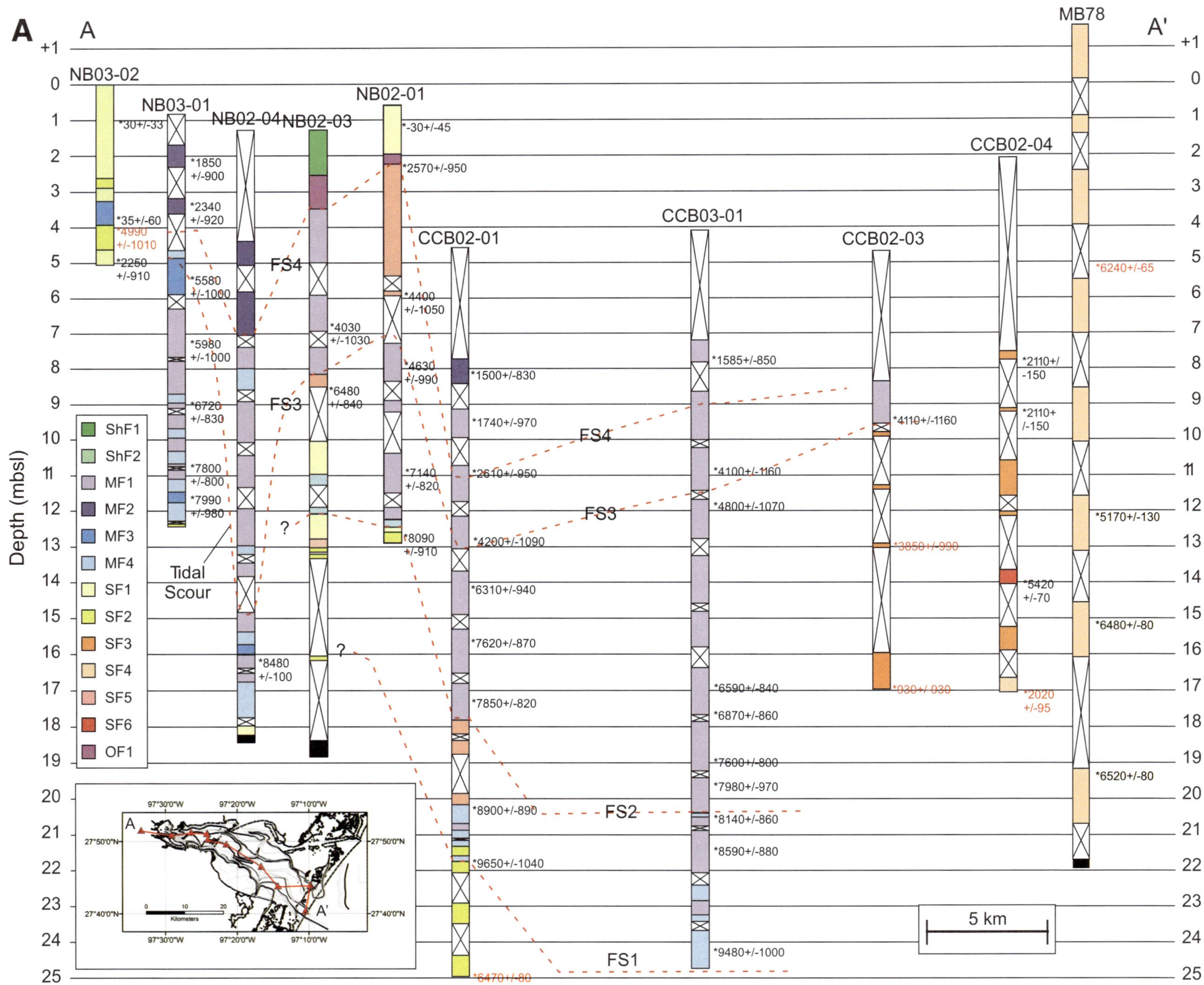

Figure 8 (*on this and following page*). (A) Dip-oriented core transect within Corpus Christi Bay illustrating lithofacies described in text. Dates in red are suspect.

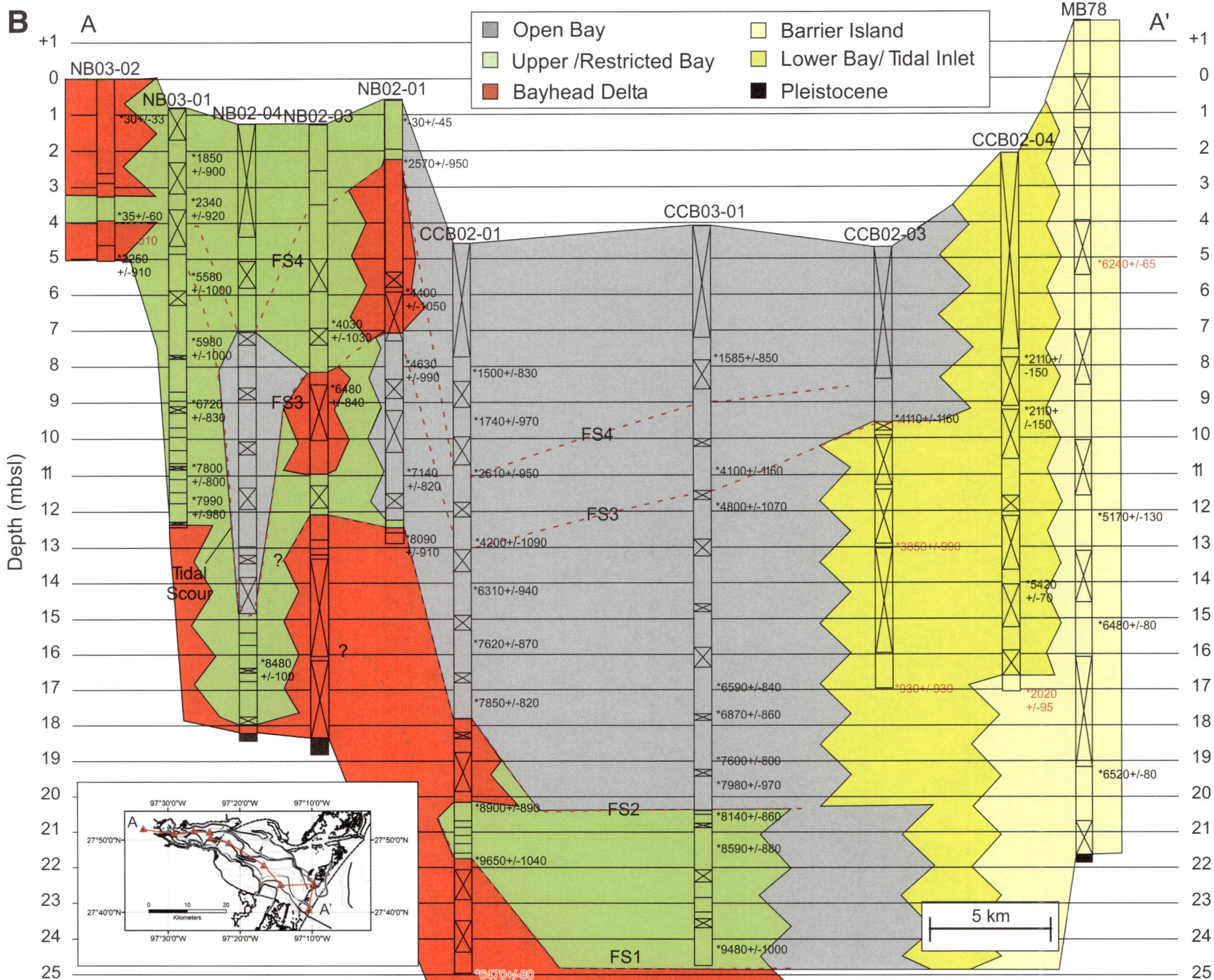

Figure 8 (*continued*). (B) Dip-oriented core transect within Corpus Christi Bay illustrating interpreted environments with isochrons. Dates in red are suspect. Dates in cal yr B.P. FS—flooding surface; MF—mud facies; SF—sand facies; ShF—shell-hash facies; OF—oyster reef facies.

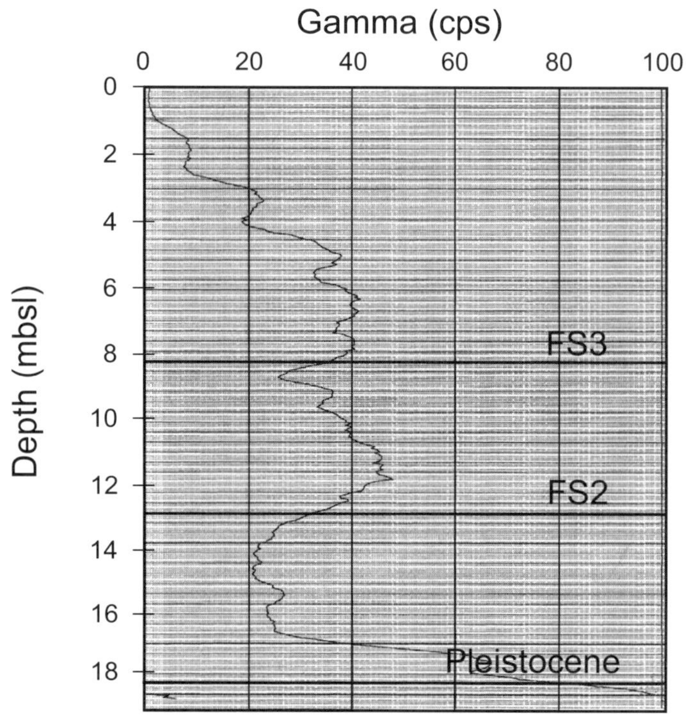

Figure 9. Gamma log from core NB02-03 illustrating the gamma response of flooding surfaces FS2 and FS3. Sands generally plot around 20 cps, and muds plot around 40–60 cps. Large values at base of hole indicate the Pleistocene.

flexuosa, *Macoma mitchelli*, etc.). In addition, core CCB94-10 sampled mouth-bar deposits (lithofacies SF2, Fig. 12A). Its coarse and lithic nature suggests proximity to a fluvial source. Core CCB94-12 contains *Lingulichnus* escape routes, which suggest rapid deposition (Fig. 12B). To account for these observations, we favor a tidally influenced bayhead delta interpretation. This interpretation would also explain the character of deposits within cores NB03-01, NB02-03, and NB02-04 at this time (Fig. 8A). A brackish lagoon would have existed landward of the bayhead delta in the vicinity of cores NB03-01, NB02-03, and NB02-04, explaining their more restricted fauna than open-bay environments but absence of sand and organic material found in a truly upper-bay and delta-influenced area. The fluvial feeder for the bayhead delta could be present beneath what is now the Port of Corpus Christi.

Tidal Delta

At some point after flooding surface FS2 formed, but prior to the formation of FS3, a tidal delta formed in the lower portions of the bay (Figs. 8B and 13). Cores sampling this feature contain SF6 and SF3 facies. This tidal delta is located 5 km from Mustang Island, which is believed to be in the same location now as it was then (Simms et al., 2006a). However, the absence of marine fauna within the delta does not suggest an open connection with the gulf, but rather connection with a shallow hypersaline bay. A possible explanation for the absence of marine fauna is a tidal delta associated with an inlet connecting a shallow lagoon landward of the open gulf with the open-bay environment of Corpus Christi Bay. This would also help to explain the tidal delta's large distance from Mustang Island.

Flooding Surface 3 (FS3)

Based on radiocarbon dates in cores NB02-03, NB02-01, CCB02-01, and CCB03-01, FS3 formed around 4.8 ka and in less than 700 yr (Fig. 8B). Within seismic profiles, this surface is marked by a shift in the location of the Nueces bayhead delta. Concurrent with that change in delta location, a large amount of erosion associated with FS3 occurred to the north of the new location of the bayhead delta (Fig. 14). In addition, a tidal-delta complex located in the lowermost portion of the bay terminates at FS3. Chronostratigraphically, the termination of the tidal delta is not as well constrained as the shift in location of the bayhead delta. Within core NB02-03, FS3 is manifest by a change from bayhead delta (SF5) to upper bay (MF1; Fig. 8A). Within core NB02-01, the contact at FS3 is not as abrupt as at FS2, but in the seismic profile through this core site, FS3 shows a prominent downlap surface associated with it (Fig. 15). The gradual lithologic change across FS3 in core NB02-01 is most likely due to the core's location in the distal portion of the bayhead delta (Figs. 8A and 15). In a more proximal location within the bayhead delta, this contact might be more abrupt. In the middle portions of the bay, such as at core locations CCB02-01 and CCB03-01, FS3 has no major lithologic changes associated with it and is often hard to map in seismic profiles. However, the location of oyster reefs shifted from the area between core sites CCB02-01 and CCB03-01 to behind present Rincon and Indian Points after the formation of FS3 (Fig. 1). At some point before FS3, but culminating after FS3, oyster reefs began to colonize the flanks of the bayhead delta (Fig. 16).

Flooding Surface 4 (FS4)

Based on radiocarbon dates in cores NB02-01, NB03-01, and CCB02-01, flooding surface FS4 formed around 2.6 ka (Fig. 8B). This event is marked by a large-scale back stepping of the Nueces bayhead delta. In core NB02-01 and NB02-03, oyster reefs replace delta-front and open-bay environments, respectively (Fig. 8A). Although the bayhead delta steps back to a location outside of the seismic coverage, lithofacies changes in core NB03-02 and reports by Brown et al. (1976) indicate that upper-bay or open-bay environments extended 10 km landward of the present bayhead delta ca. 2 ka. The locations and ages of these deposits suggest that the bayhead delta stepped back more than 20 km in less than a few centuries. Another change that occurred after FS4 is the appearance of lithological facies MF2 (Fig. 8A). MF2 is marked by discrete layers of shell material interpreted to represent storm beds. MF2 also contains a marked increase in sand without a decrease in sedimentation rates within the open-bay environments (Fig. 5). The large increase in percent sand in cores CCB02-01 and CCB03-01 is not as well constrained chronostratigraphically

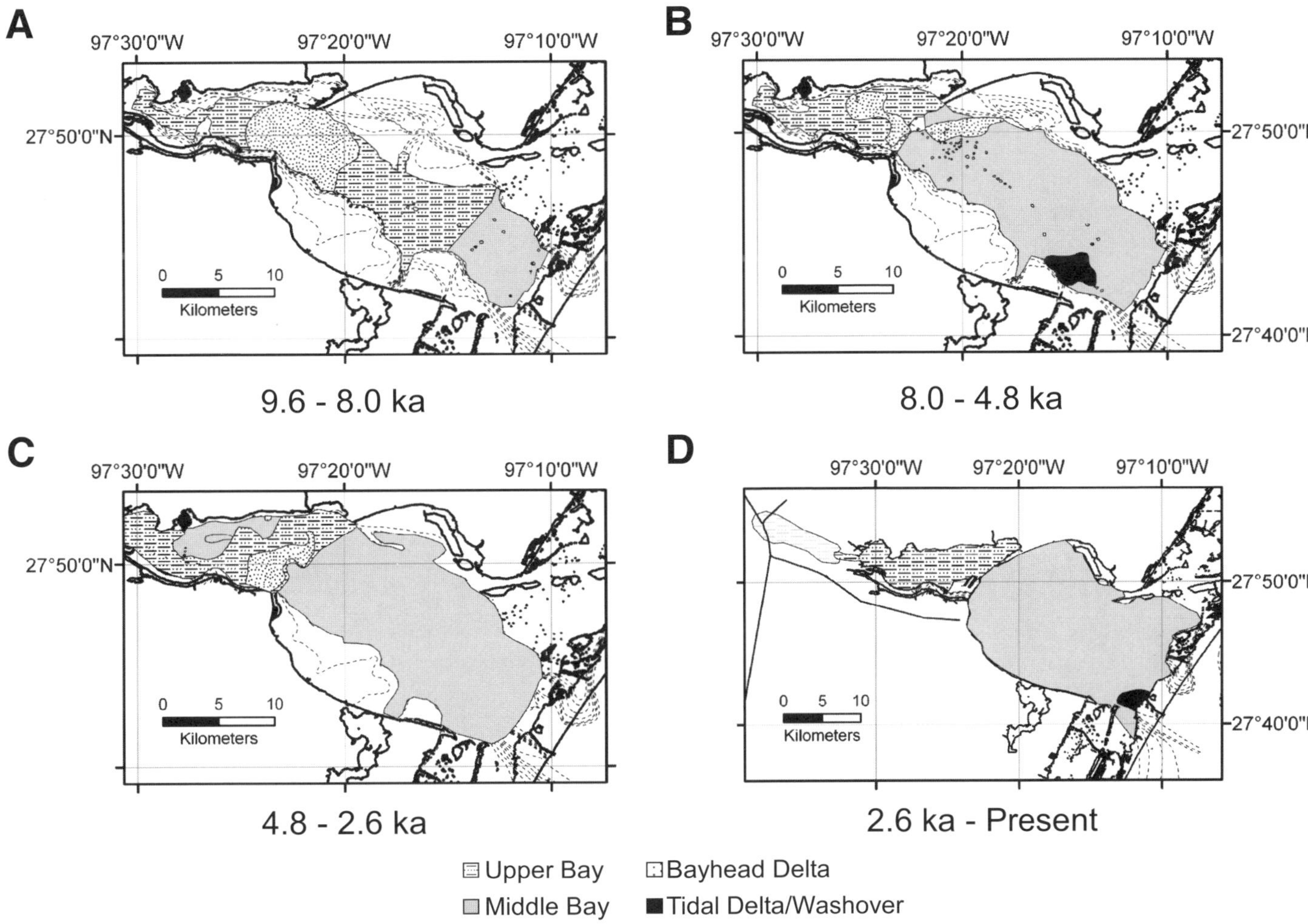

Figure 10. Paleogeography of Corpus Christi Bay illustrating environmental changes that occurred between each flooding surface. (A) Paleogeography prior to FS2 (8 ka), (B) paleogeography after FS2 but prior to FS3 (8–4.8 ka), (C) paleogeography after FS3 but prior to FS4 (4.8–2.6 ka), and (D) paleogeography after FS4 (2.6 ka).

as other changes at this time. The increase in percent sand is ubiquitous throughout the bay and does not show an increasing trend landward. This suggests a source for the sand other than the fluvial bayhead delta. This sand could possibly have been derived from wave erosion of sandy Pleistocene deposits lining the bayshore. This event does not mark any other distinct changes in the lower portions of the bay (i.e., CCB03-01).

Present Day

Over the last 2.6 ka and after FS4 formation, the bayhead delta of the Nueces River prograded to its present location (a distance of 10 km). The overstepped portions of the Nueces bayhead delta have been reworked into the present Rincon and Indian Points, effectively isolating Nueces Bay from Corpus Christi Bay, although this isolation has existed since the progradation of the delta around 7 ka. In addition, since the establishment of Indian and Rincon Points, an internal bay inlet has been established. This feature is characterized in cores by a coarse oyster shell hash (ShF1, Fig. 3). In seismic profiles, the area of the inlet is characterized by narrow cuspate ridges, possibly representing large-scale sediment waves (Fig. 11C). The processes that formed the inlet connecting Corpus Christi and Nueces Bay were more likely wind tides rather than true astronomical tides, since the tidal range within the bays is small (<0.3 m), and the winds along the central Texas coast and within Corpus Christi Bay have been shown to play the dominant role in sedimentation (Lohse, 1956; Shideler, 1984). Oyster reefs today occur only within Corpus Christi Bay, a product of the bay's salinity structure and distribution of suspended sediment (White et al., 1983).

Causes of Flooding Events

Several possible mechanisms could explain the multiple flooding events recorded in Corpus Christi Bay. These include tectonics, sea-level fluctuations, climate change, and autocyclic events, such as the flooding of antecedent topography and

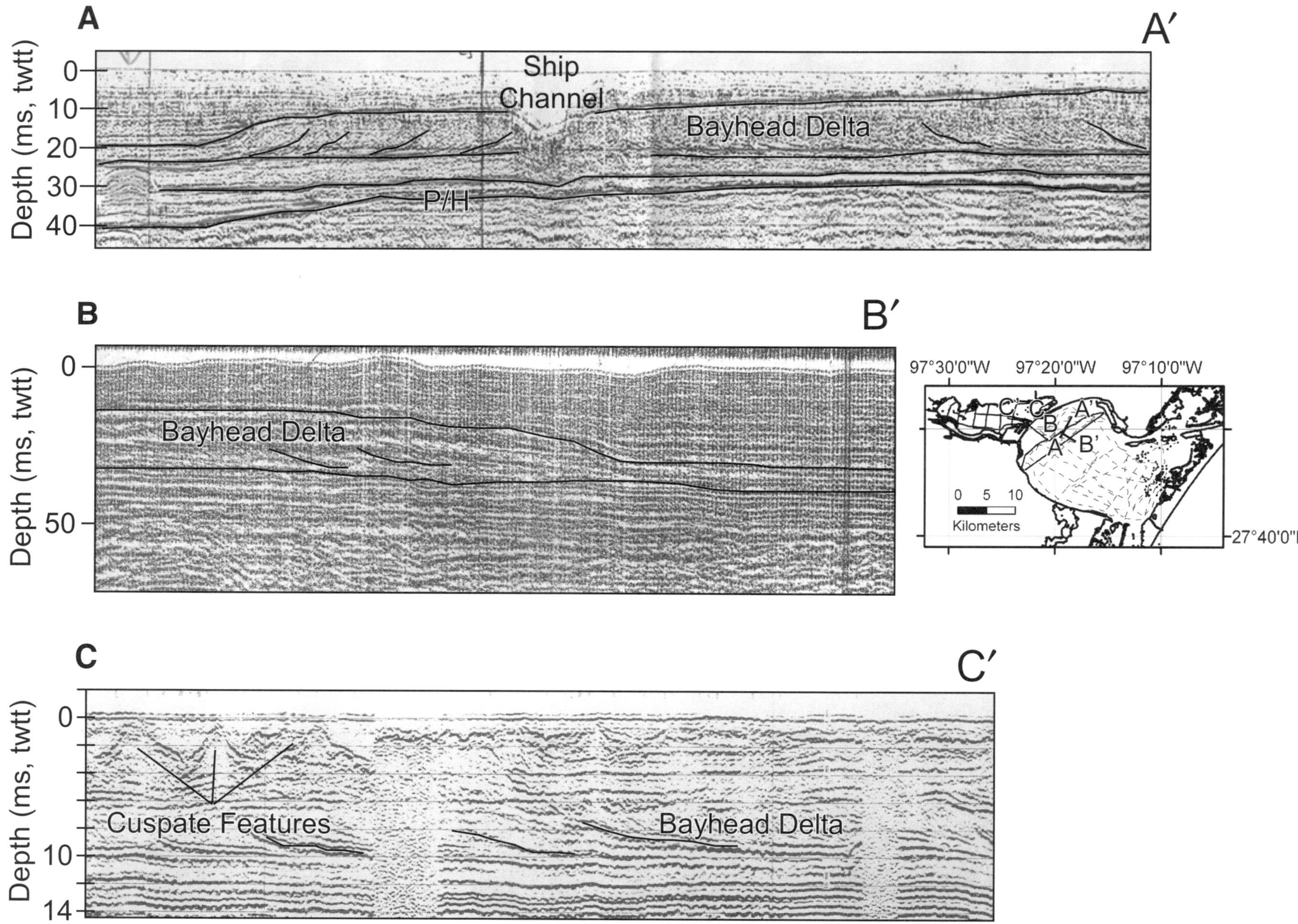

Figure 11. Seismic lines illustrating clinoforms associated with the progradation of the Nueces bayhead delta. In lines B–B′ and C–C′, the delta progrades both seaward and landward respectively, indicating a progradation direction for the bayhead delta orthogonal to the valley wall. Line C–C′ also illustrates the cuspate features associated with the inlet between Rincon and Indian Points separating Nueces and Corpus Christi Bay. P/H—Pleistocene-Holocene boundary; twtt—two-way traveltime.

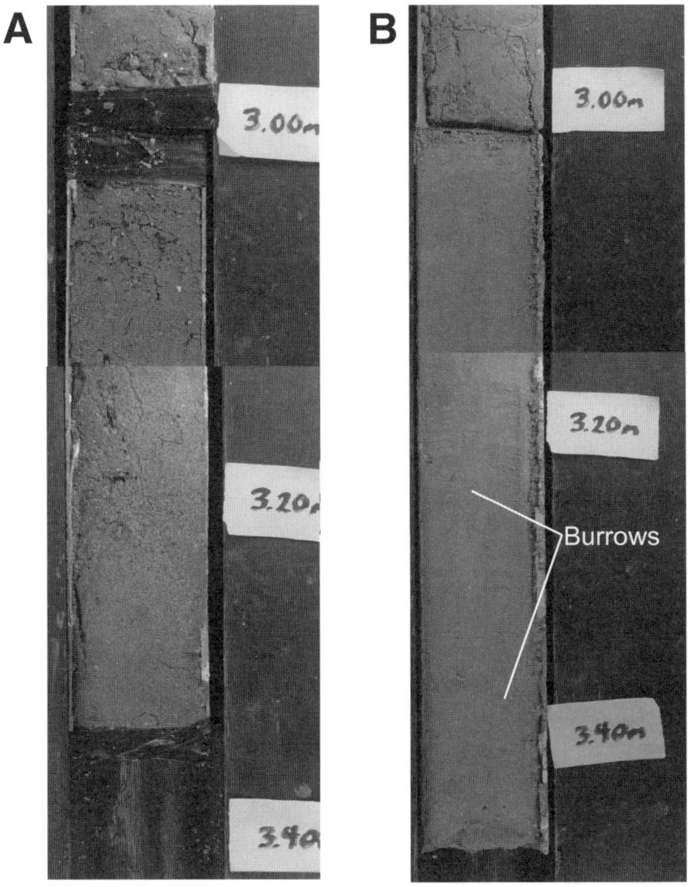

Figure 12. Photographs of (A) CCB94-10 illustrating sand facies 2 (SF2) and (B) burrowed unit in CCB94-12.

changes to the barrier-island system separating the estuary from the open gulf.

Tectonics

The only tectonic activity reported in this area is growth faulting (Price, 1933; Hillenbrand, 1985; Berryhill et al., 1986). Of these, only the Viola fault is thought to be active at the surface and not the result of hydrocarbon and groundwater withdrawal (Price, 1933; Hillenbrand, 1985). The Viola fault displaces one of the "Deweyville" terraces, named the Corpus Christi terrace (dated at 52.6 ± 5.3 ka by Durbin et al., 1997), in the upper reaches of Nueces Bay near Whites Point (Price, 1933). A seismic profile through the probable location of the Viola fault trace reveals no deformation of Holocene strata (Fig. 17). Although tectonic activity landward of the study area cannot be ruled out, it is thought that tectonic processes have not played an important role in the Holocene evolution of Corpus Christi Bay.

Changes to the Barrier-Island System

Mustang Island separates Corpus Christi Bay from the open Gulf of Mexico. It is thought to have aggraded in nearly the same location with little change for at least the last 7.5 k.y., possibly 9.5 k.y. (Shideler, 1986; Simms et al., 2006a). This is not to say changes such as the enlargement and migration of tidal inlets have not occurred. Within several of the short cores (CCB94-8, CCB94-13) from the northern margin of the bay, marine fauna (i.e., *Lucina* sp.) are more abundant than in modern deposits, suggesting a period of greater marine influence in the near past. However, Mustang Island at no point in its history disappeared altogether, and it has been near its present location during the entire time period of this study.

Flooding of Antecedent Topography

The Nueces River was in the same location during the marine isotope stage (MIS) 5e to 2 fall in sea level as it is today (Wright, 1980; Shideler, 1986; Durbin et al., 1997; Simms et al., 2006b). Terraces that formed during the fall in sea level, "Deweyville" terraces (Blum et al., 1995), are preserved along the Nueces River valley landward of Corpus Christi Bay (Durbin et al., 1997). Depositional remnants of these same terraces are preserved beneath the Holocene fill of Corpus Christi Bay (Figs. 6, 7, and 15). As sea level rose, the terraces were inundated. Each of the terraces has a gradient less than the average gradient of the base of the Nueces valley (Durbin et al., 1997). With each inundation of a terrace, the area of the estuary would increase at a higher rate, even at a constant rate of sea-level rise. For example, with the inundation of the −18 m terrace, the area of the bay increased by ~50% (260 km^2 compared to 170 km^2). These flooding events undoubtedly resulted in dramatic environmental changes caused by the reorganization of currents, salinity structures, etc. (Rodriguez et al., 2005). Given a constant rate of sediment flux spread over a larger area, slower aggradation rates would have occurred within upper-bay environments. This would help to explain the dramatic back stepping associated with FS2 (Figs. 10A and 10B). In addition, a large increase in the volume of water resulting from the flooding of a much larger bay area would increase the tidal prism coming into and out of the estuary. This could help explain the appearance of the large tidal delta after FS2. We suggest that the flooding of fluvial terraces played an important role in the flooding and formation of FS2 and FS3.

Sea-Level Changes

Sea-level rise was the first-order control on the evolution of Corpus Christi Bay. In general, the bay back stepped throughout the late Pleistocene and into the early and middle Holocene. The slowdown in the rate of sea-level rise around 5.5 ka was also coincident with the progradation of the Nueces bayhead delta. The events that caused the three major flooding surfaces, if controlled by sea level, would imply increasing rates of sea-level rise at 8.0 ka, 4.8 ka, and 2.6 ka. However, with the exception of the possible synchronous nature of FS2 with the 8.2 ka event associated with the draining of Lake Agassiz (Alley et al., 1997; Törnqvist et al., 2004b), during these time periods, the rate of sea-level rise was actually decreasing. For this reason, we interpret flooding surfaces FS3 and FS4 as resulting from other mechanisms,

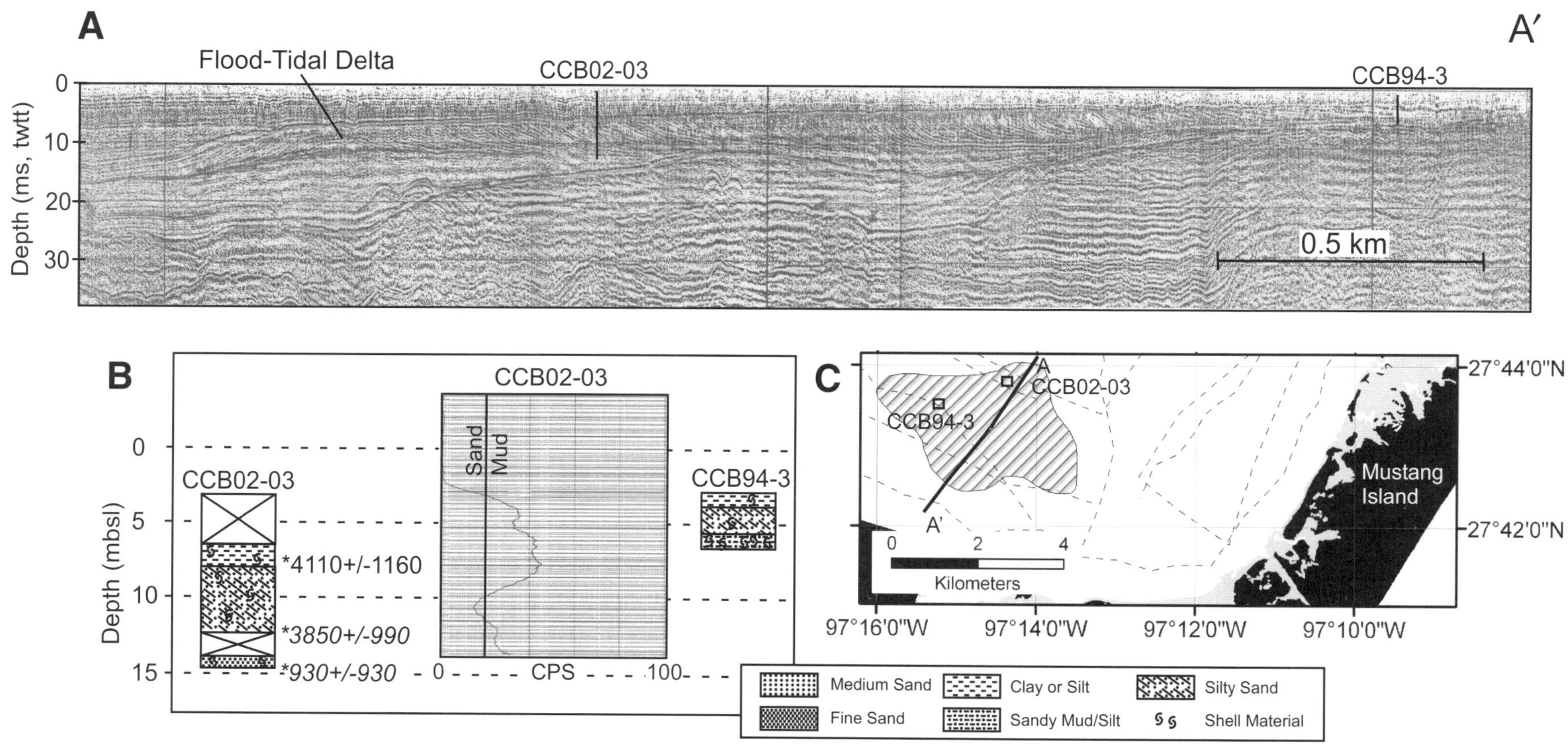

Figure 13. (A) Seismic line (A–A′) and lithology log and gamma log of cores (B) from an interpreted flood-tidal delta within Corpus Christi Bay. Seismic profile is oriented in a shore-parallel direction (C). In dip transects, the forests of the clinoforms dip landward. Core CCB02-03 sampled the more sandy flood-tidal delta and was used to constrain its age. Core CCB94-3 sampled the muddy distal edge of the delta. Within the gamma log, intervals measuring less than 20 cps are generally sandy, and those over 40 cps are muddy (from Simms et al., 2006a). Ages in cal yr B.P.

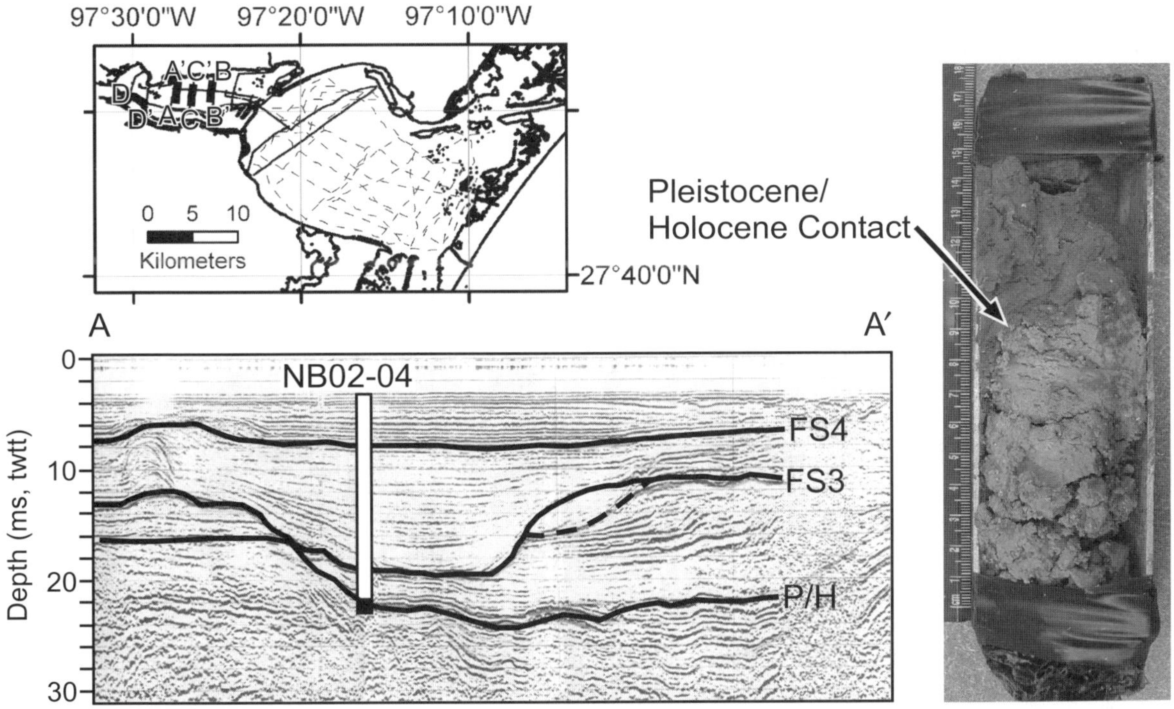

Figure 14. Seismic profile and core illustrating erosion along flooding surface FS3 that is interpreted to represent bay line ravinement possibly enhanced by a restriction between the newly avulsed delta lobe of the Nueces Delta and the edges of the Nueces paleovalley. P/H—Pleistocene-Holocene boundary; twtt—two-way traveltime.

such as climate change and the flooding of antecedent topography, rather than increases in the rate of sea-level rise. Still, it is not possible to rule out higher-order oscillations in sea level in the formation, as has been proposed by Mörner (1980), Fletcher et al. (1993), Suku and Pizzuto (1995), Donoghue and White (1995), and Froede (2002).

Climate Change

Much work has been done to reconstruct the late Quaternary climatic history of Texas (Bryant, 1977; Patton and Dibble, 1982; Bryant and Holloway, 1985; Toomey et al., 1993; Goodfriend and Ellis, 2000; Cooke et al., 2003). Unfortunately, little has been done along the Texas coast itself. The closest climate records to the study area are from cave deposits of the Edward's Plateau of central Texas (Toomey et al., 1993; Goodfriend and Ellis, 2000; Cooke et al., 2003; Ellwood and Gose, 2006). These records suggest that central Texas was cooler and effectively wetter at the end of the Pleistocene between 18 and 10.5 ka, and the early and middle Holocene experienced warmer and effectively drier conditions (Toomey et al., 1993). The effectively driest conditions occurred between 5 and 2.5 ka (Toomey et al., 1993). Recent work by Cooke et al. (2003) suggests that this event may have occurred a little earlier. Following the arid middle Holocene, the late Holocene brought higher effective moisture and somewhat cooler temperatures to the region, with a few possible higher-frequency climatic oscillations occurring around 1 ka.

An important outcome of the Toomey et al. (1993) work concerns the extent and timing of degradation of upland soils. They suggest that the soils of the Edward's Plateau began to be stripped around 8 ka and disappeared altogether by 2.5 ka. The nature and timing of these events would help to explain the progradation and ultimate back stepping of the Nueces bayhead delta. Coincident with a slowdown in the rate of sea-level rise, the stripping of soils from central Texas would provide the needed increase in sediment supply to prograde the bayhead delta. The disappearance of these soils by 2.5 ka is consistent with the dramatic back stepping of the bayhead delta around 2.6 ka. Changes in the grain size of Holocene terrace deposits along the Nueces River upstream from Corpus Christi Bay also reflect the stripping of upland soils (Durbin, 1999), although the role of fluvial storage and residence time within the Nueces River valley is unconstrained (Phillips and Slattery, 2006). In addition, the middle Holocene dry period would help to explain the presence of oysters within the Nueces bayhead delta and the appearance of *Anomalocardia cuneimeris* (a mollusk found in high-salinity bays further to the south of Corpus Christi Bay today) in lower-bay deposits during this time period.

Accompanying the climate changes around 2.5 ka, the winds across the central Texas coast could have changed to their present configuration. Models by Toomey et al. (1993) suggest that before this time, winds would have had a more southerly component and less of a southeasterly component. In conjunction with a slower rate of sea-level rise, a shift in winds around

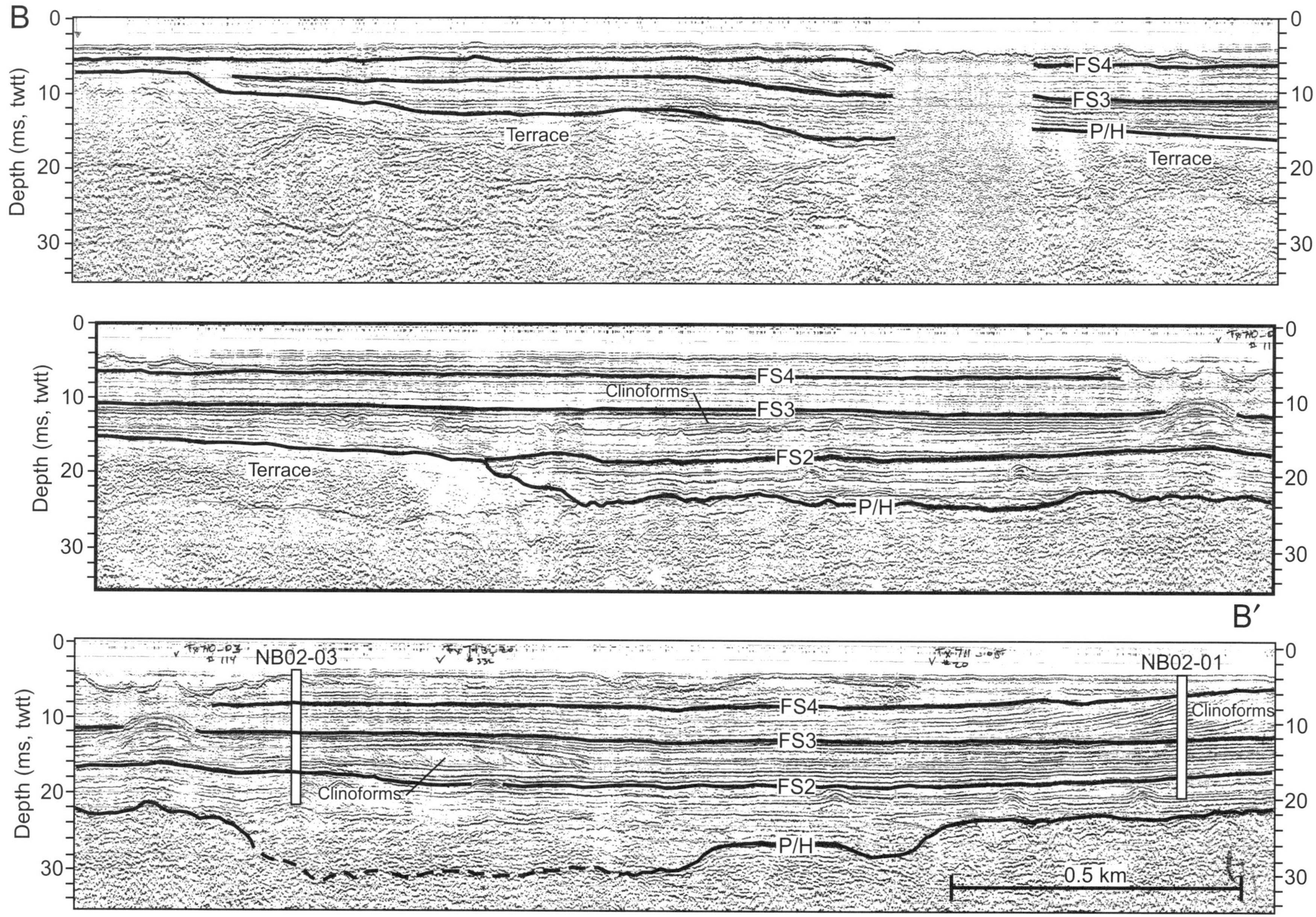

Figure 15. Seismic profile of U.S. Geological Survey 1996 line 14 illustrating the clinoforms of the bayhead delta and the location of cores NB02-01 and NB02-03 on the same seismic profile. See Figure 14 inset for line location. P/H—Pleistocene-Holocene boundary; FS—flooding surface; twtt—two-way traveltime.

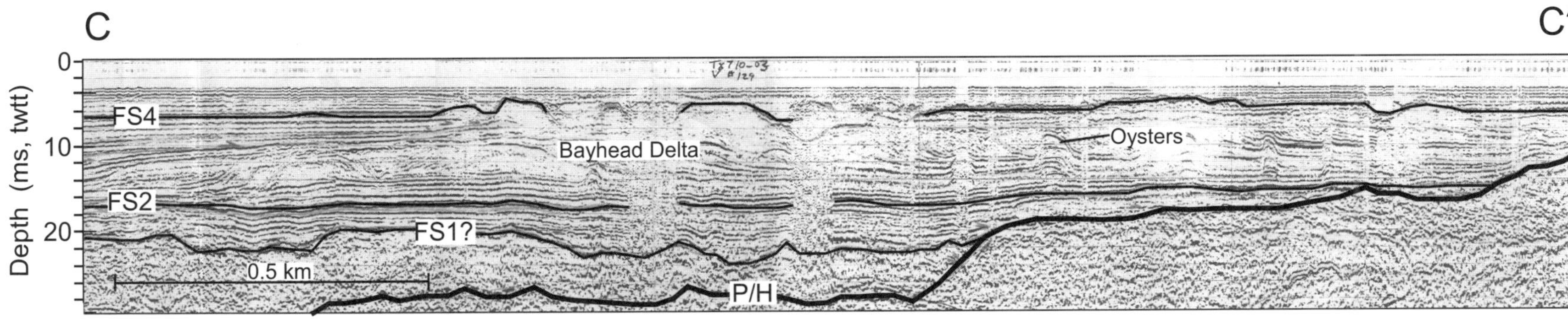

Figure 16. Seismic profile U.S. Geological Survey 1996 line 6 illustrating interpreted oyster reefs growing within the bayhead delta. See Figure 14 inset for line location. P/H—Pleistocene-Holocene boundary; FS—flooding surface; twtt—two-way traveltime.

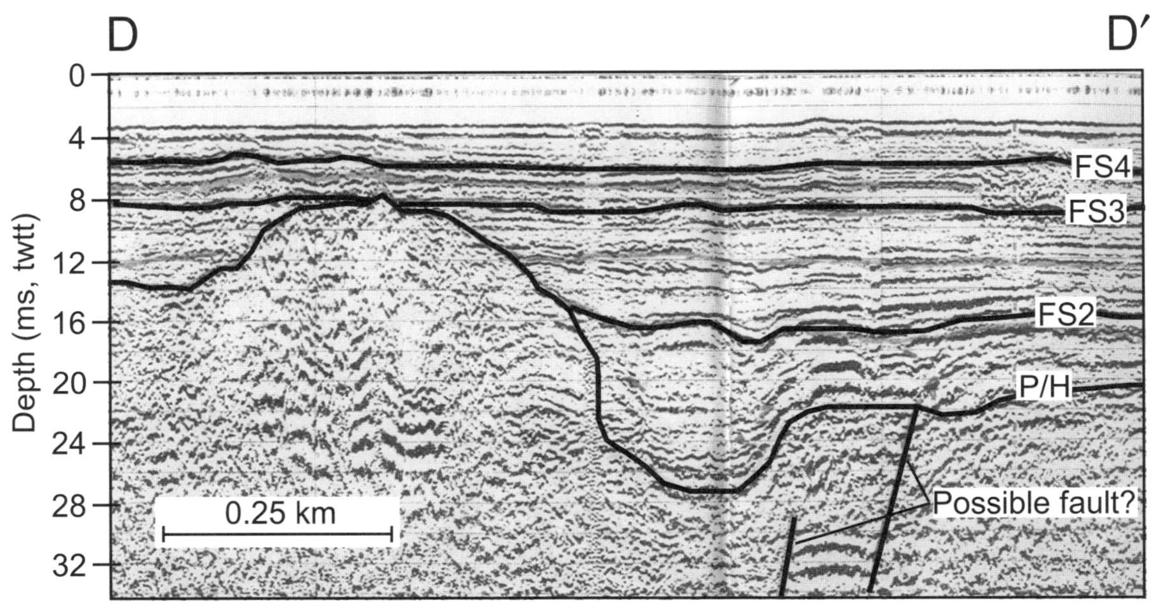

Figure 17. Seismic profile U.S. Geological Survey 1996 line 5, which passes through the fault trace of the Viola fault, is used to illustrate the lack of offset of Holocene strata. See Figure 14 inset for line location. P/H—Pleistocene-Holocene boundary; FS—flooding surface; twtt—two-way traveltime.

2.5 ka to a more southeasterly direction would allow for more erosion of sandy bay margins, explaining the ubiquitous increase in percent sand found in deposits throughout the bay since FS4 time. In addition, the climate change around 2.5 ka might have also brought about the current high hurricane frequency along the Gulf Coast, which would explain the appearance of the first discrete storm beds of lithofacies MF2.

CONCLUSIONS

Corpus Christi Bay occupies the flooded mouth of the ancestral Nueces River valley. The deposits within the bay provide an excellent record of the response of coastal environments to changes in sea level, climate, and local forcing mechanisms over the last 9.6 k.y., and they provide insights into how coastal systems may respond to future sea-level and climatic changes. These changes occurred during an overall sea-level rise but were punctuated by periods of rapid change. Three, possibly four, such rapid changes occurred at 9.6 ka, 8.0 ka, 4.8 ka, and 2.6 ka. At 8.0 ka, the upper-bay environment back stepped over 15 km, and oyster reefs were displaced 10 km to the north. At 4.8 ka, the bayhead delta of the Nueces River shifted to a new location, and a tidal inlet located in the southeast portion of the bay abruptly closed. In addition, oyster reefs started to colonize the Nueces bayhead delta. Around 2.6 ka, the Nueces bayhead delta back stepped over 20 km, well updip of its present location.

The 8.0 ka flooding event is interpreted to be the result of the flooding of fluvial terraces preserved beneath the Holocene fill of Corpus Christi Bay and possibly coinciding with the 8.2 ka eustatic event associated with the draining of Lake Agassiz. The 4.8 ka flooding event is interpreted to have resulted from a combination of a change in climate at the culmination of the middle Holocene dry period and the flooding of another fluvial terrace. The final event (2.6 ka) is thought to have been the result of a significant reduction in sediment supply following the stripping of upland soils and reestablishment of vegetation accompanying a climate change to more humid conditions.

ACKNOWLEDGMENTS

The authors would like to thank Jack Kindinger, Charles Holmes, and Nancy Soderberg at the U.S. Geological Survey for acquiring and allowing use of the two seismic surveys shot by the U.S. Geological Survey. We would also like to thank Mike Blum at Louisiana State University (LSU) for providing the nine long cores on Mustang Island. Patrick Taha, Chip Anderson, and Jessica Maddox helped in collecting cores within Corpus Christi Bay. Hermann Diaz from the Rice University geographic information systems (GIS) center helped with mapping and use of ArcView GIS. We would also like to thank the U.S. Army Corp of Engineers in Galveston, Texas, for allowing access to lithological descriptions of borings taken within Corpus Christi and Nueces Bays. James MacEachern helped to identify the trace fossils in core CCB94-12. James Durbin and Cecile Baeteman provided helpful reviews that greatly improved this manuscript. This work was funded by National Science Foundation (NSF) grant EAR-0107650, a National Science Foundation Graduate Fellowship, GCSSEPM Ed Picou Student Research Grant, ConocoPhillips, and ExxonMobil. This is Instituto di Geologia Marina (IGM) scientific contribution 1580.

REFERENCES CITED

Abdulah, K.C., 1995, The Evolution of the Brazos and Colorado Fluvial/Deltaic Systems during the Late Quaternary: An Integrated Study, Offshore Texas: Houston, Rice University, 284 p.

Alley, R.B., Mayewski, P.A., Sowers, T., Stuiver, M., Taylor, K.C., and Clark, P.U., 1997, Holocene climatic instability: A prominent, widespread event 8200 yr ago: Geology, v. 25, p. 483–486, doi: 10.1130/0091-7613(1997)025<0483:HCIAPW>2.3.CO;2.

Anderson, J.B., Rodriguez, A.B., Fletcher, C., and Fitzgerald, D., 2001, Researchers focus attention on coastal response to climate change: Eos (Transactions, American Geophysical Union), v. 82, p. 513, 519–520, doi: 10.1029/01EO00304.

Anderson, J.B., Rodriguez, A.B., Abdulah, K.C., Fillon, R.H., Banfield, L.A., McKeown, H.A., and Wellner, J.S., 2004, Late Quaternary stratigraphic evolution of the northern Gulf of Mexico margin: A synthesis, *in* Anderson, J.B., and Fillon, R.H., eds., Late Quaternary Stratigraphic Evolution of the Northern Gulf of Mexico Margin: Society for Sedimentary Geology (SEPM) Special Publication 79, p. 1–24.

Aten, L.E., 1983, Indians of the Upper Texas Coast: New York, Academic Press, 370 p.

Baeteman, C., Scott, D.B., and Van Strydonck, M., 2002, Changes in coastal zone processes at a high sea-level stand: A late Holocene example from Belgium: Journal of Quaternary Science, v. 17, p. 547–559, doi: 10.1002/jqs.707.

Bard, E., Hamelin, B., Arnold, M., Montaggioni, L., Cabioch, G., Faure, G., and Rougerie, F., 1996, Deglacial sea level record from Tahiti corals and the timing of global meltwater discharge: Nature, v. 382, p. 241–244, doi: 10.1038/382241a0.

Bartek, L.R., Cabote, B.S., Young, T., and Schroeder, W., 2004, Sequence stratigraphy of a continental margin subjected to low-energy and low-sediment-supply environmental boundary conditions: Late Pleistocene–Holocene deposition offshore Alabama, USA, *in* Anderson, J.B., and Fillon, R.H., eds., Late Quaternary Stratigraphic Evolution of the Northern Gulf of Mexico Margin: Society for Sedimentary Geology (SEPM) Special Publication 79, p. 85–109.

Belknap, D.F., and Kraft, J.C., 1981, Preservation potential of transgressive coastal lithosomes on the U.S. Atlantic shelf: Marine Geology, v. 42, p. 429–442, doi: 10.1016/0025-3227(81)90173-0.

Belknap, D.F., and Kraft, J.C., 1985, Influence of antecedent geology on stratigraphic preservation potential and evolution of Delaware's barrier systems: Marine Geology, v. 63, p. 235–262, doi: 10.1016/0025-3227(85)90085-4.

Berryhill, H.L., Suter, J.R., and Hardin, N.S., 1986, Quaternary Facies and Structure, Northern Gulf of Mexico: Interpretations from Seismic Data: American Association of Petroleum Geology Studies in Geology, v. 23, 289 p.

Blum, M.D., Morton, R.A., and Durbin, J.M., 1995, "Deweyville" terraces and deposits of the Texas Gulf Coastal Plain: Transactions of the Gulf Coast Association of Geological Societies, v. 45, p. 53–60.

Blum, M.D., Misner, T.J., Collins, E.C., Scott, D.B., Morton, R.A., and Aslam, A., 2001, Middle Holocene sea-level rise and highstand at +2 m, central Texas coast: Journal of Sedimentary Research, v. 71, no. 4, p. 581–588, doi: 10.1306/112100710581.

Blum, M.D., Carter, A.E., Zayac, T., and Goble, R., 2002, Middle Holocene sea level and evolution of the Gulf of Mexico: Journal of Coastal Research, v. 36, Special Issue, p. 65–80.

Bratton, J.F., Coleman, S.M., Thieler, E.R., and Seal, R.R., II, 2003, Birth of the modern Chesapeake Bay estuary between 7.4 and 8.2 ka and implications for global sea-level rise: Geo-Marine Letters, v. 22, p. 188–197.

Brown, L.F., Brewton, J.L., McGowen, J.H., Evans, T.J., Fisher, W.L., and Groat, C.G., 1976, Environmental Geologic Atlas of the Texas Coastal

Zone—Corpus Christi Area: Austin, Bureau of Economic Geology, University of Texas at Austin, 123 p.

Bryant, V.M., 1977, A 16,000 year pollen record of vegetation change in central Texas: Palynology, v. 1, p. 143–156.

Bryant, V.M., and Holloway, R.W., 1985, The late Quaternary paleoenvironmental record of Texas, *in* Bryant, V.M. and Holloway, R.W., eds., Pollen Records of Late-Quaternary North American Sediments: Dallas, Texas, American Association of Stratigraphic Palynologists, p. 39–70.

Carr, J.T., 1967, The Climate and Physiography of Texas: Austin, Texas, Texas Water Development Board Report 53, 27 p.

Chappell, J., and Polach, H., 1991, Post-glacial sea-level rise from a coral record at Huon Peninsula, Papua New Guinea: Nature, v. 349, p. 147–149, doi: 10.1038/349147a0.

Coleman, J.M., and Smith, W.G., 1964, Late recent rise of sea level: Geological Society of America Bulletin, v. 75, p. 833–840, doi: 10.1130/0016-7606(1964)75[833:LRROSL]2.0.CO;2.

Cooke, M.J., Stern, L.A., Banner, J.L., Mack, L.E., Stafford, T.W., Jr., and Toomey, R.S., III, 2003, Precise timing and rate of massive late Quaternary soil denudation: Geology, v. 31, no. 10, p. 853–856, doi: 10.1130/G19749.1.

Curray, J.R., 1960, Sediments and history of Holocene transgression, continental shelf, northwestern Gulf of Mexico, *in* Shepard, F.P., Phleger, F.B., and van Andel, T.H., eds., Recent Sediments, Northwestern Gulf of Mexico: Tulsa, Oklahoma, American Association of Petroleum Geologists, p. 221–266.

Donoghue, J.F., and White, N.M., 1995, Late Holocene sea-level change and delta migration, Apalachicola River region, NW Florida, USA: Journal of Coastal Research, v. 11, p. 651–663.

Durbin, J.M., 1999, Geomorphic Response to Late Quaternary Climate and Sea Level Change, Lower Nueces River, Texas [Ph.D. thesis]: Lincoln, Nebraska, University of Nebraska, 242 p.

Durbin, J.M., Blum, M., and Price, D.M., 1997, Late Pleistocene stratigraphy of the lower Nueces River, Corpus Christi, Texas: Glacio-eustatic influences on valley-fill architecture: Transactions of the Gulf Coast Association of Geological Societies, v. 47, p. 119–129.

Ellwood, B.B., and Gose, W.A., 2006, Heinrich, HI and 8200 yr B.P. climate events recorded in Hull's Cave, Texas: Geology, v. 34, p. 753–756, doi: 10.1130/G22549.1.

Eckles, B.J., Fassell, M.L., and Anderson, J.B., 2004, Late Quaternary evolution of the wave-storm-dominated central Texas shelf, *in* Anderson, J.B., and Fillon, R.H., eds., Late Quaternary Stratigraphic Evolution of the Northern Gulf of Mexico Margin: Society for Sedimentary Geology (SEPM) Special Publication 79, p. 271–287.

Fairbanks, R.G., 1989, A 17,000-year glacio-eustatic sea level record: Influence of glacial melting on the Younger Dryas event and deep-ocean circulation: Nature, v. 342, no. 6250, p. 637–642, doi: 10.1038/342637a0.

Fletcher, C.H., Van Pelt, J.E., Brush, G.S., and Sherman, J., 1993, Tidal wetland record of Holocene sea-level movements and climate history: Palaeogeography, Palaeoclimatology, Palaeoecology, v. 102, p. 177–213, doi: 10.1016/0031-0182(93)90067-S.

Froede, C.R., Jr., 2002, Rhizolith evidence in support of a late Holocene sea-level highstand at least 0.5 m higher than at present Key Biscayne, Florida: Geology, v. 30, p. 203–206, doi: 10.1130/0091-7613(2002)030<0203:REISOA>2.0.CO;2.

Goodfriend, G.A., and Ellis, G.L., 2000, Stable carbon isotope record of Middle to Late Holocene climate changes from land snail shells at Hinds Cave, Texas: Quaternary International, v. 67, p. 47–60, doi: 10.1016/S1040-6182(00)00008-2.

Henley, D.E., and Rauschuber, D.G., 1981, Freshwater Needs of Fish and Wildlife Resources in the Nueces–Corpus Christi Bay Area, Texas: A Literature Synthesis: Washington, D.C., U.S. Fish and Wildlife Service, Office of Biological Services Report FWS/OBS-80/10, 410 p.

Hillenbrand, C.J., III, 1985, Subsidence and Fault Activation Related to Fluid Extraction Saxet Field, Nueces County, Texas: Houston, University of Houston, 144 p.

Lambeck, K., and Chappell, J., 2001, Sea level change through the last glacial cycle: Science, v. 292, p. 679–686, doi: 10.1126/science.1059549.

Lighty, R.G., Mcintyre, I.G., and Stuckenrath, R., 1982, *Acropora palmata* reef framework; a reliable indicator of sea level in the Western Atlantic for the past 10,000 years: Coral Reefs, v. 1, p. 125–130, doi: 10.1007/BF00301694.

Lohse, E.A., 1956, Dynamic geology of the modern coastal region, northwestern Gulf of Mexico, *in* Hough, J.L., ed., Finding Ancient Shorelines: A Symposium with Discussions: Society of Economic Mineralogists and Paleontologists (SEPM) Special Publication 3, p. 99–104.

Mannino, A., and Montagna, P.A., 1996, Fine-scale spatial variation of sediment composition and salinity in Nueces Bay of south Texas: The Texas Journal of Science, v. 48, no. 1, p. 35–47.

Marmer, H.A., 1954, Tides and sea level in the Gulf of Mexico, *in* Gulf of Mexico, Its Origin, Waters, and Marine Life: U.S. Department of Interior Fish Wildlife Service Bulletin, v. 89, p. 101–118.

McFarlan, E., Jr., 1961, Radiocarbon dating of late Quaternary deposits, South Louisiana: Geological Society of America Bulletin, v. 72, p. 129–158, doi: 10.1130/0016-7606(1961)72[129:RDOLQD]2.0.CO;2.

Mitchum, R.M., Jr., Vail, P.R., and Sangree, J.B., 1977, Seismic stratigraphy and global changes of sea level: Part 6. Stratigraphic interpretation of seismic reflection patterns in depositional sequences, *in* Payton, C.E., ed., Seismic Stratigraphy, Applications to Hydrocarbon Exploration: American Association of Petroleum Geology Memoir 26, p. 117–133.

Moore, D.G., and Scruton, P.C., 1957, Minor internal structures of some recent unconsolidated sediments: American Association of Petroleum Geologists Bulletin, v. 41, p. 2723–2751.

Mörner, N.A., 1980, The northwestern European "sea-level laboratory" and regional Holocene eustasy: Palaeogeography, Palaeoclimatology, and Palaeoecology, v. 29, p. 281–300, doi: 10.1016/0031-0182(79)90086-5.

Morton, R.A., and Kindinger, J.L., 1998, Responses of coastal systems to late Pleistocene and Holocene sea-level fluctuations, Corpus Christi Region: Corpus Christi, Texas, Gulf Coast Association of Geological Societies, Field Trip Guidebook, 24 p.

Morton, R.A., and McGowen, J.H., 1980, Modern Depositional Environments of the Texas Coast: Austin, Bureau of Economic Geology, The University of Texas at Austin, Guidebook 20, 167 p.

Morton, R.A., Paine, J.G., and Blum, M.D., 2000, Responses of stable bay-margin and barrier-island systems to Holocene sea-level highstands, western Gulf of Mexico: Journal of Sedimentary Research, v. 70, no. 3, p. 478–490, doi: 10.1306/2DC40921-0E47-11D7-8643000102C1865D.

Nelson, H.F., and Bray, E.E., 1970, Stratigraphy and history of the Holocene sediments in the Sabine-High Island area, Gulf of Mexico, *in* Morgan, J.P., eds., Deltaic Sedimentation Modern and Ancient: Society of Economic Paleontologists and Mineralogists (SEPM) Special Publication 15, p. 48–77.

Paine, J.G., 1993, Subsidence of the Texas Coast: Inferences from historical and late Pleistocene sea levels: Tectonophysics, v. 222, p. 445–458.

Patton, P.C., and Dibble, D.S., 1982, Archeologic and geomorphic evidence for the paleohydrologic record of the Pecos River in West Texas: American Journal of Science, v. 282, p. 97–121.

Peltier, W.R., Farrell, W.E., and Clark, J.A., 1978, Glacial isostasy and relative sea-level: A global finite element model: Tectonophysics, v. 50, p. 81–110, doi: 10.1016/0040-1951(78)90129-4.

Penland, S., Boyd, R., and Suter, J.R., 1988, Transgressive depositional systems of the Mississippi Delta Plain: A model for barrier shoreline and shelf sand movement: Journal of Sedimentary Petrology, v. 58, p. 932–949.

Phillips, J.D., and Slattery, M.C., 2006, Sediment storage, sea level, and sediment delivery to the ocean by coastal plain rivers: Progress in Physical Geography, v. 30, p. 513–530, doi: 10.1191/0309133306pp494ra.

Price, W.A., 1933, Role of diastrophism in topography of Corpus Christi Area: South Texas: American Association of Petroleum Geology Bulletin, v. 17, no. 8, p. 907–962.

Raymond, P.A., and Bauer, J.E., 2001, Riverine export of aged terrestrial organic matter to the North Atlantic Ocean: Nature, v. 409, p. 497–499, doi: 10.1038/35054034.

Rehkemper, J.L., 1969, Sedimentology of Holocene Estuarine Deposits, Galveston Bay, Texas: Houston, Rice University, 61 p.

Rodriguez, A.B., Anderson, J.B., Siringan, F.P., and Taviani, M., 2004, Holocene evolution of the east Texas coast and inner shelf: Along-strike variability in coastal retreat rates: Journal of Sedimentary Research, v. 74, p. 405–422, doi: 10.1306/092403740405.

Rodriguez, A.B., Anderson, J.B., and Simms, A.R., 2005, Terrace inundation as an autocyclic mechanism for parasequences formation, Galveston estuary, Texas: Journal of Sedimentary Research, v. 75, p. 608–618, doi: 10.2110/jsr.2005.050.

Shepard, F., 1953, Sedimentation rates in Texas estuaries and lagoons: American Association of Petroleum Geologists Bulletin, v. 37, p. 1919–1934.

Shepard, F., and Moore, D.G., 1954, Sedimentary environments differentiated by coarse-fraction studies: American Association of Petroleum Geology Bulletin, v. 38, no. 8, p. 1792–1802.

Shepard, F., and Moore, D.G., 1955, Central Texas coast sedimentation: Characteristics of sedimentary environment, recent history, and diagenesis: American Association of Petroleum Geology Bulletin, v. 39, no. 8, p. 1463–1593.

Shepard, F.P., and Suess, H.E., 1956, Rate of postglacial rise of sea level: Science, v. 123, p. 1082–1083, doi: 10.1126/science.123.3207.1082-a.

Shideler, G.L., 1984, Suspended sediment responses in a wind-dominated estuary of the Texas Gulf Coast: Journal of Sedimentary Petrology, v. 54, no. 3, p. 731–745.

Shideler, G.L., 1986, Seismic and physical stratigraphy of late Quaternary deposits, south Texas coastal complex, *in* Shideler, G.L., ed., Stratigraphic Studies of a Late Quaternary Barrier-Type Coastal Complex, Mustang Island–Corpus Christi Area, South Texas Gulf Coast: U.S. Geological Survey Professional Paper 1328-B, p. 9–31.

Simms, A.R., Anderson, J.B., and Rodriguez, A.B., 2003, The role of antecedent topography in shaping coastal change caused by sea-level rise: XVI International Union for Quaternary Research (INQUA) Congress, Programs and Abstracts, p. 239.

Simms, A.R., Anderson, J.B., and Blum, M.D., 2006a, Barrier-island aggradation via inlet migration, Mustang Island Texas: Sedimentary Geology, v. 187, p. 105–125, doi: 10.1016/j.sedgeo.2005.12.023.

Simms, A.R., Anderson, J.B., Taha, Z.P., and Rodriguez, A.B., 2006b, Overfilled versus underfilled incised valleys: Lessons from the Gulf of Mexico, *in* Dalrymple, R., Leckie, S., and Tillman, R., eds., Incised Valleys through Space and Time: Society for Sedimentary Geology (SEPM) Special Publication 85, p. 117–139.

Simms, A.R., Lambeck, K., Purcell, A., Anderson, J.B., and Rodriguez, A.B., 2007, Sea-level history of the Gulf of Mexico since the Last Glacial Maximum with implications for the melting history of the Laurentide Ice Sheet: Quaternary Science Reviews, v. 26, p. 920–940.

Simpson, R.H., and Lawrence, M.B., 1971, Atlantic hurricane frequencies along the US coastline: U.S. Department of Commerce, National Oceanic and Atmospheric Administration Technical Memorandum NWS SR-58.

Snedden, J.W., Nummedal, D., and Amos, A.F., 1988, Storm- and fair-weather combined flow on the central Texas shelf: Journal of Sedimentary Petrology, v. 58, no. 4, p. 580–595.

Sperazza, M., Moore, J.N., and Hendrix, M.S., 2004, High-resolution particle size analysis of naturally occurring very fine–grained sediment through laser diffractometry: Journal of Sedimentary Research, v. 74, p. 736–743, doi: 10.1306/031104740736.

Stapor, F.W., Jr., and Stone, F.W., 2004, A new depositional model for the buried 4000 yr BP New Orleans barrier: Implications for sea-level fluctuations and onshore transport from a nearshore shelf source: Marine Geology, v. 204, p. 215–234, doi: 10.1016/S0025-3227(03)00350-5.

Streif, H., 2004, Sedimentary record of Pleistocene and Holocene marine inundations along the North Sea coast of Lower Saxony, Germany: Quaternary International, v. 112, p. 3–28, doi: 10.1016/S1040-6182(03)00062-4.

Stuiver, M., and Braziunas, T.F., 1993, Modeling atmospheric ^{14}C influences and ^{14}C ages of marine samples to 10,000 B.C.: Radiocarbon, v. 35, no. 1, p. 137–189.

Stuiver, M., Reimer, P.J., Bard, E., Beck, J.W., Burr, G.S., Hughen, K.A., Kromer, B., McCormac, G., van der Plicht, J., and Spurk, M., 1998, INTCAL98 radiocarbon age calibration, 24,000–0 cal B.P.: Radiocarbon, v. 40, no. 3, p. 1041–1083.

Suku, J.J., and Pizzuto, J.E., 1995, Accelerated sea-level rise 2,000 years BP in the Delaware Bay and stratigraphic evidence from the Leipsic River valley, Delaware, USA: Journal of Coastal Research, v. 11, p. 573–582.

Sydow, J., and Roberts, H.H., 1994, Stratigraphic framework of a late Pleistocene shelf-edge delta, northeast Gulf of Mexico: The American Association of Petroleum Geologists Bulletin, v. 78, p. 1276–1312.

Thomas, M.A., and Anderson, J.B., 1994, Sea-level controls on the facies architecture of the Trinity/Sabine incised-valley system, Texas continental shelf, *in* Dalrymple, R.W., Boyd, R., and Zaitlin, B.A., eds., Incised-Valley Systems: Origins and Sedimentary Sequences: Society for Sedimentary Geology (SEPM) Special Publication 51, p. 63–82.

Thornthwaite, C.W., 1948, An approach toward a rational classification of climate: Geographical Review, v. 38, no. 1, p. 55–94, doi: 10.2307/210739.

Toomey, R.S., III, Blum, M.D., and Valastro, S.V., Jr., 1993, Late Quaternary climates and environments of the Edwards Plateau, Texas: Global and Planetary Change, v. 7, p. 299–320, doi: 10.1016/0921-8181(93)90003-7.

Törnqvist, T.E., Gonzalez, J.L., Newsom, L.E., van der Borg, K., de Jong, A.F.M., and Kurnik, C.W., 2004a, Deciphering Holocene sea-level history on the US Gulf Coast: A high-resolution record from the Mississippi Delta: Geological Society of America Bulletin, v. 116, p. 1026–1039, doi: 10.1130/B2525478.1.

Törnqvist, T.E., Bick, S.J., Gonzalez, J.L., van der Borg, K., and de Jong, A.F.M., 2004b, Tracking the sea-level signature of the 8.2 ka cooling event: New constraints from the Mississippi Delta: Geophysical Research Letters, v. 31, doi: 10.1029/2004GL021429.

van de Plassche, O., 1990, Mid-Holocene sea-level change on the eastern shore of Virginia: Marine Geology, v. 91, p. 149–154, doi: 10.1016/0025-3227(90)90138-A.

van de Plassche, O., Mook, W.G., and Bloom, A.L., 1989, Submergence of coastal Connecticut 6,000–3,000 (^{14}C) years B.P.: Marine Geology, v. 86, p. 349–354, doi: 10.1016/0025-3227(89)90093-5.

Walker, K.J., Stapor, F.W., Jr., and Marquardt, W.H., 1995, Archeological evidence for a 1750–1450 B.P. higher-than-present sea level along Florida's Gulf Coast: Holocene Cycles, Sea Levels, and Sedimentation: Journal of Coastal Research, v. 17, Special Issue, p. 205–218.

White, W.A., Calnan, T.R., Morton, R.A., Kimble, R.S., Littleton, T.G., McGowen, J.H., Nance, H.S., and Schmedes, K.E., 1983, Submerged Lands of Texas, Corpus Christi Area: Sediments, Geochemistry, Benthic Macroinvertebrates, and Associated Wetlands: Austin, Texas, Bureau of Economic Geology, University of Texas at Austin, 154 p.

Wright, S.S., 1980, Seismic Stratigraphy and Depositional History of Holocene Sediments on the Central Texas Gulf Coast: Austin, Texas, University of Texas at Austin, 123 p.

Yeager, K.M., Santschi, P.H., Schindler, K.J., Andres, M.J., and Weaver, E.A., 2006, The relative importance of terrestrial versus marine sediment sources to the Nueces–Corpus Christi estuary, Texas: An isotopic approach: Estuaries and Coasts, v. 29, p. 443–454.

Manuscript Accepted by the Society 10 March 2008

Printed in the USA